Panspermia & The Tardigrade

" Lifeforms That Can Live In Space "

Edited by Paul F. Kisak

Contents

Chapter 1

Tardigrade

For members of the taxonomic suborder *Folivora/Phyllophaga*, historically sometimes designated *Tardigrada* ("slow-walking"), see Sloth (animal).

Tardigrades (also known as **water bears** or **moss piglets**)[2][3][4] are water-dwelling, segmented micro-animals, with eight legs.[2] They were first discovered by the German pastor Johann August Ephraim Goeze in 1773. The name ***Tardigrada*** (meaning "slow stepper") was given three years later by the Italian biologist Lazzaro Spallanzani.[5] The phylum has been sighted from mountaintops to the deep sea, from tropical rain forests to the Antarctic[6] Tardigrades can survive in extreme environments. For example, they can withstand temperatures from just above absolute zero to well above the boiling point of water (100 °C), pressures about six times greater than those found in the deepest ocean trenches, ionizing radiation at doses hundreds of times higher than the lethal dose for a human, and the vacuum of outer space.[7] They can go without food or water for more than 10 years, drying out to the point where they are 3% or less water, only to rehydrate, forage, and reproduce.[3][8][9][10]

They are not considered extremophilic because they are not adapted to exploit these conditions. This means that their chances of dying increase the longer they are exposed to the extreme environments,[5] whereas true extremophiles thrive in a physically or geochemically extreme condition that would harm most other organisms.[3][11][12]

Usually, tardigrades are about 0.5 mm (0.020 in) long when they are fully grown.[2] They are short and plump with four pairs of legs, each with four to eight claws also known as "disks".[2] The first three pairs of legs are directed ventrolaterally and are the primary means of locomotion, while the fourth pair is directed posteriorly on the terminal segment of the trunk and is used primarily for grasping the substrate.[13]

The animals are prevalent in mosses and lichens and feed on plant cells, algae, and small invertebrates. When collected, they may be viewed under a very-low-power microscope, making them accessible to students and amateur scientists.[14]

Tardigrades form the phylum *Tardigrada*, part of the superphylum Ecdysozoa. It is an ancient group, with fossils dating from 530 million years ago, in the Cambrian period.[15]

1.1 Description

Johann August Ephraim Goeze originally named the tardigrade *kleiner Wasserbär* (*Bärtierchen* today), meaning 'little water bear' in German. The name *Tardigrada* means "slow walker" and was given by Lazzaro Spallanzani in 1776.[16] The name *water bear* comes from the way they walk, reminiscent of a bear's gait. The biggest adults may reach a body length of 1.5 mm (0.059 in), the smallest below 0.1 mm. Newly hatched tardigrades may be smaller than 0.05 mm.

About 1,150 species of tardigrades have been described.[17][18] Tardigrades can be found throughout the world, from the Himalayas[19] (above 6,000 m (20,000 ft)), to the deep sea (below 4,000 m (13,000 ft)) and from the polar regions to the equator.

The most convenient place to find tardigrades is on lichens and mosses. Other environments are dunes, beaches, soil, and marine or freshwater sediments, where they may occur quite frequently (up to 25,000 animals per liter). Tardigrades, in the case of *Echiniscoides Wyethi*,[20] may be found on barnacles. Often, tardigrades can be found by soaking a piece of moss in water.[21]

1.2 Anatomy and morphology

Tardigrades have barrel-shaped bodies with four pairs of stubby legs. Most range from 0.3 to 0.5 mm (0.012 to 0.020 in) in length, although the largest species may reach 1.2 mm (0.047 in). The body consists of a head, three body segments with a pair of legs each, and a caudal segment with a fourth pair of legs. The legs are without joints while the feet have four to eight claws each. The cuticle contains chitin and protein and is moulted periodically.

Tardigrades are eutelic, meaning all adult tardigrades of the same species have the same number of cells. Some species have as many as 40,000 cells in each adult, while others have far fewer.[22][23]

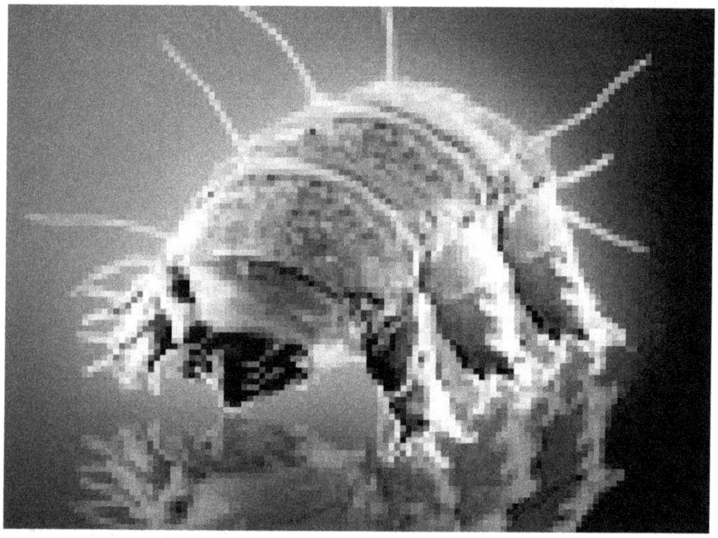

Echiniscus testudo

The body cavity consists of a haemocoel, but the only place where a true coelom can be found is around the gonad. There are no respiratory organs, with gas exchange able to occur across the whole of the body. Some tardigrades have three tubular glands associated with the rectum; these may be excretory organs similar to the Malpighian tubules of arthropods, although the details remain unclear.[24]

The tubular mouth is armed with stylets, which are used to pierce the plant cells, algae, or small invertebrates on which the tardigrades feed, releasing the body fluids or cell contents. The mouth opens into a triradiate, muscular, sucking pharynx. The stylets are lost when the animal molts, and a new pair is secreted from a pair of glands that lie on either side of the mouth. The pharynx connects to a short esophagus, and then to an intestine that occupies much of the length of the body, which is the main site of digestion. The intestine opens, via a short rectum, to an anus located at the terminal end of the body. Some species only defecate when they molt, leaving the feces behind with the shed cuticle.[24]

The brain includes multiple lobes, mostly consisting of three bilaterally paired clusters of neurons.[25] The brain is attached to a large ganglion below the esophagus, from which a double ventral nerve cord runs the length of the body. The cord possesses one ganglion per segment, each of which produces lateral nerve fibres that run into the limbs. Many species possess a pair of rhabdomeric pigment-cup eyes, and there are numerous sensory bristles on the head and body.[26]

Tardigrades all possess a buccopharyngeal apparatus, which, along with the claws, is used to differentiate among species.

1.3 Reproduction

Although some species are parthenogenic, both males and females are usually present, each with a single gonad located above the intestine. Two ducts run from the testis in males, opening through a single pore in front of the anus. In contrast, females have a single duct opening either just above the anus or directly into the rectum, which thus forms a cloaca.[24]

Tardigrades are oviparous, and fertilization is usually external. Mating occurs during the molt with the eggs being laid inside the shed cuticle of the female and then covered with sperm. A few species have internal fertilization, with mating occurring before the female fully sheds her cuticle. In most cases, the eggs are left inside the shed cuticle to develop, but some attach them to nearby substrate.[24]

The eggs hatch after no more than 14 days, with the young already possessing their full complement of adult cells. Growth to the adult size therefore occurs by enlargement of the individual cells (hypertrophy), rather than by cell division. Tardigrades may molt up to 12 times.[24]

1.4 Ecology and life history

Most tardigrades are phytophagous (plant eaters) or bacteriophagous (bacteria eaters), but some are predatory (e.g., *Milnesium tardigradum*).[27][28]

1.5 Physiology

Scientists have reported tardigrades in hot springs, on top of the Himalayas, under layers of solid ice, and in ocean sediments. Many species can be found in milder environments such as lakes, ponds, and meadows, while others can be found in stone walls and roofs. Tardigrades are most common in moist environments, but can stay active wherever they can retain at least some moisture.

Tardigrades are one of the few groups of species that are capable of reversibly suspending their metabolism and going into a state of cryptobiosis. Several species of tardigrade regularly survive in a dehydrated state for nearly 10 years. Depending on the environment, they may enter this state via anhydrobiosis, cryobiosis, osmobiosis, or anoxybiosis. While in this state, their metabolism lowers to less than 0.01% of normal and their water content can drop to 1% of normal.[7] Their ability to remain desiccated for such a long period is largely dependent on the high levels of the nonreducing sugar trehalose, which protects their membranes. In this cryptobiotic state, the tardigrade is known as a tun.[29]

Tardigrades are able to survive in extreme environments that would kill almost any other animal. Extremes at which tardigrades can survive include those of:

- Temperature – tardigrades can survive being heated for a few minutes to 151 °C (304 °F),[30] or being chilled for days at −200 °C (−328 °F).[30] Some can even survive cooling to −272 °C (~1 degree above absolute zero or −458 °F)[31] for a few minutes.
- Pressure – they can withstand the extremely low pressure of a vacuum and also very high pressures, more than 1,200 times atmospheric pressure. Tardigrades can survive the vacuum of open space and solar radiation combined

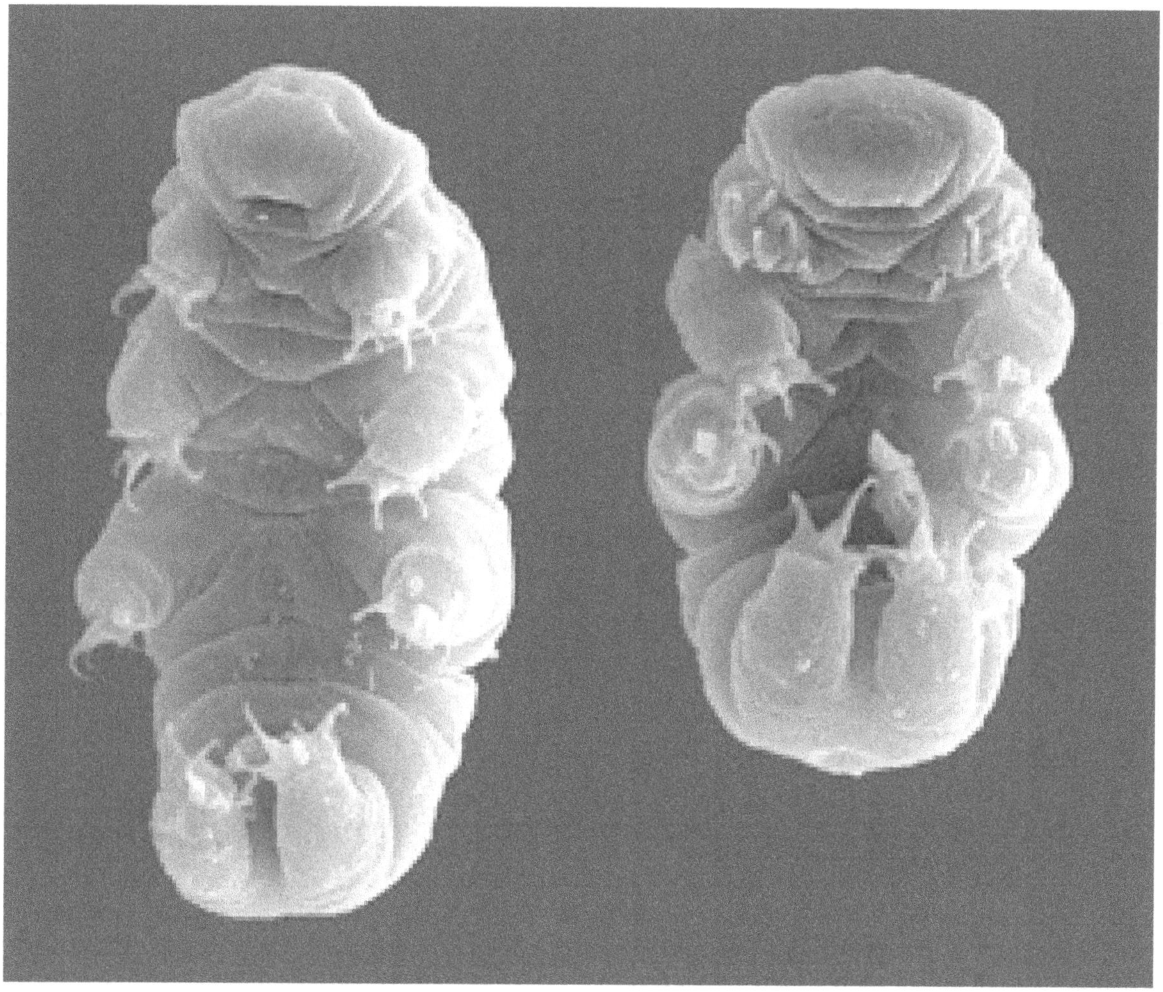

Hypsibius dujardini *imaged with a scanning electron microscope*

for at least 10 days.[32] Some species can also withstand pressure of 6,000 atmospheres, which is nearly six times the pressure of water in the deepest ocean trench, the Mariana trench.[22]

- Dehydration – the longest that living tardigrades have been shown to survive in a dry state is nearly 10 years,[10][33] although there is one report of leg movement, not generally considered "survival",[34] in a 120-year-old specimen from dried moss.[35] When exposed to extremely low temperatures, their body composition goes from 85% water to only 3%. As water expands upon freezing, dehydration ensures the tardigrades do not get ripped apart by the freezing ice.[36]
- Radiation – tardigrades can withstand 1,000 times more radiation than other animals,[37] median lethal doses of 5,000 Gy (of gamma rays) and 6,200 Gy (of heavy ions) in hydrated animals (5 to 10 Gy could be fatal to a human).[38] The only explanation found in earlier experiments for this ability was that their lowered water state provides fewer reactants for the ionizing radiation.[39] However, subsequent research found that tardigrades, when hydrated, still remain highly resistant to shortwave UV radiation in comparison to other animals, and that one factor for this is their ability to efficiently repair damage to their DNA resulting from that exposure.[40]

Irradiation of tardigrade eggs collected directly from a natural substrate (moss) showed a clear dose-related response, with a steep decline in hatchability at doses up to 4 kGy, above which no eggs hatched.[41] The eggs were more tolerant to radiation late in development. No eggs irradiated at the early developmental stage hatched, and only one egg at middle stage hatched, while eggs irradiated in the late stage hatched at a rate indistinguishable from controls.[41]

- Environmental toxins – tardigrades can undergo chemobiosis, a cryptobiotic response to high levels of environmental toxins. However, as of 2001, these laboratory results have yet to be verified.[34][35]

- Outer space – tardigrades are the first known animal to survive in space. On September 2007, dehydrated tardigrades were taken into low Earth orbit on the FOTON-M3 mission carrying the BIOPAN astrobiology payload. For 10 days, groups of tardigrades were exposed to the hard vacuum of outer space, or vacuum and solar UV radiation.[3][42][43] After being rehydrated back on Earth, over 68% of the subjects protected from high-energy UV radiation revived within 30 minutes following rehydration, but subsequent mortality was high; many of these produced viable embryos.[32][44] In contrast, hydrated samples exposed to the combined effect of vacuum and full solar UV radiation had significantly reduced survival, with only three subjects of *Milnesium tardigradum* surviving.[32] In May 2011, Italian scientists sent tardigrades on board the International Space Station along with extremophiles on STS-134, the final flight of Space Shuttle *Endeavour*.[45][46][47] Their conclusion was that microgravity and cosmic radiation "did not significantly affect survival of tardigrades in flight, confirming that tardigrades represent a useful animal for space research."[48] In November 2011, they were among the organisms to be sent by the US-based Planetary Society on the Russian Fobos-Grunt mission's Living Interplanetary Flight Experiment to Phobos; however, the launch failed. It remains unclear whether tardigrade specimens survived the failed launch.

1.6 Taxonomy

See also: List of bilaterial animal orders

Scientists have conducted morphological and molecular studies to understand how tardigrades relate to other lineages of ecdysozoan animals. Two plausible placements have been proposed: tardigrades are either most closely related to Arthropoda ± Onychophora, or tardigrades are most closely related to nematodes. Evidence for the former is a common result of morphological studies; evidence of the latter is found in some molecular analyses.

The latter hypothesis has been rejected by recent microRNA and expressed sequence tag analyses.[49] Apparently, the grouping of tardigrades with nematodes found in a number of molecular studies is a long branch attraction artifact. Within the arthropod group (called panarthropoda and comprising onychophora, tardigrades and euarthropoda), three patterns of relationship are possible: tardigrades sister to onychophora plus arthropods (the lobopodia hypothesis); onychophora sister to tardigrades plus arthropods (the tactopoda hypothesis); and onychophora sister to tardigrades.[50] Recent analyses indicate that the panarthropoda group is monophyletic, and that tardigrades are a sister group of Lobopodia, the lineage consisting of arthropods and Onychophora.[49][51]

The minute sizes of tardigrades and their membranous integuments make their fossilization both difficult to detect and highly unusual. The only known fossil specimens are those from mid-Cambrian deposits in Siberia and a few rare specimens from Cretaceous amber.[52]

The Siberian tardigrade fossils differ from living tardigrades in several ways. They have three pairs of legs rather than four, they have a simplified head morphology, and they have no posterior head appendages. But they share with modern tardigrades their columnar cuticle construction.[53] Scientists think they represent a stem group of living tardigrades.[52]

Rare specimens in Cretaceous amber have been found in two North American locations. *Milnesium swolenskyi*, from New Jersey, is the older of the two; its claws and mouthparts are indistinguishable from the living *M. tardigradum*. The other specimens from amber are from western Canada, some 15–20 million years younger than *M. swolenskyi*. One of the two specimens from Canada has been given its own genus and family, *Beorn leggi* (the genus named by Cooper after the character Beorn from *The Hobbit* by J. R. R. Tolkien and the species named after his student William M. Legg); however, it bears a strong resemblance to many living specimens in the family Hypsibiidae.[52][54]

Aysheaia from the middle Cambrian Burgess shale has been proposed as a sister-taxon to an arthropod-tardigrade clade.[55]

Tardigrades have been proposed to be among the closest living relatives of the Burgess Shale oddity *Opabinia*.[56]

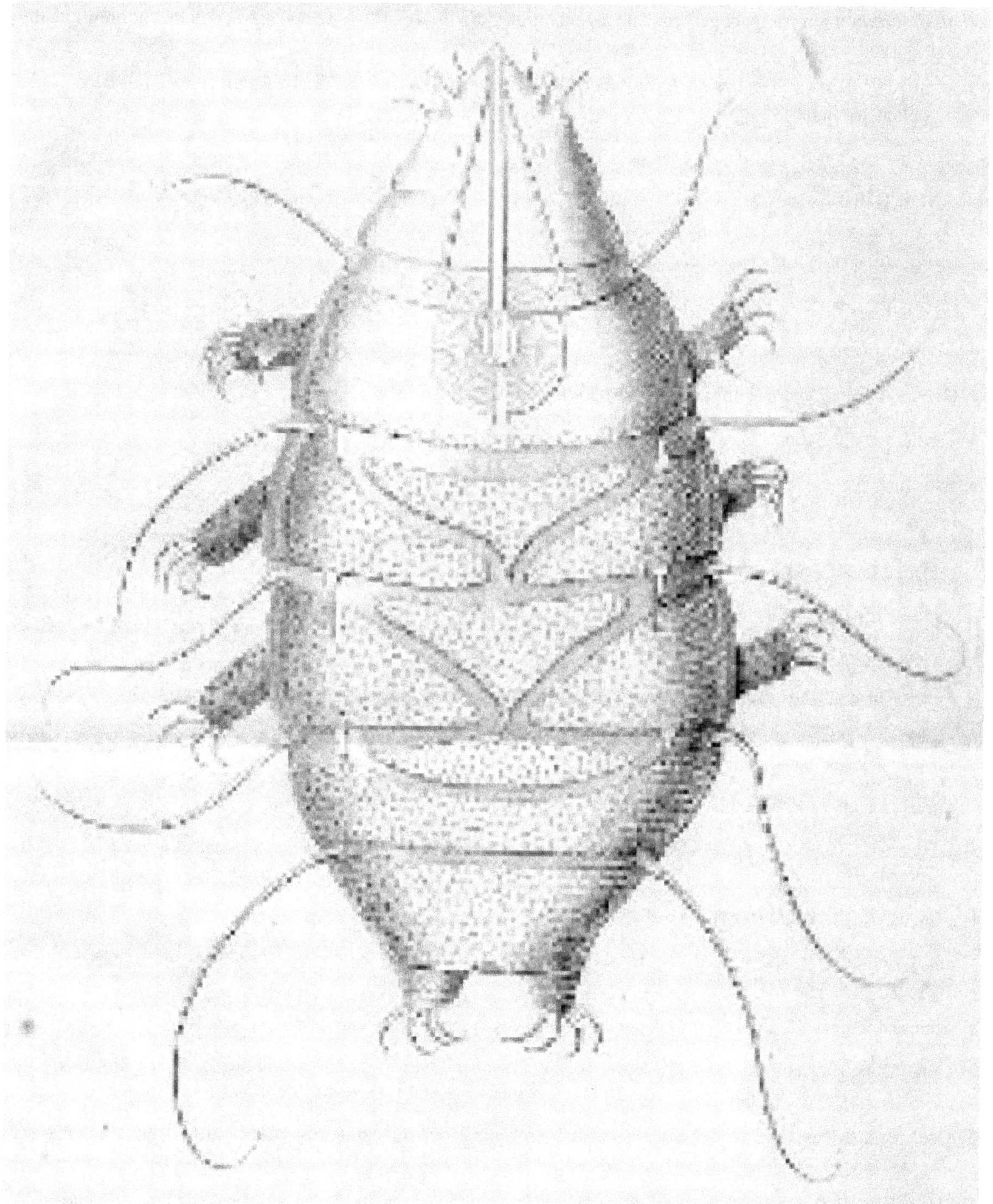

Illustration of Echiniscus *sp. from 1861*

1.7 Genomes and genome sequencing

Tardigrade genomes vary in size, from about 75 to 800 megabase pairs of DNA.[57] The genome of a tardigrade species, *Hypsibius dujardini*, is being sequenced at the Broad Institute.[58]

Hypsibius dujardini has a compact genome and a generation time of about two weeks, and it can be cultured indefinitely and cryopreserved.[59]

The genome of *R. varieornatus* has been reported to be sequenced but the results of this effort have not been published or made publicly available.[60]

1.8 See also

- List of microorganisms tested in outer space
- Living Interplanetary Flight Experiment - study of selected microorganisms in outer space
- *Mopsechiniscus franciscae* - tardigrade found in Victoria Land Antarctica
- Oncopoda - hypothetical animal group which includes Tardigrada
- Tardigrade animation - *Cosmos: A Spacetime Odyssey* (eps 2 & 6)
- Extremophile

1.9 References

[1] Budd, G.E. (2001). "Tardigrades as 'stem-group arthropods': the evidence from the Cambrian fauna". *Zool. Anz.* **240** (3–4): 265–279. doi:10.1078/0044-5231-00034.

[2] Miller William. "Tardigrades". American Scientist. Retrieved 2013-12-02.

[3] Simon, Matt (March 21, 2014). "Absurd Creature of the Week: The Incredible Critter That's Tough Enough to Survive in the vacuum of Space". *Wired*. Retrieved 2014-03-21.

[4] Copley, Jon (1999-10-23). "Indestructible". *New Scientist* (2209). Retrieved 2010-02-06.

[5] Bordenstein, Sarah. "Tardigrades (Water Bears)". *Microbial Life Educational Resources*. National Science Digital Library. Retrieved 2014-01-24.

[6] "Tardigrades". *Tardigrade*. Retrieved 2015-09-21.

[7] Dean, Cornelia (September 7, 2015). "The Tardigrade: Practically Invisible, Indestructible 'Water Bears'". *New York Times*. Retrieved September 7, 2015.

[8] Brennand, Emma (2011-05-17). "Tardigrades: Water bears in space". BBC. Retrieved 2013-05-31.

[9] Crowe, John H.; Carpenter, John F.; Crowe, Lois M. (October 1998). "The role of vitrification in anhydrobiosis". *Annual Review of Physiology* **60**. pp. 73–103. doi:10.1146/annurev.physiol.60.1.73. PMID 9558455.

[10] Guidetti, R. & Jönsson, K.I. (2002). "Long-term anhydrobiotic survival in semi-terrestrial micrometazoans". *Journal of Zoology* **257** (2): 181–187. doi:10.1017/S095283690200078X.

[11] Rampelotto, P. H. (2010). "Resistance of microorganisms to extreme environmental conditions and its contribution to Astrobiology". *Sustainability* **2**: 1602–1623. doi:10.3390/su2061602.

[12] Rothschild, L.J.; Mancinelli, R.L. (2001). "Life in extreme environments".*Nature***409**(6823): 1092–1101. doi:10.1038/3505. PMID 11234023.

[13] Florida Entomologist: On Water Bears

[14] Shaw, Michael W. "How to Find Tardigrades". tardigrades.us. Retrieved 2013-01-14.

[15] "Tardigrada (water bears, tardigrades)". biodiversity explorer. Retrieved 2013-05-31.

[16] Bordenstein, Sarah (December 17, 2008). "Tardigrades (Water Bears)". Carleton College. Retrieved 2012-09-16.

[17] Zhang, Z.-Q. (2011). “Animal biodiversity: An introduction to higher-level classification and taxonomic richness” (PDF). *Zootaxa* **3148**: 7–12.

[18] Degma, P., Bertolani, R. & Guidetti, R. 2009–2011. Actual checklist of Tardigrada species. Ver. 18: 27-04-2011. tardigrada

[19] Hogan, C. Michael. 2010. “Extremophile”. eds. E.Monosson and C.Cleveland. *Encyclopedia of Earth.* National Council for Science and the Environment, washington DC

[20] Staff (September 29, 2015). “Researchers discover new tiny organism, name it for Wyeths”. *AP News.* Retrieved September 29, 2015.

[21] Goldstein, B. & Blaxter, M. (2002). “Quick Guide: Tardigrades”. *Current Biology* **12** (14): R475. doi:10.1016/S0960-9822(02)00959-4.

[22] Seki, Kunihiro; Toyoshima, Masato (1998-10-29). “Preserving tardigrades under pressure”. *Nature* **395** (6705): 853–854. doi:10.1038/27576.

[23] Kinchin, Ian M. (1994) *The Biology of Tardigrades*, Ashgate Publishing

[24] Barnes, Robert D. (1982). *Invertebrate Zoology*. Philadelphia, PA: Holt-Saunders International. pp. 877–880. ISBN 0-03-056747-5.

[25] Zantke, Juliane; Wolff, Carsten; Scholtz, Gerhard (2008). “Three-dimensional reconstruction of the central nervous system of Macrobiotus hufelandi (Eutardigrada, Parachela): implications for the phylogenetic position of Tardigrada” (PDF). *Zoomorphology* **127** (1): 21–26. doi:10.1007/s00435-007-0045-1.

[26] Greven, H. (Dec 2007). “Comments on the eyes of tardigrades”. *Arthropod structure & development* **36** (4): 401–407. doi:10.1016/j.asd.2007.06.003. ISSN 1467-8039. PMID 18089118.

[27] Morgan, Clive I. (1977). “Population Dynamics of two Species of Tardigrada, *Macrobiotus hufelandii* (Schultze) and *Echiniscus (Echiniscus) testudo* (Doyere), in Roof Moss from Swansea”. *The Journal of Animal Ecology* (British Ecological Society) **46** (1): 263–279. doi:10.2307/3960. JSTOR 3960.

[28] Lindahl, K. (2008-03-15). “Tardigrade Facts”.

[29] Piper, Ross (2007), *Extraordinary Animals: An Encyclopedia of Curious and Unusual Animals*, Greenwood Press.

[30] Horikawa, Daiki D. (2012). Alexander V. Altenbach; Joan M. Bernhard; Joseph Seckbach, eds. *Anoxia Evidence for Eukaryote Survival and Paleontological Strategies.* (21st ed.). Springer Netherlands. pp. 205–217. ISBN 978-94-007-1895-1. Retrieved 2012-01-21.

[31] Becquerel, P. (1950). “La suspension de la vie au dessous de 1/20 K absolu par demagnetization adiabatique de l'alun de fer dans le vide les plus eléve”. *C. R. Hebd. Séances Acad. Sci. Paris* (in French) **231**: 261–263.

[32] Jönsson, K. Ingemar; Rabbow, Elke; Schill, Ralph O.; Harms-Ringdahl, Mats & Rettberg, Petra (9 September 2008). “Tardigrades survive exposure to space in low Earth orbit”. *Current Biology* **18** (17): R729–R731. doi:10.1016/j.cub.2008.06.048. PMID 18786368.

[33] Crowe, John H.; Carpenter, John F.; Crowe, Lois M. (October 1998). “The role of vitrification in anhydrobiosis”. *Annual Review of Physiology* **60**. pp. 73–103. doi:10.1146/annurev.physiol.60.1.73. PMID 9558455.

[34] Jönsson, K. Ingemar & Bertolani, R. (2001). “Facts and fiction about long-term survival in tardigrades”. *Journal of Zoology* **255**: 121–123. doi:10.1017/S0952836901001169.

[35] Franceschi, T. (1948). “Anabiosi nei tardigradi”. *Bolletino dei Musei e degli Istituti Biologici dell'Università di Genova* **22**: 47–49.

[36] Kent, Michael (2000), *Advanced Biology*, Oxford University Press

[37] Radiation tolerance in the tardigrade Milnesium tardigradum

[38] Horikawa DD; Sakashita T; Katagiri C; Watanabe M; Kikawada T; Nakahara Y; Hamada N; Wada S; et al. (2006). “Radiation tolerance in the tardigrade Milnesium tardigradum”. *International Journal of Radiation Biology* **82** (12): 843–8. doi:10.1080/09553000600972956. PMID 17178624.

[39] Horikawa, Daiki D.; Sakashita, Tetsuya; Katagiri, Chihiro; Watanabe, Masahiko; Kikawada, Takahiro; Nakahara, Yuichi; Hamada, Nobuyuki; Wada, Seiichi; et al. (1 January 2006). “Radiation tolerance in the tardigrade”. *International Journal of Radiation Biology* **82** (12): 843–848. doi:10.1080/09553000600972956. PMID 17178624.

[40] Horikawa, Daiki D. “UV Radiation Tolerance of Tardigrades”. NASA.com. Retrieved 2013-01-15.

[41] Jönsson, Ingemar; Beltran-Pardo, Eliana; Haghdoost, Siamak; Wojcik, Andrzej; Bermúdez-Cruz, Rosa Maria; Bernal Villegas, Jaime E.; Harms-Ringdahl, Mats (2013). “Tolerance to gamma-irradiation in eggs of the tardigrade Richtersius coronifer depends on stage of development”. *Journal of Limnology* **71** (12th International Symposium on Tardigrada). Retrieved 2013-08-05.

[42] “Creature Survives Naked in Space”. Space.com. 8 September 2008. Retrieved 2011-12-22.

[43] Mustain, Andrea (22 December 2011). “Weird wildlife: The real land animals of Antarctica”. MSNBC. Retrieved 2011-12-22.

[44] Courtland, Rachel (2008-09-08). "'Water bears' are first animal to survive space vacuum”. *New Scientist*. Retrieved 2011-05-22.

[45] NASA Staff (2011-05-17). “BIOKon In Space (BIOKIS)". NASA. Retrieved 2011-05-24.

[46] Brennard, Emma (2011-05-17). “Tardigrades: Water bears in space”. BBC. Retrieved 2011-05-24.

[47] “Tardigrades: Water bears in space”. BBC Nature. 2011-05-17.

[48] Rebecchi, L.; et al. “Two Tardigrade Species On Board the STS-134 Space Flight” in “International Symposium on Tardigrada, 23-26 July 2012” (PDF). p. 89. Retrieved 2013-01-14.

[49] Campbell, Lahcen; Omar Rota-Stabelli; Gregory D. Edgecombe; Trevor Marchioro; Stuart J. Longhorn; Maximilian J. Telford; Hervé Philippe; Lorena Rebecchi; et al. (2011). “MicroRNAs and phylogenomics resolve the relationships of Tardigrada and suggest that velvet worms are the sister group of Arthropoda”. *PNAS Early Edition* **108** (38): 15920–4. doi:10.1073/pnas.1105 499108.PMC3179045. PMID 21896763.

[50] Telford, Maximilian; Sarah J Bourlat; Andrew Economou; Daniel Papillon & Omar Rota-Stabelli (April 2008). “The evolution of the Ecdysozoa”. *Phil. Trans. R. Soc. B* **363** (1496): 1529–1537. doi:10.1098/rstb.2007.2243. PMC 2614232. PMID 18192181. Retrieved 2013-09-09.

[51] “Sequencing of Tardigrade Genome” (PDF). The Royal Society. 2003. Retrieved 2013-05-31.

[52] Grimaldi, David A.; Engel, Michael S. (2005). *Evolution of the Insects*. Cambridge University Press. pp. 96–97. ISBN 0-521-82149-5.

[53] Budd, G. (2001). “Tardigrades as 'Stem-Group Arthropods': The Evidence from the Cambrian Fauna”. *Zoologischer Anzeiger - A Journal of Comparative Zoology* **240** (3–4): 265–279. doi:10.1078/0044-5231-00034. ISSN 0044-5231.

[54] Cooper, Kenneth W. (1964). “The first fossil tardigrade: *Beorn leggi*, from Cretaceous Amber”. *Psyche – Journal of Entomology* **71** (2): 41. doi:10.1155/1964/48418.

[55] Fortey, Richard A.; Thomas, Richard H. (2001). *Arthropod Relationships*. Chapman & Hall. p. 383. ISBN 0412754207.

[56] Budd, G.E. (1996). “The morphology of *Opabinia regalis* and the reconstruction of the arthropod stem-group”. *Lethaia* **29** (1): 1–14. doi:10.1111/j.1502-3931.1996.tb01831.x.

[57] “Genome Size of Tardigrades”.

[58] Entrez. “Genome Projects for Hypsibius dujardini”.

[59] Gabriel, W; McNuff, Robert; Patel, Sapna K.; Gregory, T. Ryan; Jeck, William R.; Jones, Corbin D.; Goldstein, Bob (2007). “The tardigrade Hypsibius dujardini, a new model for studying the evolution of development”. *Developmental Biology* **312** (2): 545–559. doi:10.1016/j.ydbio.2007.09.055. PMID 17996863.

[60] Horikawa D.; et al. (2013). “Analysis of DNA Repair and Protection in the Tardigrade Ramazzottius varieornatus and Hypsibius dujardini after Exposure to UVC Radiation.”. *PLoS ONE* **8** (8(6): e64793.): e64793. doi:10.1371/journal.pone.0064793. PMC 3675078. PMID 23762256.

1.10 External links

- Tardigrade water bear NYTimes, 2015
- Tardigrada Newsletter
- Tardigrades – Pictures and Movies
- The Edinburgh Tardigrade project
- Instructions for finding tardigrades
- The incredible water bear!
- Tardigrade Reference Center
- Tardigrades in space
- Tardigrade data and analysis
- A short film about tardigrade research from NPR's Science Friday
- Tardigrada at the Tree of Life Web Project
- Swiss Center of Tardigrade Research – Ecology, Physiology and Evolutionary Biology of Tardigrades
- Tardigrade in Moss
- Video (07:54) - First Animal to Survive in Space
- Video (00:38) - Tardigrade Movement in Water
- Tardigrades are so tough, they can survive outer space (March 2015). *BBC*
- The International Society of Tardigrade Hunters

Chapter 2

Micro-animal

Micro-animals are animals so small that they can only be visually observed under a microscope. Mostly these microorganisms are multicellular but none are vertebrates. Microscopic arthropods include dust mites, spider mites, and some crustaceans such as copepods and the cladocera.Another common group of microscopic animals are the rotifers, which are filter feeders that are usually found in fresh water. Some nematode species are microscopic,[1] as well as many loricifera, including the recently discovered anaerobic species, which spend their entire lives in an anoxic environment.[2][3] The tardigrades, water-dwelling micro-animals, can survive extreme living conditions that they are not evolved for and have survived the hard vacuum of space, solar and UVradiation in an astrobiology experiment into space.[4]

2.1 References

[1] http://www.mt.nrcs.usda.gov/technical/ecs/agronomy/technotes/agtechnoteMT73-150.3/microscopic.html

[2] "Animals thrive without oxygen at sea bottom". *nature.com*.

[3] "Briny deep basin may be home to animals thriving without oxygen". *Science News*.

[4] Simon, Matt (March 21, 2014). "Absurd Creature of the Week: The Incredible Critter That's Tough Enough to Survive in Space". *Wired*. Retrieved 2014-03-21.

A microscopic arachnid Lorryia formosa

Chapter 3

Ecdysozoa

Ecdysozoa /ˌɛkdɪsəˈzoʊ.ə/ is a group of protostome animals,[1] including Arthropoda (insects, chelicerata, crustaceans, and myriapods), nematoda, and several smaller phyla. They were first defined by Aguinaldo *et al.* in 1997, based mainly on phylogenetic trees constructed using 18S ribosomal RNA genes.[2] A large study in 2008 by Dunn *et al.* strongly supported the Ecdysozoa as a clade, that is, a group consisting of a common ancestor and all its descendants.[3]

The group is also supported by morphological characters, and includes all animals that shed their exoskeleton (see ecdysis).

The group was initially contested by a significant minority of biologists. Some argued for groupings based on more traditional taxonomic techniques,[4] while others contested the interpretation of the molecular data.[5][6]

3.1 Group characters

See also: List of bilaterial animal orders

The most notable characteristic shared by ecdysozoans is a three-layered cuticle (four in Tardigrada[7]) composed of organic material, which is periodically molted as the animal grows. This process of molting is called ecdysis, and gives the group its name. The ecdysozoans lack locomotory cilia and produce mostly amoeboid sperm, and their embryos do not undergo spiral cleavage as in most other protostomes. Ancestrally, the group exhibited sclerotized teeth within the foregut, and a ring of spines around the mouth opening, though these features have been secondarily lost in certain groups.[8]

3.2 Group membership

The Ecdysozoa include the following phyla: Arthropoda, Onychophora, Tardigrada, Kinorhyncha, Priapulida, Loricifera, Nematoda, and Nematomorpha. A few other groups, such as the gastrotrichs, have been considered possible members but lack the main characters of the group, and are now placed elsewhere. The Arthropoda, Onychophora, and Tardigrada have been grouped together as the Panarthropoda because they are distinguished by segmented body plans.[9] Dunn *et al.* in 2008 suggested that the tardigrada could be grouped along with the nematodes, leaving Onychophora as the sister group to the arthropods.[3]

The non-panarthropod members of Ecdysozoa have been grouped as Cycloneuralia but they are more usually considered paraphyletic.

3.3 Alternative groupings that used to be favoured before Ecdysozoa gained widespread acceptance

3.3.1 Articulata hypothesis

The grouping proposed by Aguinaldo *et al.* is almost universally accepted, although Wikipedia alleges that some zoologists still hold to the original view that Panarthropoda should be classified with Annelida in a group called the Articulata, and that Ecdysozoa are polyphyletic. Nielsen has suggested that a possible solution is to regard Ecdysozoa as a sister-group of Annelida.,[10] though later considered them unrelated.[11] Inclusion of the roundworms within the Ecdysozoa was initially contested[5][12] but since 2003, a broad consensus has formed supporting the Ecdysozoa [13] and in 2011 the Darwin–Wallace Medal was awarded for the discovery of the New Animal Phylogeny consisting of the *Ecdysozoa*, the *Lophotrochozoa*, and the *Deuterostomia*.

3.3.2 Coelomata hypothesis

Before Ecdysozoa, one of the prevailing theories for the evolution of the bilateral animals was based on the morphology of their body cavities. There were three types, or grades of organization: the Acoelomata (no coelom), the Pseudocoelomata (partial coelom), and the Eucoelomata (true coelom). Adoutte and coworkers were among the first to strongly support the Ecdysozoa.[14] With the introduction of molecular phylogenetics, the coelomate hypothesis was abandoned, although some molecular, phylogenetic support for the Coelomata continued until as late as 2005.[15]

3.4 References

[1] Telford MJ, Bourlat SJ, Economou A, Papillon D, Rota-Stabelli O (April 2008). "The evolution of the Ecdysozoa". *Philos. Trans. R. Soc. Lond., B, Biol. Sci.* **363** (1496): 1529–37. doi:10.1098/rstb.2007.2243. PMC 2614232. PMID 18192181.

[2] Aguinaldo, A. M. A.; J. M. Turbeville; L. S. Linford; M. C. Rivera; J. R. Garey; R. A. Raff; J. A. Lake (1997). "Evidence for a clade of nematodes, arthropods, and other moulting animals". *Nature* **387** (6632): 489–493. Bibcode:1997Natur.387R.489A. doi:10.1038/387489a0. PMID 9168109.

[3] Dunn, CW; Hejnol, A; Matus, DQ; Pang, K; Browne, WE; Smith, SA; Seaver, E; Rouse, GW; et al. (2008). "Broad phylogenomic sampling improves resolution of the animal tree of life". *Nature* **452** (7188): 745–749. Bibcode:2008Natur.452..745D. doi:10.1038/nature06614. PMID 18322464.

[4] Nielsen, Claus (1995). *Animal Evolution: Interrelationships of the Living Phyla*. Oxford University Press. ISBN 978-0-19-850682-9.

[5] Blair, J. E.; Kazuho Ikeo; Takashi Gojobori; S. Blair Hedges (2002). "The evolutionary position of nematodes". *BMC Evolutionary Biology* **2**: 7. doi:10.1186/1471-2148-2-7. PMC 102755. PMID 11985779.

[6] Wägele, J. W.; T. Erikson; P. Lockhart; B. Misof (1999). "The Ecdysozoa: Artifact or monophylum?". *Journal of Zoological Systematics and Evolutionary Research* **37** (4): 211–223. doi:10.1111/j.1439-0469.1999.tb00985.x.

[7] Barnes, Robert D. (1982). *Invertebrate Zoology*. Philadelphia, PA: Holt-Saunders International. pp. 877–880. ISBN 0-03-056747-5.

[8] Smith, Martin R.; Caron, Jean-Bernard (2015). "Hallucigenia's head and the pharyngeal armature of early ecdysozoans". *Nature* **523** (7558): 75. Bibcode:2015Natur.523...75S. doi:10.1038/nature14573. PMID 26106857.

[9] Paleos Invertebrates: Panarthropoda – URL retrieved February 17, 2007

[10] Nielsen, C. (2003) Proposing a solution to the Articulata–Ecdysozoa controversy. *Zoologica Scripta* 32:5, 475-482

[11] Nielsen, Claus (2012). *Animal Evolution: Interrelationships of the Living Phyla 3rd ed.* Oxford University Press. ISBN 978-0-19-960603-0.

[12] Wägele, J. W.; B. Misof (2001). "On quality of evidence in phylogeny reconstruction: a reply to Zrzavý's defence of the 'Ecdysozoa' hypothesis". *J. Zool. Syst. Evol. Research* **39** (3): 165–176. doi:10.1046/j.1439-0469.2001.00177.x.

[13] Maximilian J Telford and D. Timothy J Littlewood (2008). "The evolution of the animals: introduction to a Linnean tercentenary celebration". *Phil. Trans. R. Soc. B* **363**: 1421–1424. doi:10.1098/rstb.2007.2231.

[14] Adoutte, A.; Balavoine, G.; Lartillot, N.; Lespinet, O.; Prud'homme, B.; de Rosa, R. (25 April 2000). "Special Feature: The new animal phylogeny: Reliability and implications". *Proceedings of the National Academy of Sciences* **97** (9): 4453–4456. doi:10.1073/pnas.97.9.4453. PMC 34321. PMID 10781043. Retrieved 5 April 2013.

[15] Philip, G.K.; C.J. Creevey; J.O. McInerney (9 February 2005). "The Opisthokonta and the Ecdysozoa May Not Be Clades: Stronger Support for the Grouping of Plant and Animal than for Animal and Fungi and Stronger Support for the Coelomata than Ecdysozoa" (PDF). *Molecular Biology and Evolution* **22** (5): 1175–1184. doi:10.1093/molbev/msi102. Retrieved 5 April 2013.

3.5 External links

- UCMP-Ecdysozoa introduction
- http://www.palaeos.com/Kingdoms/Animalia/Ecdysozoa.html
- http://www.nematodes.org/tardigrades/Tardigrades_and_Ecdysozoa.html
- http://chuma.cas.usf.edu/~{ }garey/articulata.html
- http://chuma.cas.usf.edu/~{ }garey/essential.html
- http://www.nematomorpha.net

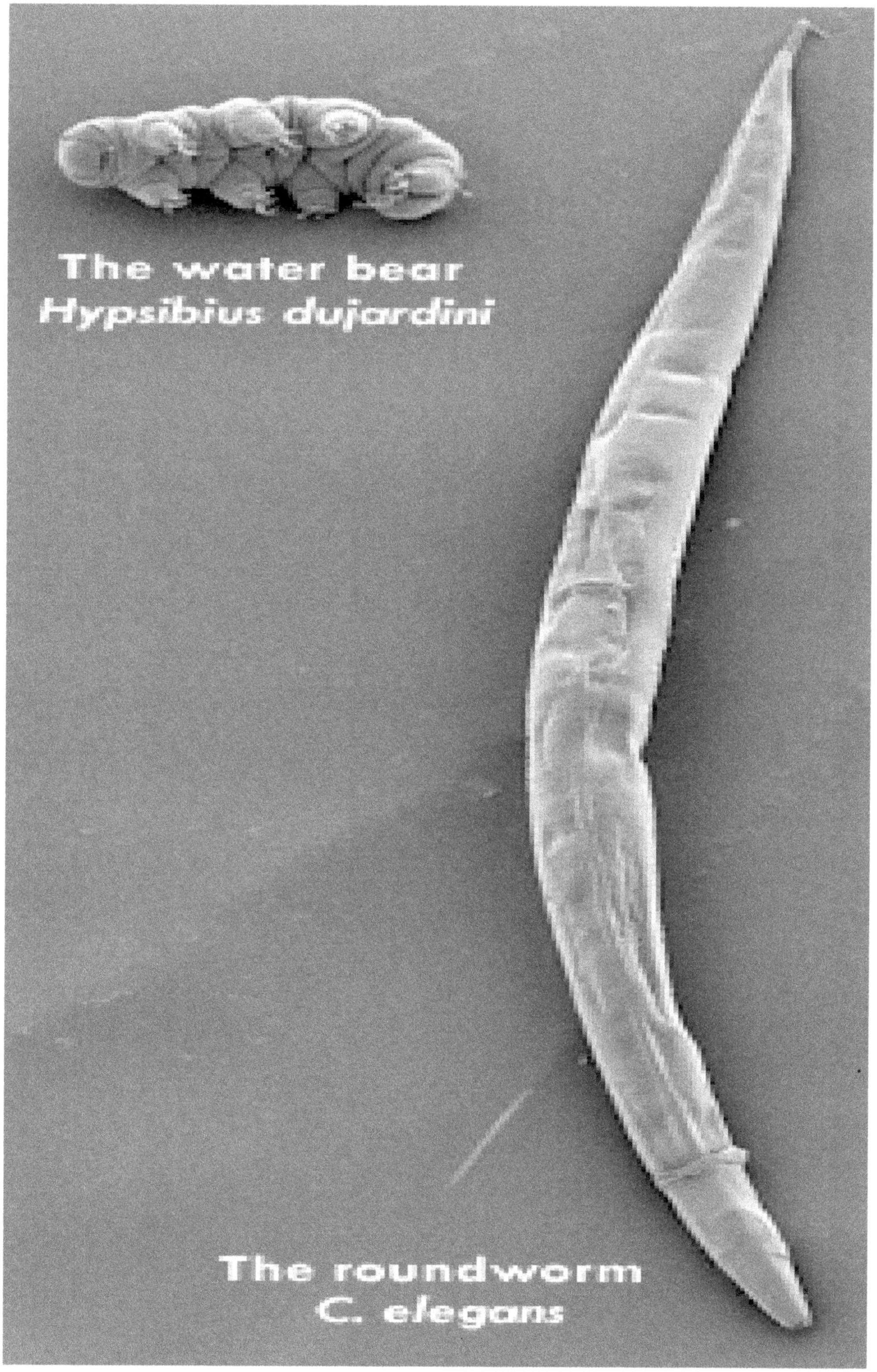

A tardigrade (water bear) and a nematode (roundworm)

Chapter 4

Cambrian

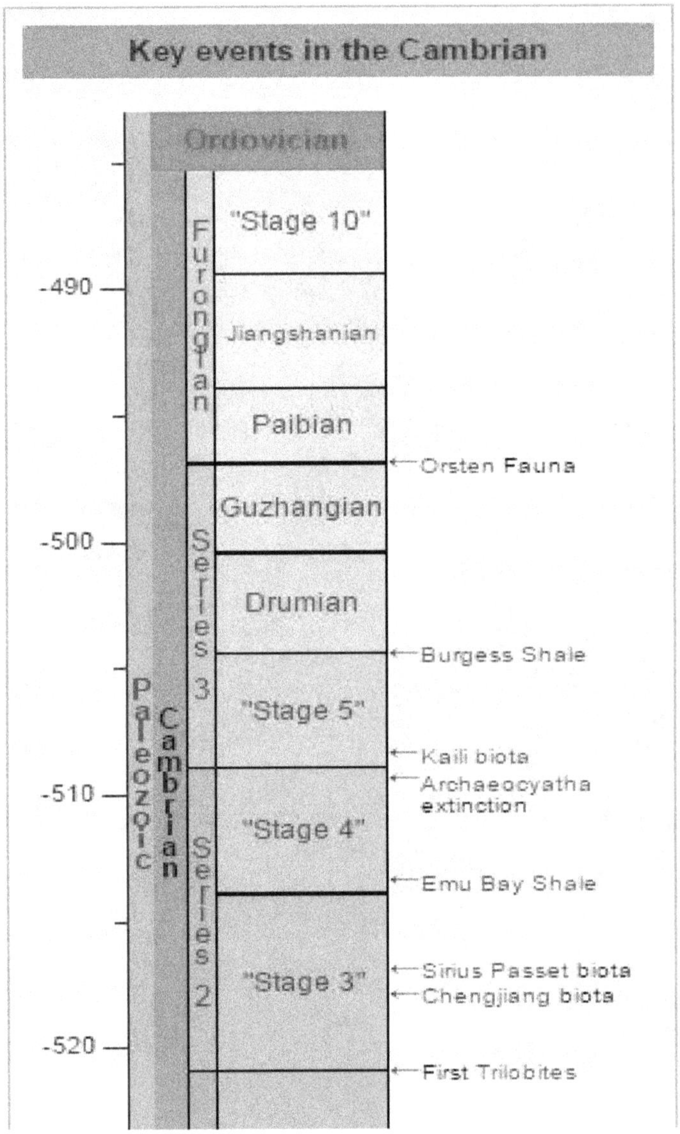

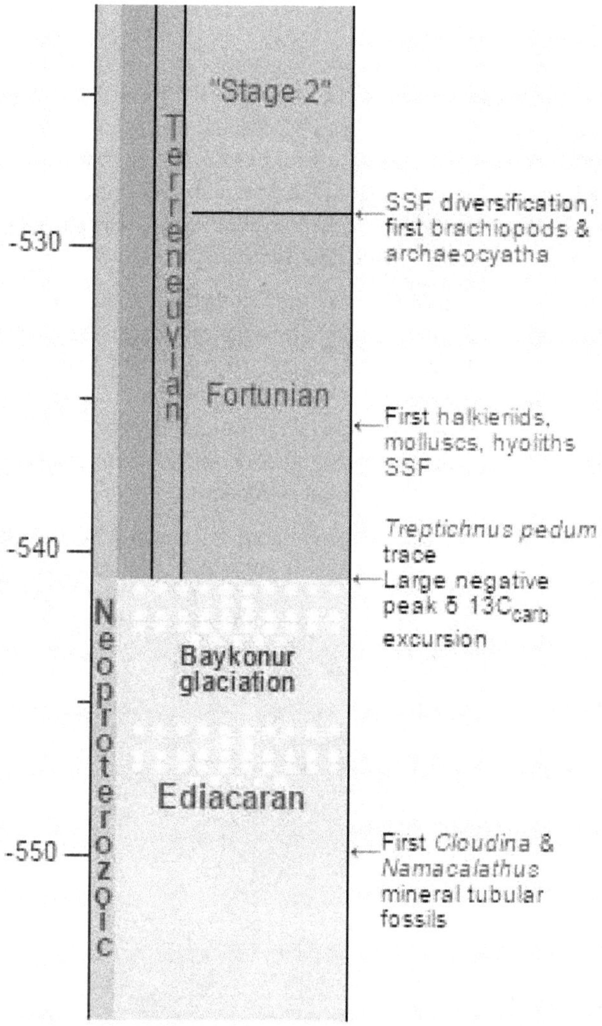

Stratigraphic scale of the ICS subdivisions and Precambrian/Cambrian boundary.

The **Cambrian** (/ˈkæmbriən/ or /ˈkeɪmbriən/) is the first geological period of the Paleozoic Era,[5] lasting from 541.0 ± 1.0 to 485.4 ± 1.9 million years ago (mya) and is succeeded by the Ordovician.[6] Its subdivisions, and indeed its base, are somewhat in flux. The period was established (as "Cambrian series") by Adam Sedgwick,[5] who named it after Cambria, the Latinised form of *Cymru*, the Welsh name for Wales, where Britain's Cambrian rocks are best exposed.[7][8][9] The Cambrian is unique in its unusually high proportion of lagerstätte sedimentary deposits. These are sites of exceptional preservation, where "soft" parts of organisms are preserved as well as their more resistant shells. This means that our understanding of the Cambrian biology surpasses that of some later periods.[10]

The Cambrian marked a profound change in life on Earth; prior to the Cambrian, the majority of living organisms on the whole were small, unicellular and simple; the Precambrian *Charnia* being exceptional. Complex, multicellular organisms gradually became more common in the millions of years immediately preceding the Cambrian, but it was not until this period that mineralized – hence readily fossilized – organisms became common.[11] The rapid diversification of lifeforms in the Cambrian, known as the Cambrian explosion, produced the first representatives of all modern animal phyla. Phylogenetic analysis has supported the view that during the Cambrian radiation, metazoa (animals) evolved monophyletically from a single common ancestor: flagellated colonial protists similar to modern choanoflagellates.

While diverse life forms prospered in the oceans, the land was comparatively barren – with nothing more complex than a microbial soil crust[12] and a few molluscs that emerged to browse on the microbial biofilm[13] Most of the continents were probably dry and rocky due to a lack of vegetation. Shallow seas flanked the margins of several continents created during the breakup of the supercontinent Pannotia. The seas were relatively warm, and polar ice was absent for much of the period.

The United States Federal Geographic Data Committee uses a "barred capital C" ⟨Є⟩ character similar to the capital letter Ukrainian Ye ⟨Є⟩ to represent the Cambrian Period.[14] The proper[15] Unicode character is U+A792 ⍰ latin capital letter c with bar.[16]

4.1 Stratigraphy

Further information: Stratigraphy of the Cambrian

Despite the long recognition of its distinction from younger Ordovician rocks and older Precambrian rocks, it was not until 1994 that this time period was internationally ratified. The base of the Cambrian is defined on a complex assemblage of trace fossils known as the *Treptichnus pedum* assemblage.[17] Nevertheless, the usage of *Treptichnus pedum*, a reference ichnofossil for the lower boundary of the Cambrian, for the stratigraphic detection of this boundary is always risky because of occurrence of very similar trace fossils belonging to the Treptichnids group well below the *T. pedum* in Namibia, Spain and Newfoundland, and possibly, in the western USA. The stratigraphic range of *T. pedum* overlaps the range of the Ediacaran fossils in Namibia, and probably in Spain.[18][19]

4.1.1 Subdivisions

The Cambrian period follows the Ediacaran and is followed by the Ordovician period. The Cambrian is divided into four epochs or series and ten ages or stages. Currently only two series and five stages are named and have a GSSP.

Because the international stratigraphic subdivision is not yet complete, many local subdivisions are still widely used. In some of these subdivisions the Cambrian is divided into three epochs with locally differing names – the Early Cambrian (Caerfai or Waucoban, 541 ± 0.3 to 509 ± 1.7 mya), Middle Cambrian (St Davids or Albertan, 509 ± 0.3 to 497 ± 1.7 mya) and Furongian (497 ± 0.3 to 485.4 ± 1.7 mya; also known as Late Cambrian, Merioneth or Croixan). Rocks of these epochs are referred to as belonging to the Lower, Middle, or Upper Cambrian.

Trilobite zones allow biostratigraphic correlation in the Cambrian.

Each of the local epochs is divided into several stages. The Cambrian is divided into several regional faunal stages of which the Russian-Kazakhian system is most used in international parlance:

*In Russian scientific thought the lower boundary of the Cambrian is suggested to be defined at the base of the Tommotian Stage which is characterized by diversification and global distribution of organisms with mineral skeletons and the appearance of the first Archaeocyath bioherms.[20][21][22]

4.1.2 Cambrian dating

The time range for the Cambrian has classically been thought to have been from about 542 million-years-ago (mya) to about 488 mya. The lower boundary of the Cambrian was traditionally set at the earliest appearance of trilobites and also unusual forms known as archeocyathids (literally "ancient cup") that are thought to be the earliest sponges and also the first non-microbial reef builders.

The end of the period was eventually set at a fairly definite faunal change now identified as an extinction event. Fossil discoveries and radiometric dating in the last quarter of the 20th century have called these dates into question. Date inconsistencies as large as 20 million years are common between authors. Framing dates of *ca.* 545 to 490 mya were proposed by the International Subcommission on Global Stratigraphy as recently as 2002.

A radiometric date from New Brunswick puts the end of the Lower Cambrian around 511 mya. This leaves 21 mya for the other two series/epochs of the Cambrian.

Archeocyathids from the Poleta formation in the Death Valley area

A more precise date of 542 ± 0.3 mya for the extinction event at the beginning of the Cambrian has recently been submitted.[23] The rationale for this precise dating is interesting in itself as an example of paleological deductive reasoning. Exactly at the Cambrian boundary there is a marked fall in the abundance of carbon-13, a "reverse spike" that paleontologists call an *excursion*. It is so widespread that it is the best indicator of the position of the Precambrian-Cambrian boundary in stratigraphic sequences of roughly this age. One of the places that this well-established carbon-13 excursion occurs is in Oman. Amthor (2003) describes evidence from Oman that indicates the carbon-isotope excursion relates to a mass extinction: the disappearance of distinctive fossils from the Precambrian coincides exactly with the carbon-13 anomaly. Fortunately, in the Oman sequence, so too does a volcanic ash horizon from which zircons provide a very precise age of 542 ± 0.3 mya (calculated on the decay rate of uranium to lead). This new and precise date tallies with the less precise dates for the carbon-13 anomaly, derived from sequences in Siberia and Namibia.

4.2 Paleogeography

Plate reconstructions suggest a global supercontinent, Pannotia, was in the process of breaking up early in the period,[24][25] with Laurentia (North America), Baltica, and Siberia having separated from the main supercontinent of Gondwana to form isolated land masses.[26] Most continental land was clustered in the Southern Hemisphere at this time, but was gradually drifting north.[26] Large, high-velocity rotational movement of Gondwana appears to have occurred in the Early Cambrian.[27]

With a lack of sea ice – the great glaciers of the Marinoan Snowball Earth were long melted[28] – the sea level was high, which led to large areas of the continents being flooded in warm, shallow seas ideal for thriving life. The sea levels fluctuated somewhat, suggesting there were 'ice ages', associated with pulses of expansion and contraction of a south polar

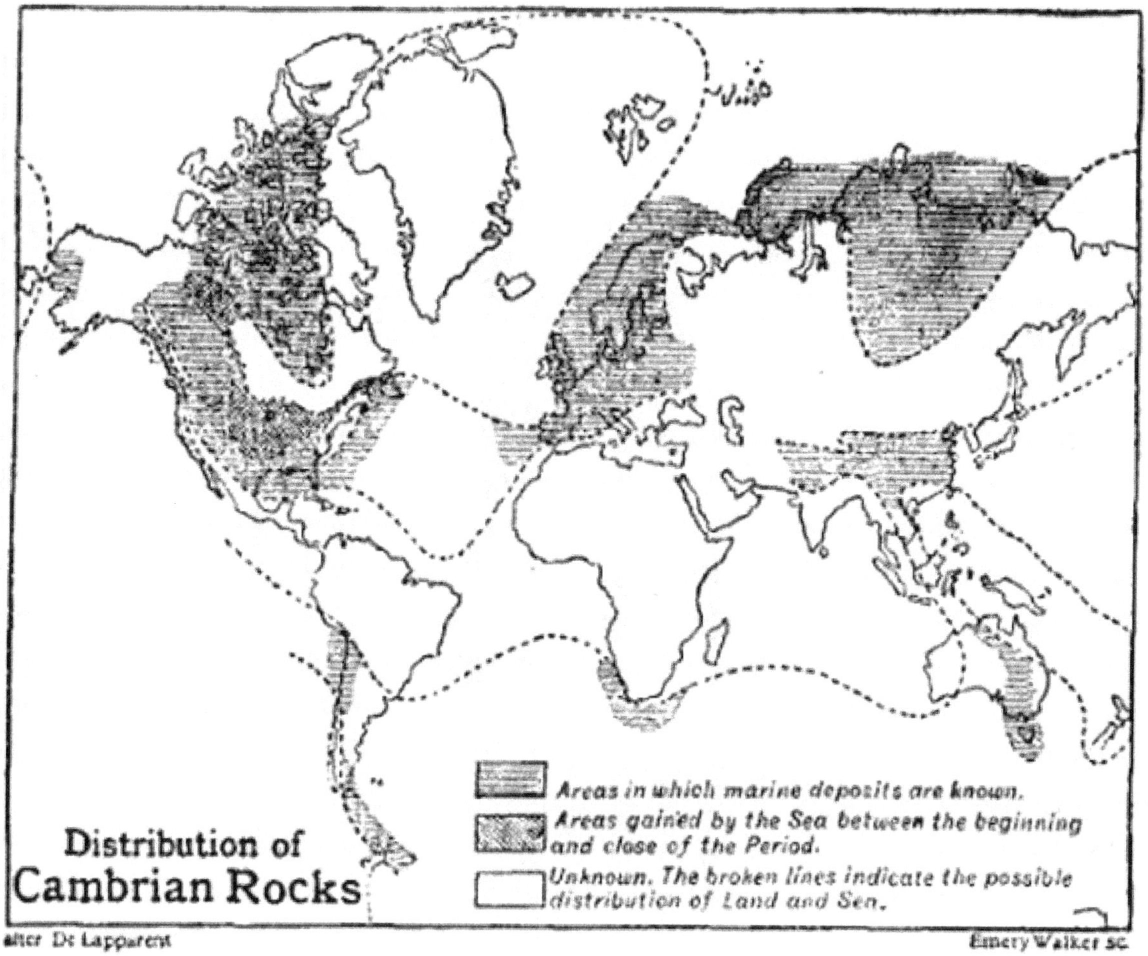

Distribution of Cambrian Rocks

ice cap.[29]

4.3 Climate

The Earth was generally cold during the early Cambrian, probably due to the ancient continent of Gondwana covering the South Pole and cutting off polar ocean currents. There were likely polar ice caps and a series of glaciations, as the planet was still recovering from an earlier Snowball Earth. It became warmer towards the end of the period; the glaciers receded and eventually disappeared, and sea levels rose dramatically. This trend would continue into the Ordovician period.

4.4 Flora

Although there were a variety of macroscopic marine plants (e.g. *Margaretia* and *Dalyia*), no true land plant (embryophyte) fossils are known from the Cambrian. However, biofilms and microbial mats were well developed on Cambrian tidal flats and beaches.,[30] and further inland were a variety of lichens, fungi and microbes forming microbial earth ecosystems, comparable with modern soil crust of desert regions, contributing to soil formation.[31][32]

A reconstruction of Margaretia dorus *from the Burgess Shale, which are believed to be green algae*

4.5 Fauna

Main article: Cambrian explosion

Most animal life during the Cambrian was aquatic, with trilobites assumed to be the dominant life form,[33] which has since proven to be incorrect. Arthropods in general were by far the most dominating animals in the ocean, but at the time, trilobites were only a minor part of the total arthropod diversity. What made them different from their relatives was their heavy armor, which fossilized far more easily than the fragile exoskeleton of other arthropods, leaving behind numerous preserved remains which give the false impression that they were the most abundant part of the fauna.[34] The period marked a steep change in the diversity and composition of Earth's biosphere. The incumbent Ediacaran biota suffered a mass extinction at the base of the period, which corresponds to an increase in the abundance and complexity of burrowing behaviour. This behaviour had a profound and irreversible effect on the substrate which transformed the seabed ecosystems. Before the Cambrian, the sea floor was covered by microbial mats. By the end of the period, burrowing animals had destroyed the mats through bioturbation, and gradually turned the seabeds into what they are today. As a consequence, many of those organisms that were dependent on the mats went extinct, while the other species adapted to the changed environment that now offered new ecological niches.[35] Around the same time there was a seemingly rapid appearance of representatives of all the mineralized phyla except the Bryozoa, which appear in the Lower Ordovician.[36] However, many of these phyla were represented only by stem-group forms; and since mineralized phyla generally have a benthic origin, they may not be a good proxy for (more abundant) non-mineralized phyla.[37]

While the early Cambrian showed such diversification that it has been named the Cambrian Explosion, this changed later in the period, when it was exposed to a sharp drop in biodiversity. About 515 million years ago, the number of species going extinct exceeded the amount of new species appearing. Five million years later, the number of genera had dropped from an earlier peak of about 600 to just 450. Also the speciation rate in many groups was reduced to between a fifth and a third of previous levels. 500 million years ago, oxygen levels fell dramatically in the oceans, leading to hypoxia, while the levels of poisonous hydrogen sulfide simultaneously increased, causing another extinction. The later half of Cambrian was surprisingly barren and show evidence of several rapid extinction events; the stromatolites which had been replaced by reef building sponges known as Archaeocyatha, returned once more as the archaeocyathids went extinct. This declining trend did not change before the Ordovician.[38][39]

Some Cambrian organisms ventured onto land, producing the trace fossils *Protichnites* and *Climactichnites*. Fossil evidence suggests that euthycarcinoids, an extinct group of arthropods, produced at least some of the *Protichnites*.[40][41] Fossils of the maker of *Climactichnites* have not been found; however, fossil trackways and resting traces suggest a large, slug-like mollusk.[42][43]

In contrast to later periods, the Cambrian fauna was somewhat restricted; free-floating organisms were rare, with the majority living on or close to the sea floor;[44] and mineralizing animals were rarer than in future periods, in part due to the unfavourable ocean chemistry.[44]

Many modes of preservation are unique to the Cambrian, resulting in an abundance of *Lagerstätten*.

- Stromatolites of the Pika Formation (Middle Cambrian) near Helen Lake, Banff National Park, Canada
- Trilobites were very common during this time
- *Anomalocaris* was an early marine predator, among the various arthropods of the time.
- *Pikaia* was an early chordate from the Middle Cambrian
- *Opabinia* was a creature with an unusual body plan; it was probably related to arthropods
- *Protichnites* were the trackways of arthropods that walked Cambrian beaches.

4.6 See also

- Cambro-Ordovician extinction event – circa 488 mya

- Dresbachian extinction event—circa 502 mya
- End Botomian extinction event—circa 517 mya
- List of fossil sites *(with link directory)*
- Type locality (geology), the locality where a particular rock type, stratigraphic unit, fossil or mineral species is first identified

4.7 References

[1] Image:Sauerstoffgehalt-1000mj.svg

[2] Image:Phanerozoic Carbon Dioxide.png

[3] Image:All palaeotemps.png

[4] Haq, B. U.; Schutter, SR (2008). "A Chronology of Paleozoic Sea-Level Changes".*Science***322**(5898): 64–8. Bibcode:2008Sci doi:10.1126/science.1161648. PMID 18832639.

[5] Chisholm, Hugh, ed. (1911). "Cambrian System". *Encyclopædia Britannica* (11th ed.). Cambridge University Press.

[6] "Stratigraphic Chart 2012" (PDF). International Stratigraphic Commission. Retrieved 9 November 2012.

[7] Sedgwick and R. I. Murchison (1835) "On the Silurian and Cambrian systems, exhibiting the order in which the older sedimentary strata succeed each other in England and Wales," Notices and Abstracts of Communications to the British Association for the Advancement of Science at the Dublin meeting, August 1835, pp. 59-61, in: *Report of the Fifth Meeting of the British Association for the Advancement of Science; held in Dublin in 1835* (1836). From p. 60: "Professor Sedgwick then described in descending order the groups of slate rocks, as they are seen in Wales and Cumberland. To the highest he gave the name of *Upper Cambrian* group. ... To the next inferior group he gave the name of *Middle Cambrian*. ... The *Lower Cambrian* group occupies the S.W. coast of Cærnarvonshire, ... "

[8] Sedgwick, A. (1852). "On the classification and nomenclature of the Lower Paleozoic rocks of England and Wales". *Q. J. Geol. Soc. Land.* **8**: 136–138. doi:10.1144/GSL.JGS.1852.008.01-02.20.

[9] "Chambers 21st Century Dictionary". *Chambers Dictionary* (Revised ed.). New Dehli: Allied Publishers. 2008. p. 203. ISBN 978-81-8424-329-1.

[10] Orr, P. J.; Benton, M. J.; Briggs, D. E. G. (2003). "Post-Cambrian closure of the deep-water slope-basin taphonomic window". *Geology* **31** (9): 769–772. Bibcode:2003Geo....31..769O. doi:10.1130/G19193.1. Retrieved 2008-06-28.

[11] Butterfield, N. J. (2007). "Macroevolution and macroecology through deep time".*Palaeontology***50**(1): 41–55. doi:10.1111/j 4983.2006.00613.x.

[12] Schieber, 2007, pp. 53–71.

[13] Seilacher, A.; Hagadorn, J.W. (2010). "Early Molluscan evolution: evidence from the trace fossil record". *Palaios* **25**: 565–575.

[14] Federal Geographic Data Committee, ed. (August 2006). *FGDC Digital Cartographic Standard for Geologic Map Symbolization FGDC-STD-013-2006* (PDF). U.S. Geological Survey for the Federal Geographic Data Committee. p. A–32–1. Retrieved 23 August 2010.

[15] Priest, Lorna A.; Iancu, Laurentiu; Everson, Michael (October 2010). "Proposal to Encode C WITH BAR" (PDF). Retrieved 6 April 2011.

[16] Unicode Character 'LATIN CAPITAL LETTER C WITH BAR' (U+A792). fileformat.info. Accessed 15 Jun 2015

[17] A. Knoll, M. Walter, G. Narbonne, and N. Christie-Blick (2004) "The Ediacaran Period: A New Addition to the Geologic Time Scale." Submitted on Behalf of the Terminal Proterozoic Subcommission of the International Commission on Stratigraphy.

[18] M.A. Fedonkin, B.S. Sokolov, M.A. Semikhatov, N.M.Chumakov (2007). "Vendian versus Ediacaran: priorities, contents, prospectives." In: edited by M. A. Semikhatov "The Rise and Fall of the Vendian (Ediacaran) Biota. Origin of the Modern Biosphere. Transactions of the International Conference on the IGCP Project 493, August 20–31, 2007, Moscow." Moscow: GEOS.

[19] A. Ragozina, D. Dorjnamjaa, A. Krayushkin, E. Serezhnikova (2008). "*Treptichnus pedum* and the Vendian-Cambrian boundary". 33 Intern. Geol. Congr. 6–14 August 2008, Oslo, Norway. Abstracts. Section HPF 07 Rise and fall of the Ediacaran (Vendian) biota. P. 183.

[20] A.Yu. Rozanov, V.V. Khomentovsky, Yu.Ya. Shabanov, G.A. Karlova, A.I. Varlamov, V.A. Luchinina, T.V. Pegel', Yu.E. Demidenko, P.Yu. Parkhaev, I.V. Korovnikov, N.A. Skorlotova (2008). "To the problem of stage subdivision of the Lower Cambrian". *Stratigraphy and Geological Correlation* **16** (1): 1–19. Bibcode:2008SGC....16....1R. doi:10.1007/s11506-008-1001-3.

[21] B. S. Sokolov, M. A. Fedonkin (1984). "The Vendian as the Terminal System of the Precambrian" (PDF). *Episodes* **7** (1): 12–20.

[22] V. V. Khomentovskii and G. A. Karlova (2005). "The Tommotian Stage Base as the Cambrian Lower Boundary in Siberia". *Stratigraphy and Geological Correlation* **13** (1): 21–34.

[23] Gradstein, F.M.; Ogg, J.G.; Smith, A.G.; et al. (2004). *A Geologic Time Scale 2004*. Cambridge University Press.

[24] Powell, C.M.; Dalziel, I.W.D.; Li, Z.X.; McElhinny, M.W. (1995). "Did Pannotia, the latest Neoproterozoic southern supercontinent, really exist". *Eos, Transactions, American Geophysical Union* **76**: 46–72.

[25] Scotese, C.R. (1998). "A tale of two supercontinents: the assembly of Rodinia, its break-up, and the formation of Pannotia during the Pan-African event". *Journal of African Earth Sciences* **27** (1A): 171. Bibcode:1998JAfES..27....1A. doi:10.1016/S0899-5362(98)00028-1.

[26] Mckerrow, W. S.; Scotese, C. R.; Brasier, M. D. (1992). "Early Cambrian continental reconstructions". *Journal of the Geological Society* **149** (4): 599–593. doi:10.1144/gsjgs.149.4.0599.

[27] Mitchell, R. N.; Evans, D. A. D.; Kilian, T. M. (2010). "Rapid Early Cambrian rotation of Gondwana". *Geology* **38** (8): 755. Bibcode:2010Geo....38..755M. doi:10.1130/G30910.1.

[28] Smith, A.G. (2008). "Neoproterozoic time scales and stratigraphy". *Geol. Soc.* (Special publication).

[29] Brett, C. E.; Allison, P. A.; Desantis, M. K.; Liddell, W. D.; Kramer, A. (2009). "Sequence stratigraphy, cyclic facies, and lagerstätten in the Middle Cambrian Wheeler and Marjum Formations, Great Basin, Utah". *Palaeogeography Palaeoclimatology Palaeoecology* **277**: 9–33. doi:10.1016/j.palaeo.2009.02.010.

[30] Schieber et al., 2007, pp. 53–71.

[31] Retallack, G.J. (2008). "Cambrian palaeosols and landscapes of South Australia".*Alcheringa***55**: 1083–1106. Bibcode:2008A doi:10.1080/08120090802266568.

[32] Greening of the Earth pushed way back in time

[33] Cambrian HSU NHM

[34] Out of Thin Air: Dinosaurs, Birds, and Earth's Ancient Atmosphere

[35] As the worms churn

[36] Taylor, P.D.; Berning, B.; Wilson, M.A. (2013). "Reinterpretation of the Cambrian 'bryozoan' *Pywackia* as an octocoral". *Journal of Paleontology* **87** (6): 984–990. doi:10.1666/13-029.

[37] Budd, G. E.; Jensen, S. (2000). "A critical reappraisal of the fossil record of the bilaterian phyla". *Biological Reviews of the Cambridge Philosophical Society* **75** (2): 253–95. doi:10.1111/j.1469-185X.1999.tb00046.x. PMID 10881389.

[38] The Ordovician: Life's second big bang

[39] Oxygen crash led to Cambrian mass extinction

[40] Collette & Hagadorn, 2010.

[41] Collette, Gass & Hagadorn, 2012

[42] Yochelson & Fedonkin, 1993.

[43] Getty & Hagadorn, 2008.

[44] Munnecke, A.; Calner, M.; Harper, D. A. T.; Servais, T. (2010). "Ordovician and Silurian sea-water chemistry, sea level, and climate: A synopsis". *Palaeogeography, Palaeoclimatology, Palaeoecology* **296** (3–4): 389–413. doi:10.1016/j.palaeo.2010.08.001.

4.8 Further reading

- Amthor, J. E.; Grotzinger, John P.; Schröder, Stefan; Bowring, Samuel A.; Ramezani, Jahandar; Martin, Mark W.; Matter, Albert (2003). "Extinction of *Cloudina* and *Namacalathus* at the Precambrian-Cambrian boundary in Oman".*Geology***31**(5): 431–434. Bibcode:2003Geo....31..431A.doi:10.1130/0091-7613(2003)031<0431:EO

- Collette, J. H., Gass, K. C. & Hagadorn, J. W. (2012). "*Protichnites eremita* unshelled? Experimental model-based neoichnology and new evidence for a euthycarcinoid affinity for this ichnospecies". *Journal of Paleontology* **86** (3): 442–454. doi:10.1666/11-056.1.

- Collette, J. H. & Hagadorn, J. W. (2010). "Three-dimensionally preserved arthropods from Cambrian Lagerstatten of Quebec and Wisconsin". *Journal of Paleontology* **84** (4): 646–667. doi:10.1666/09-075.1.

- Getty, P. R. & Hagadorn, J. W. (2008). "Reinterpretation of *Climactichnites* Logan 1860 to include subsurface burrows, and erection of *Musculopodus* for resting traces of the trailmaker". *Journal of Paleontology* **82** (6): 1161–1172. doi:10.1666/08-004.1.

- Gould, S. J.; Wonderful Life: the Burgess Shale and the Nature of Life (New York: Norton, 1989)

- Ogg, J.; June 2004, *Overview of Global Boundary Stratotype Sections and Points (GSSPs)* http://www.stratigraphy.org/gssp.htm Accessed 30 April 2006.

- Owen, R. (1852). "Description of the impressions and footprints of the *Protichnites* from the Potsdam sandstone of Canada". *Geological Society of London Quarterly Journal* **8**: 214–225. doi:10.1144/GSL.JGS.1852.008.01-02.26.

- Peng, S.; Babcock, L.E.; Cooper, R.A. (2012). "The Cambrian Period". *The Geologic Time Scale* (PDF).

- Schieber, J.; Bose, P. K.; Eriksson, P. G.; Banerjee, S.; Sarkar, S.; Altermann, W.; Catuneau, O. (2007). *Atlas of Microbial Mat Features Preserved within the Clastic Rock Record.* Elsevier. pp. 53–71.

- Yochelson, E. L. and M. A. Fedonkin (1993). "Paleobiology of *Climactichnites*, and Enigmatic Late Cambrian Fossil" (Free full text). *Smithsonian Contributions to Paleobiology* **74**: 1–74. doi:10.5479/si.00810266.74.1.

4.9 External links

-
- Cambrian period on *In Our Time* at the BBC. (listen now)
- Biostratigraphy – includes information on Cambrian trilobite biostratigraphy
- Dr. Sam Gon's trilobite pages (contains numerous Cambrian trilobites)
- Examples of Cambrian Fossils
- Paleomap Project
- Report on the web on Amthor and others from *Geology* vol. 31
- Weird Life on the Mats

Chapter 5

Parthenogenesis

For a similar process in plants, rather than animals, see Apomixis.
Not to be confused with Pathogenesis.

Parthenogenesis /ˌpɑrθənəˈdʒɛnɨsɨs/ (from the Greek παρθένος parthenos, "virgin", + γένεσις genesis, "creation"[1]) is a natural form of asexual reproduction in which growth and development of embryos occur without fertilization. In animals, parthenogenesis means development of an embryo from an unfertilized egg cell and is a component process of apomixis.

Gynogenesis and pseudogamy are closely related phenomena in which a sperm or pollen triggers the development of the egg cell into an embryo but makes no genetic contribution to the embryo. The rest of the cytology and genetics of these phenomena are mostly identical to that of parthenogenesis.

The term is sometimes used inaccurately to describe reproduction modes in hermaphroditic species that can reproduce by themselves because they contain reproductive organs of both sexes in a single individual's body.

Parthenogenesis occurs naturally in many plants, some invertebrate animal species (including nematodes, water fleas, some scorpions, aphids, some bees, some Phasmida and parasitic wasps) and a few vertebrates (such as some fish,[2] amphibians, reptiles[3][4] and very rarely birds[5]). This type of reproduction has been induced artificially in a few species including fish and amphibians.[6]

Normal egg cells form after meiosis and are haploid, with half as many chromosomes as their mother's body cells. Haploid individuals, however, are usually non-viable, and parthenogenetic offspring usually have the diploid chromosome number. Depending on the mechanism involved in restoring the diploid number of chromosomes, parthenogenetic offspring may have anywhere between all and half of the mother's alleles. The offspring having all of the mother's genetic material are called full clones and those having only half are called half clones. Full clones are usually formed without meiosis. If meiosis occurs, the offspring will get only a fraction of the mother's alleles.

Parthenogenetic offspring in species that use either the XY or the X0 sex-determination system have two X chromosomes and are female. In species that use the ZW sex-determination system, they have either two Z chromosomes (male) or two W chromosomes (mostly non-viable but rarely a female), or they could have one Z and one W chromosome (female).

5.1 Life history types

Some species reproduce exclusively by parthenogenesis (such as the Bdelloid rotifers), while others can switch between sexual reproduction and parthenogenesis. This is called facultative parthenogenesis (other terms are cyclical parthenogenesis, heterogamy[7][8] or heterogony[9][10]). The switch between sexuality and parthenogenesis in such species may be triggered by the season (aphid, some gall wasps), or by a lack of males or by conditions that favour rapid population growth (rotifers and cladocerans like daphnia). In these species asexual reproduction occurs either in summer (aphids) or as long as conditions are favourable. This is because in asexual reproduction a successful genotype can spread quickly without being modified by sex or wasting resources on male offspring who won't give birth. In times of stress, offspring

The asexual, all-female whiptail species Cnemidophorus neomexicanus *(center), which reproduces via parthenogenesis, is shown flanked by two sexual species having males* C. inornatus *(left) and* C. tigris *(right), which hybridized naturally to form the* C. neomexicanus *species.*

produced by sexual reproduction may be fitter as they have new, possibly beneficial gene combinations. In addition, sexual reproduction provides the benefit of meiotic recombination between non-sister chromosomes, a process associated with repair of DNA double-strand breaks and other DNA damages that may be induced by stressful conditions.[11][12] (See also Meiosis section: Origin and function of meiosis.)

Many taxa with heterogony have within them species that have lost the sexual phase and are now completely asexual. Many other cases of obligate parthenogenesis (or gynogenesis) are found among polyploids and hybrids where the chromosomes cannot pair for meiosis.

The production of female offspring by parthenogenesis is referred to as thelytoky (e.g., aphids) while the production of males by parthenogenesis is referred to as arrhenotoky (e.g., bees). When unfertilized eggs develop into both males and females, the phenomenon is called deuterotoky.[13]

5.2 Types and mechanisms

Parthenogenesis can occur without meiosis through mitotic oogenesis. This is called **apomictic parthenogenesis**. Mature egg cells are produced by mitotic divisions, and these cells directly develop into embryos. In flowering plants, cells of the gametophyte can undergo this process. The offspring produced by apomictic parthenogenesis are *full clones* of their mother. Examples include aphids.

Parthenogenesis involving meiosis is more complicated. In some cases, the offspring are haploid (e.g., male ants). In other cases, collectively called **automictic parthenogenesis**, the ploidy is restored to diploidy by various means. This is because haploid individuals are not viable in most species. In automictic parthenogenesis the offspring differ from one another and from their mother. They are called *half clones* of their mother.

5.2.1 Automixis

Automixis is a term that covers several reproductive mechanisms, some of which are parthenogenetic.[14]

Diploidy might be restored by the doubling of the chromosomes without cell division before meiosis begins or after meiosis is completed. This is referred to as an *endomitotic* cycle. This may also happen by the fusion of the first two blastomeres. Other species restore their ploidy by the fusion of the meiotic products. The chromosomes may not separate at one of the two anaphases (called **restitutional meiosis**), or the nuclei produced may fuse or one of the polar bodies may fuse with the egg cell at some stage during its maturation.

Some authors consider all forms of automixis sexual as they involve recombination. Many others classify the endomitotic variants as asexual, and consider the resulting embryos parthenogenetic. Among these authors the threshold for classifying automixis as a sexual process depends on when the products of anaphase I or of anaphase II are joined together. The criterion for "sexuality" varies from all cases of restitutional meiosis,[15] to those where the nuclei fuse or to only those where gametes are mature at the time of fusion.[14] Those cases of automixis that are classified as sexual reproduction are compared to self-fertilization in their mechanism and consequences.

The genetic composition of the offspring depends on what type of apomixis takes place. When endomitosis occurs before meiosis[16][17] or when *central fusion* occurs (restitutional meiosis of anaphase I or the fusion of its products), the offspring get all[16][18] to more than half of the mother's genetic material and heterozygosity is mostly preserved[19] (if the mother has two alleles for a locus, it is likely that the offspring will get both). This is because in anaphase I the homologous chromosomes are separated. Heterozygosity is not completely preserved when crossing over occurs in central fusion.[20] In the case of pre-meiotic doubling, recombination -if it happens- occurs between identical sister chromatids.[16]

If *terminal fusion* (restitutional meiosis of anaphase II or the fusion of its products) occurs, a little over half the mother's genetic material is present in the offspring and the offspring are mostly homozygous.[21] This is because at anaphase II the sister chromatids are separated and whatever heterozygosity is present is due to crossing over. In the case of endomitosis after meiosis the offspring is completely homozygous and has only half the mother's genetic material.

This can result in parthenogenetic offspring being unique from each other and from their mother.

5.2.2 Sex of the offspring

In apomictic parthenogenesis, the offspring are clones of the mother and hence are usually (except for aphids) female. In the case of aphids, parthenogenetically produced males and females are clones of their mother except that the males lack one of the X chromosomes (XO).[22]

When meiosis is involved, the sex of the offspring will depend on the type of sex determination system and the type of apomixis. In species that use the XY sex-determination system, parthenogenetic offspring will have two X chromosomes and are female. In species that use the ZW sex-determination system the offspring genotype may be one of ZW (female),[18][19] ZZ (male), or WW (non-viable in most species[21] but a fertile, viable female in a few (e.g., boas)).[21] ZW offspring are produced by endoreplication before meiosis or by central fusion.[18][19] ZZ and WW offspring occur either by terminal fusion[21] or by endomitosis in the egg cell.

In polyploid obligate parthenogens like the whiptail lizard, all the offspring are female.[17]

In many hymenopteran insects such as honeybees, female eggs are produced sexually, using sperm from a drone father, while the production of further drones (males) depends on the queen (and occasionally workers) producing unfertilised eggs. This means that females (workers and queens) are always diploid, while males (drones) are always haploid, and produced parthenogenetically.

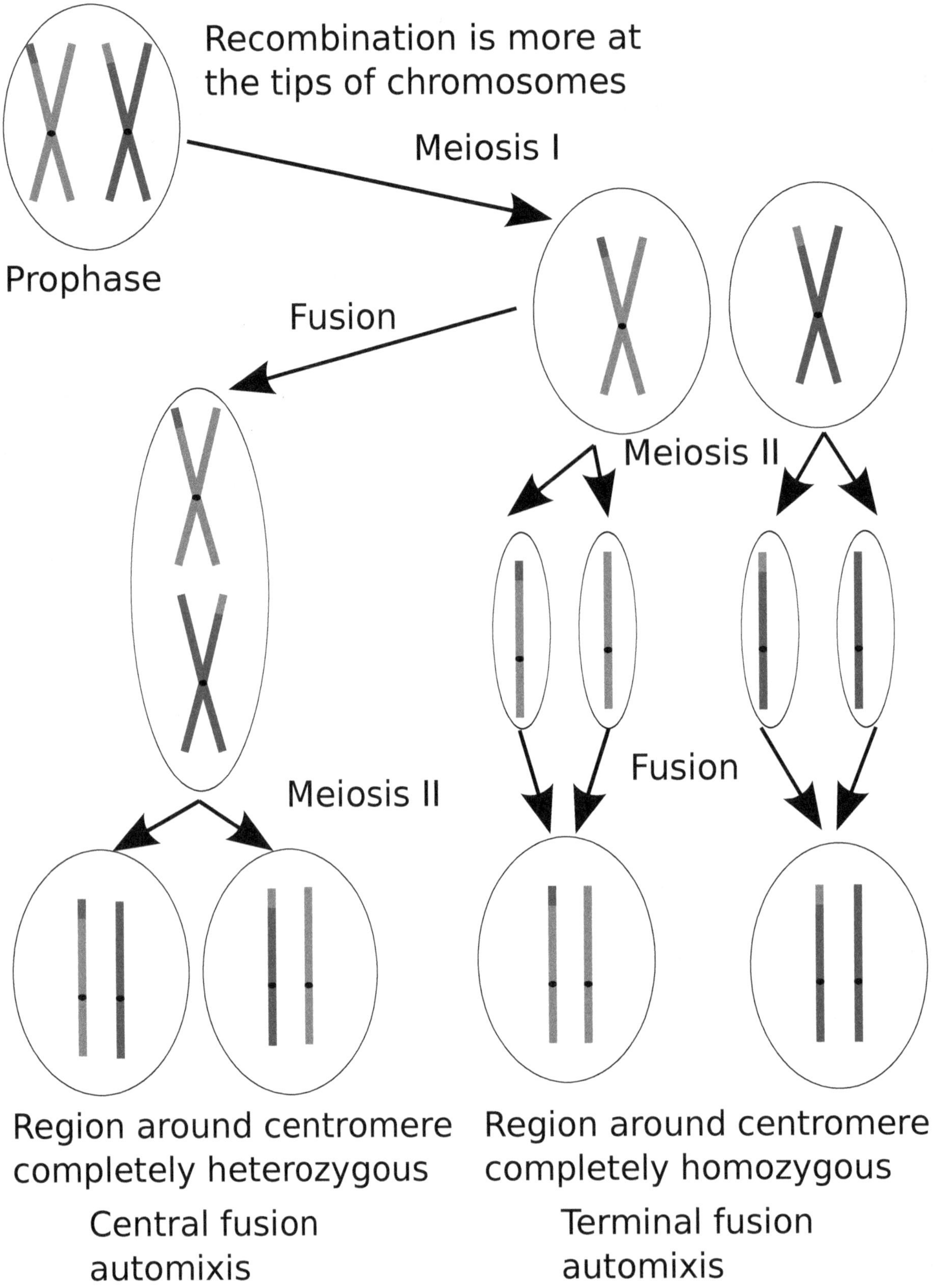

The effects of central fusion and terminal fusion on heterozygosity

5.2.3 Facultative parthenogenesis

Facultative parthenogenesis is the term for when a female can produce offspring either sexually or via asexual reproduction.[23] Facultative parthenogenesis is extremely rare in nature, with only a few examples of animal taxa capable of facultative parthenogenesis.[23] One of the best known examples of taxa exhibiting facultative parthenogenesis are mayflies; presumably this is the default reproductive mode of all species in this insect order.[24] Facultative parthenogenesis is believed to be a response to a lack of a viable male. A female may undergo facultative parthenogenesis if a male is absent from the habitat or if it is unable to produce viable offspring.

Facultative parthenogenesis is often incorrectly used to describe cases of accidental or spontaneous parthenogenesis in normally sexual animals.[25] For example, many cases of accidental parthenogenesis in sharks, some snakes, Komodo dragons and a variety of domesticated birds were widely perpetuated as facultative parthenogenesis.[26] These cases are, however, examples of accidental parthenogenesis, given the frequency of asexually produced eggs and their hatching rates are extremely low, in contrast to 'true' facultative parthenogenesis where the majority of asexually produced eggs hatch.[23][25] In addition, asexually produced offspring in vertebrates are virtually always sterile, highlighting that this mode of reproduction is not adaptive. The occurrence of such asexually produced eggs in sexual animals can be explained by a meiotic error, leading to automictically produced eggs.[25][27]

5.2.4 Obligate parthenogenesis

Obligate parthenogenesis is the process in which organisms exclusively reproduce through asexual means.[28] Many species have been shown to transition to obligate parthenogenesis over evolutionary time. Among these species, one of the most well documented transitions to obligate parthenogenesis was found in almost all metazoan taxa, albeit through highly diverse mechanisms. These transitions often occur as a result of inbreeding or mutation within large populations.[29] There are a number of documented species, specifically salamanders and geckos, that rely on obligate parthenogenesis as their major method of reproduction. As such, there are over 80 species of unisex reptiles, mostly lizards but including a single snake species, amphibians and fishes in nature for which males are no longer a part of the reproductive process.[30] A female will produce an ovum with a full set (two sets of genes) provided solely by the mother. Thus, a male is not needed to provide sperm to fertilize the egg. This form of asexual reproduction is thought in some cases to be a serious threat to biodiversity for the subsequent lack of gene variation and potentially decreased fitness of the offspring.[28]

5.3 Natural occurrence

Parthenogenesis is seen to occur naturally in aphids, *Daphnia*, rotifers, nematodes and some other invertebrates, as well as in many plants. Among vertebrates, strict parthenogenesis is only known to occur in lizards, snakes,[31] birds[32] and sharks,[33] with fish, amphibians and reptiles exhibiting various forms of gynogenesis and hybridogenesis (an incomplete form of parthenogenesis).[34] The first all-female (unisexual) reproduction in vertebrates was described in the fish *Poecilia formosa* in 1932.[35] Since then at least 50 species of unisexual vertebrate have been described, including at least 20 fish, 25 lizards, a single snake species, frogs, and salamanders.[34] Other, usually sexual, species may occasionally reproduce parthenogenetically and Komodo dragons; the hammerhead and blacktip sharks are recent additions to the known list of spontaneous parthenogenetic vertebrates. As with all types of asexual reproduction, there are both costs (low genetic diversity and therefore susceptibility to adverse mutations that might occur) and benefits (reproduction without the need for a male) associated with parthenogenesis.

Parthenogenesis is distinct from artificial animal cloning, a process where the new organism is necessarily genetically identical to the cell donor. In cloning, the nucleus of a diploid cell from a donor organism is inserted into an enucleated egg cell and the cell is then stimulated to undergo continued mitosis, resulting in an organism that is genetically identical to the donor. Parthenogenesis is different, in that it originates from the genetic material contained within an egg cell and the new organism is not necessarily genetically identical to the parent.

Parthenogenesis may be achieved through an artificial process as described below under the discussion of mammals.

5.3.1 Insects

Parthenogenesis in insects can cover a wide range of mechanisms.[36] The offspring produced by parthenogenesis may be of both sexes, only female (thelytoky, e.g. aphids) or only male (arrhenotoky, e.g. most hymenopterans). Both true parthenogenesis and pseudogamy (**gynogenesis** or **sperm-dependent parthenogenesis**) are known to occur.[23] The egg cells, depending on the species may be produced without meiosis (apomictically) or by one of the several automictic mechanisms.

A related phenomenon, polyembryony is a process that produces multiple clonal offspring from a single egg cell. This is known in some hymenopteran parasitoids and in Strepsiptera.[36]

In automictic species the offspring can be haploid or diploid. Diploids are produced by doubling or fusion of gametes after meiosis. Fusion is seen in the Phasmatodea, Hemiptera (Aleurodids and Coccidae), Diptera, and some Hymenoptera.[36]

In addition to these forms is hermaphroditism, where both the eggs and sperm are produced by the same individual, but is not a type of parthenogenesis. This is seen in three species of *Icerya* scale insects.[36]

Parasitic bacteria like *Wolbachia* have been noted to induce automictic thelytoky in many insect species with haplodiploid systems. They also cause gamete duplication in unfertilized eggs causing them to develop into female offspring.[36]

Honey Bee on a plum blossom

Among species with the haplo-diploid sex-determination system, such as hymenopterans (ants, bees and wasps) and thysanopterans (thrips), haploid males are produced from unfertilized eggs. Usually eggs are laid only by the queen, but the unmated workers may also lay haploid, male eggs either regularly (e.g. stingless bees) or under special circumstances. An example of non-viable parthenogenesis is common among domesticated honey bees. The queen bee is the only fertile female in the hive; if she dies without the possibility for a viable replacement queen, it is not uncommon for the worker bees to lay eggs. This is a result of the lack of the queen's pheromones and the pheromones secreted by uncapped brood, which normally suppress ovarian development in workers. Worker bees are unable to mate, and the unfertilized eggs produce only drones (males), which can mate only with a queen. Thus, in a relatively short period, all the worker bees die off, and the new drones follow if they have not been able to mate before the collapse of the colony. This behaviour is believed to have evolved to allow a doomed colony to produce drones which may mate with a virgin queen and thus preserve the colony's genetic progeny.

A few ants and bees are capable of producing diploid female offspring parthenogenetically. These include a honey bee subspecies from South Africa, *Apis mellifera capensis*, where workers are capable of producing diploid eggs parthenogenetically, and replacing the queen if she dies; other examples include some species of small carpenter bee, (genus *Ceratina*). Many parasitic wasps are known to be parthenogenetic, sometimes due to infections by *Wolbachia*.

The workers in five[20] ant species and the queens in some ants are known to reproduce by parthenogenesis. In *Cataglyphis cursor*, a European formicine ant, the queens and workers can produce new queens by parthenogenesis. The workers are produced sexually.[20]

In Central and South American electric ants, *Wasmannia auropunctata*, queens produce more queens through automictic parthenogenesis with central fusion. Sterile workers usually are produced from eggs fertilized by males. In some of the eggs fertilized by males, however, the fertilization can cause the female genetic material to be ablated from the zygote. In this way, males pass on only their genes to become fertile male offspring. This is the first recognized example of an animal species where both females and males can reproduce clonally resulting in a complete separation of male and female gene pools.[37] As a consequence, the males will only have fathers and the queens only mothers, while the sterile workers are the only ones with both parents of both genders.

These ants get both the benefits of both asexual and sexual reproduction[20][37] — the daughters who can reproduce (the queens) have all of the mother's genes, while the sterile workers whose physical strength and disease resistance are important are produced sexually.

Other examples of insect parthenogenesis can be found in gall-forming aphids (e.g. *Pemphigus betae*), where females reproduce parthenogenetically during the gall-forming phase of their life cycle and in grass thrips. In the grass thrips genus *Aptinothrips* there have been, despite the very limited number of species in the genus, several transitions to asexuality.[38]

5.3.2 Crustaceans

Crustacean reproduction varies both across and within species. The water flea *Daphnia pulex* alternates between sexual and parthenogenetic reproduction.[39] Among the better-known large decapod crustaceans, some crayfish reproduce by parthenogensis. "Marmorkrebs" are parthenogenetic crayfish that were discovered in the pet trade in the 1990s.[40] Offspring are genetically identical to the parent, indicating it reproduces by apomixis, i.e. parthenogenesis in which the eggs did not undergo meiosis.[41] Spinycheek crayfish (*Orconectes limosus*) can reproduce both sexually and by parthenogenesis.[42] The Louisiana red swamp crayfish (*Procambarus clarkii*), which normally reproduces sexually, has also been suggested to reproduce by parthenogenesis,[43] although no individuals of this species have been reared this way in the lab.

5.3.3 Rotifers

In bdelloid rotifers, females reproduce exclusively by parthenogenesis (obligate parthenogenesis),[44] while in monogonont rotifers, females can alternate between sexual and asexual reproduction (cyclical parthenogenesis). At least in one normally cyclical parthenogenetic species obligate parthenogenesis can be inherited: a recessive allele leads to loss of sexual reproduction in homozygous offspring.[45]

5.3.4 Flatworms

At least two species in the genus *Dugesia*, flatworms in the Turbellaria sub-division of the phylum Platyhelminthes, include polyploid individuals that reproduce by parthenogenesis.[46] This type of parthenogenesis requires mating, but the sperm does not contribute to the genetics of the offspring (the parthenogenesis is pseudogamous, alternatively referred to as gynogenetic). A complex cycle of matings between diploid sexual and polyploid parthenogenetic individuals produces new parthenogenetic lines.

5.3.5 Snails

Several species of parthenogenetic gastropods have been studied, especially with respect to their status as invasive species. Such species include the New Zealand mud snail (*Potamopyrgus antipodarum*),[47] the red-rimmed melania (*Melanoides tuberculata*),[48] and the Quilted melania (*Tarebia granifera*).[49]

5.3.6 Squamata

Main article: Parthenogenesis in squamata

Most reptiles of the squamata order (lizards and snakes) reproduce sexually, but parthenogenesis has been observed to

Komodo dragon, Varanus komodoensis*, rarely reproduces offspring via parthenogenesis.*

occur naturally in certain species of whiptails, some geckos, rock lizards,[3][50][51]Komodo dragons and snakes.[52] Some of these like the mourning gecko *Lepidodactylus lugubris*, Indo-Pacific house gecko *Hemidactylus garnotii*, the hybrid whiptails *Cnemidophorus*, Caucasian rock lizards *Darevskia*, and the brahminy blindsnake, *Indotyphlops braminus* are unisexual and obligately parthenogenetic. Others reptiles, such as the Komodo dragon, other monitor lizards,[53] and some species of boas,[6][21][54] pythons,[19] filesnakes,[55][56] gartersnakes[57] and rattlesnakes[58][59] were previously considered as cases of facultative parthenogenesis, but are in fact cases of accidental parthenogenesis.[25]

In 2012, facultative parthenogenesis was reported in wild vertebrates for the first time by US researchers amongst captured pregnant copperhead and cottonmouth female pit-vipers.[60] The Komodo dragon, which normally reproduces sexually, has also been found able to reproduce asexually by parthenogenesis.[61] A case has been documented of a Komodo dragon reproducing via sexual reproduction after a known parthenogenetic event,[62] highlighting that these cases of parthenogenesis are reproductive accidents, rather than adaptive, facultative parthenogenesis.[25]

Some reptile species use a ZW chromosome system, which produces either males (ZZ) or females (ZW). Until 2010, it was thought that the ZW chromosome system used by reptiles was incapable of producing viable WW offspring, but a (ZW) female boa constrictor was discovered to have produced viable female offspring with WW chromosomes.[63]

Parthenogenesis has been studied extensively in the New Mexico whiptail in the genus *Cnemidophorus* (also known as *Aspidoscelis*) of which 15 species reproduce exclusively by parthenogenesis. These lizards live in the dry and sometimes harsh climate of the southwestern United States and northern Mexico. All these asexual species appear to have arisen through the hybridization of two or three of the sexual species in the genus leading to polyploid individuals. The mechanism by which the mixing of chromosomes from two or three species can lead to parthenogenetic reproduction is unknown. Recently, a hybrid parthenogenetic whiptail lizard was bred in the laboratory from a cross between an asexual and a sexual whiptail[64] Because multiple hybridization events can occur, individual parthenogenetic whiptail species can consist of multiple independent asexual lineages. Within lineages, there is very little genetic diversity, but different lineages may have quite different genotypes.

An interesting aspect to reproduction in these asexual lizards is that mating behaviors are still seen, although the populations are all female. One female plays the role played by the male in closely related species, and mounts the female that is about to lay eggs. This behaviour is due to the hormonal cycles of the females, which cause them to behave like males shortly after laying eggs, when levels of progesterone are high, and to take the female role in mating before laying eggs, when estrogen dominates. Lizards who act out the courtship ritual have greater fecundity than those kept in isolation, due to the increase in hormones that accompanies the mounting. So, although the populations lack males, they still require sexual behavioral stimuli for *maximum* reproductive success.[65]

Some lizard parthenogens show a pattern of geographic parthenogenesis, occupying high mountain areas where their ancestral forms have an inferior competition ability.[66] In Caucasian rock lizards of genus *Darevskia*, which have six parthenogenetic forms of hybrid origin [50][51][67] hybrid parthenogenetic form *D. "dahli"* has a broader niche then either of its bisexual ancestors and its expansion throughout the Central Lesser Caucasus caused decline of the ranges of both its maternal and paternal species [68]

5.3.7 Amphibians

Main article: Parthenogenesis in amphibians

5.3.8 Sharks

Parthenogenesis in sharks has been confirmed in at least three species, the bonnethead,[33] the blacktip shark,[69] and the zebra shark,[70] and reported in others.

A bonnethead, a type of small hammerhead shark, was found to have produced a pup, born live on 14 December 2001 at Henry Doorly Zoo in Nebraska, in a tank containing three female hammerheads, but no males. The pup was thought to have been conceived through parthenogenic means. The shark pup was apparently killed by a stingray within days of birth.[71] The investigation of the birth was conducted by the research team from Queen's University Belfast, Southeastern University in Florida, and Henry Doorly Zoo itself, and it was concluded after DNA testing that the reproduction was parthenogenic. The testing showed the female pup's DNA matched only one female who lived in the tank, and that no male DNA was present in the pup. The pup was not a twin or clone of her mother, but rather, contained only half of her mother's DNA ("automictic parthenogenesis"). This type of reproduction had been seen before in bony fish, but never in cartilaginous fish such as sharks, until this documentation.

In the same year, a female Atlantic blacktip shark in Virginia reproduced via parthenogenesis.[72] On 10 October 2008 scientists confirmed the second case of a virgin birth in a shark. The Journal of Fish Biology reported a study in which scientists said DNA testing proved that a pup carried by a female Atlantic blacktip shark in the Virginia Aquarium & Marine Science Center contained no genetic material from a male.[69]

In 2002, two white-spotted bamboo sharks were born at the Belle Isle Aquarium in Detroit. They hatched 15 weeks after being laid. The births baffled experts as the mother shared an aquarium with only one other shark, which was female. The female bamboo sharks had laid eggs in the past. This is not unexpected, as many animals will lay eggs even if there is not a male to fertilize them. Normally, the eggs are assumed to be inviable and are discarded. This batch of eggs was left undisturbed by the curator as he had heard about the previous birth in 2001 in Nebraska and wanted to observe whether they would hatch. Other possibilities had been considered for the birth of the Detroit bamboo sharks including thoughts that the sharks had been fertilized by a male and stored the sperm for a period of time, as well as the possibility that the Belle Isle bamboo shark is a hermaphrodite, harboring both male and female sex organs, and capable of fertilizing its own eggs, but that is not confirmed.[73]

In 2008, a Hungarian aquarium had another case of parthenogenesis after its lone female shark produced a pup without ever having come into contact with a male shark.

The repercussions of parthenogenesis in sharks, which fails to increase the genetic diversity of the offspring, is a matter of concern for shark experts, taking into consideration conservation management strategies for this species, particularly in areas where there may be a shortage of males due to fishing or environmental pressures. Although parthenogenesis may help females who cannot find mates, it does reduce genetic diversity.

In 2011, recurring shark parthenogenesis over several years was demonstrated in a captive zebra shark, a type of carpet shark.[70][74]

5.3.9 Birds

Parthenogenesis in birds is known mainly from studies of domesticated turkeys and chickens, although it has also been noted in the domestic pigeon.[32] In most cases the egg fails to develop normally or completely to hatching.[32][75] The first description of parthenogenetic development in a passerine was demonstrated in captive zebra finches, although the dividing cells exhibited irregular nuclei and the eggs did not hatch.[32]

Parthenogenesis in turkeys appears to result from a conversion of haploid cells to diploid;[75] most embryos produced in this way die early in development. Rarely, viable birds result from this process, and the rate at which this occurs in turkeys can be increased by selective breeding,[76] however male turkeys produced from parthenogenesis exhibit smaller testes and reduced fertility.[77]

5.3.10 Mammals

There are no known cases of naturally occurring mammalian parthenogenesis in the wild. Parthenogenetic progeny of mammals would have two X chromosomes, and would therefore be female.

In 1936, Gregory Goodwin Pincus reported successfully inducing parthenogenesis in a rabbit.[78] In April 2004, scientists at Tokyo University of Agriculture used parthenogenesis successfully to create a fatherless mouse. Using gene targeting, they were able to manipulate two imprinted loci H19/IGF2 and DLK1/MEG3 to produce bi-maternal mice at high frequency[79] and subsequently show that fatherless mice have enhanced longevity.[80]

Induced parthenogenesis in mice and monkeys often results in abnormal development. This is because mammals have imprinted genetic regions, where either the maternal or the paternal chromosome is inactivated in the offspring in order for development to proceed normally. A mammal created by parthenogenesis would have double doses of maternally imprinted genes and lack paternally imprinted genes, leading to developmental abnormalities. It has been suggested[81] that defects in placental folding or interdigitation are one cause of swine parthenote abortive development. As a consequence, research on human parthenogenesis is focused on the production of embryonic stem cells for use in medical treatment, not as a reproductive strategy.

Use of an electrical or chemical stimulus can produce the beginning of the process of parthenogenesis in the asexual development of viable offspring.[82]

During oocyte development, high metaphase promoting factor (MPF) activity causes mammalian oocytes to arrest at the metaphase II stage until fertilization by a sperm. The fertilization event causes intracellular calcium oscillations, and targeted degradation of cyclin B, a regulatory subunit of MPF, thus permitting the MII-arrested oocyte to proceed through

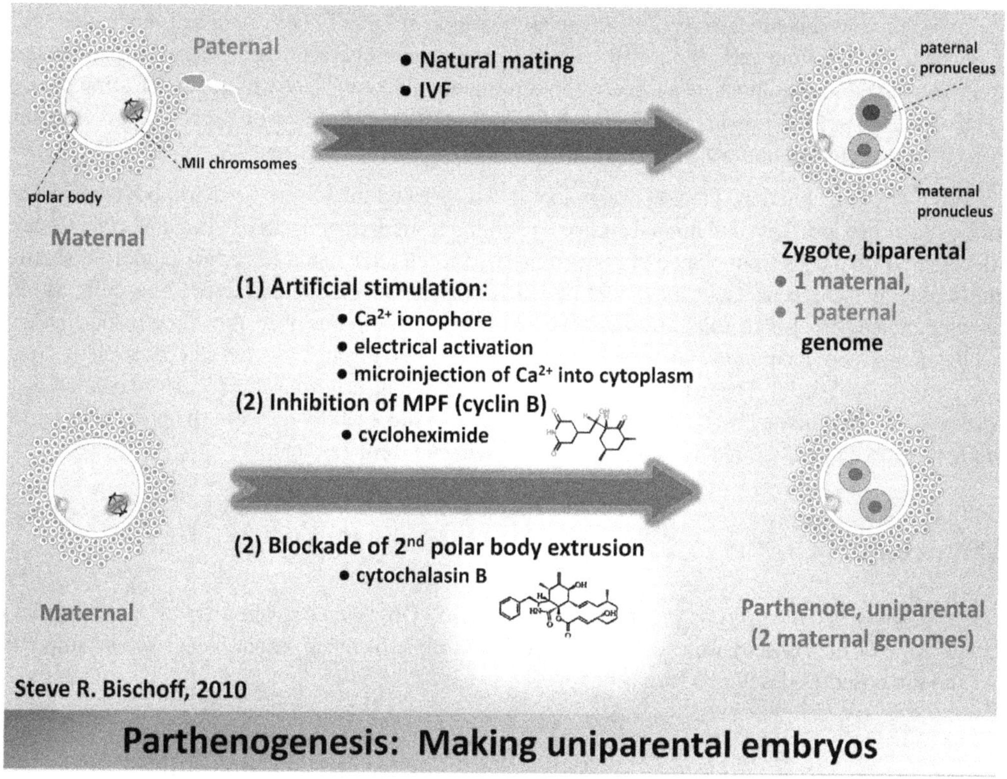

Induction of parthenogenesis in swine. *Parthenogenetic development of swine oocytes.*[81] *High metaphase promoting factor (MPF) activity causes mammalian oocytes to arrest at the metaphase II stage until fertilization by a sperm. The fertilization event causes intracellular calcium oscillations, and targeted degradation of cyclin B, a regulatory subunit of MPF, thus permitting the MII-arrested oocyte to proceed through meiosis. To initiate parthenogenesis of swine oocytes, various methods exist to induce an artificial activation that mimics sperm entry, such as calcium ionophore treatment, microinjection of calcium ions, or electrical stimulation. Treatment with cycloheximide, a non-specific protein synthesis inhibitor, enhances parthenote development in swine presumably by continual inhibition of MPF/cyclin B.*[83] *As meiosis proceeds, extrusion of the second polar is blocked by exposure to cytochalasin B. This treatment results in a diploid (2 maternal genomes) parthenote. Parthenotes can be surgically transferred to a recipient oviduct for further development, but will succumb by developmental failure after ~30 days of gestation. The swine parthenote placentae often appears hypo-vascular and is approximately 50% smaller than biparental offspring placentae: see free image (Figure 1) in linked reference.*[81]

meiosis.

To initiate parthenogenesis of swine oocytes, various methods exist to induce an artificial activation that mimics sperm entry, such as calcium ionophore treatment, microinjection of calcium ions, or electrical stimulation. Treatment with cycloheximide, a non-specific protein synthesis inhibitor, enhances parthenote development in swine presumably by continual inhibition of MPF/cyclin B.[83] As meiosis proceeds, extrusion of the second polar is blocked by exposure to cytochalasin B. This treatment results in a diploid (2 maternal genomes) parthenote [(Bischoff et al., 2009), PMID 19571260] Parthenotes can be surgically transferred to a recipient oviduct for further development, but will succumb to developmental failure after ~30 days of gestation. The swine parthenote placentae often appears hypo-vascular: see free image (Figure 1) in linked reference.[81]

Humans

On June 26, 2007, International Stem Cell Corporation (ISCC), a California-based stem cell research company, announced that their lead scientist, Dr. Elena Revazova, and her research team were the first to intentionally create human stem cells from unfertilized human eggs using parthenogenesis. The process may offer a way for creating stem cells that are genetically matched to a particular woman for the treatment of degenerative diseases that might affect her. In December 2007, Dr. Revazova and ISCC published an article[84] illustrating a breakthrough in the use of parthenogenesis to produce human stem cells that are homozygous in the HLA region of DNA. These stem cells are called HLA homozygous parthenogenetic human stem cells (hpSC-Hhom) and have unique characteristics that would allow derivatives of these cells to be implanted into millions of people without immune rejection.[85] With proper selection of oocyte donors according to HLA haplotype, it is possible to generate a bank of cell lines whose tissue derivatives, collectively, could be MHC-matched with a significant number of individuals within the human population.

On August 2, 2007, after much independent investigation, it was revealed that discredited South Korean scientist Hwang Woo-Suk unknowingly produced the first human embryos resulting from parthenogenesis. Initially, Hwang claimed he and his team had extracted stem cells from cloned human embryos, a result later found to be fabricated. Further examination of the chromosomes of these cells show indicators of parthenogenesis in those extracted stem cells, similar to those found in the mice created by Tokyo scientists in 2004. Although Hwang deceived the world about being the first to create artificially cloned human embryos, he did contribute a major breakthrough to stem cell research by creating human embryos using parthenogenesis.[86] The truth was discovered in 2007, long after the embryos were created by him and his team in February 2004. This made Hwang the first, unknowingly, to successfully perform the process of parthenogenesis to create a human embryon and, ultimately, a human parthenogenetic stem cell line.

5.3.11 Oomycetes

Apomixis can apparently occur in *Phytophthora*,[87] an Oomycete. Oospores derived after an experimental cross were germinated, and some of the progeny were genetically identical to one or other parent, which would imply that meiosis did not occur and the oospores developed by parthenogenesis.

5.4 Similar phenomena

5.4.1 Gynogenesis

A form of asexual reproduction related to parthenogenesis is gynogenesis. Here, offspring are produced by the same mechanism as in parthenogenesis, but with the requirement that the egg merely be stimulated by the *presence* of sperm in order to develop. However, the sperm cell does not contribute any genetic material to the offspring. Since gynogenetic species are all female, activation of their eggs requires mating with males of a closely related species for the needed stimulus. Some salamanders of the genus *Ambystoma* are gynogenetic and appear to have been so for over a million years. It is believed that the success of those salamanders may be due to rare fertilization of eggs by males, introducing new material to the gene pool, which may result from perhaps only one mating out of a million. In addition, the amazon molly is known to reproduce by gynogenesis.

5.4.2 Hybridogenesis

See also: Hybridogenesis in water frogs

Hyridogenesis is a mode of reproduction of hybrids. Hybridogenetic hybrids (for example AB genome), usually females, during gametogenesis exclude one of parental genomes (A) and produce gametes with unrecombined[88] genome of second parental species (B), instead of containing mixed recombined parental genomes. First genome (A) is restored by fertilization of these gametes with gametes from the first species (AA, sexual host,[88] usually male).[88][89][90]

So hybridogenesis is not completely asexual, but instead hemiclonal: half of genome is passed to the next generation clonally, unrecombined, intact (B), other half sexually, recombined (A).[88][91]

This process continues, so that each generation is half (or hemi-) clonal on the mother's side and has half new genetic material from the father's side.

This form of reproduction is seen in some live-bearing fish of the genus *Poeciliopsis*[89][92] as well as in some of the *Pelophylax* spp. ("green frogs" or "waterfrogs"):

- *P. kl. esculentus* (edible frog): *P. lessonae* × *P. ridibundus*,[88][93][94]
- *P. kl. grafi* (Graf's hybrid frog): *P. perezi* × *P. ridibundus*[88]
- *P. kl. hispanicus* (Italian edible frog) - unknown origin: *P. bergeri* × *P. ridibundus* or *P. kl. esculentus*[88]

and perhaps in *P. demarchii*.

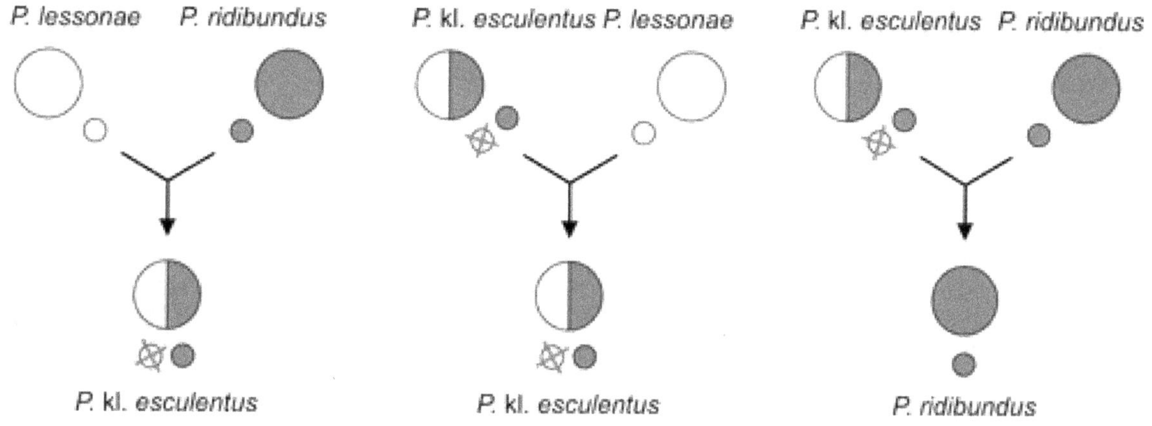

*Example crosses between pool frog (*Pelophylax lessonae*), marsh frog (*P. ridibundus*) and their hybrid - edible frog (*P. *kl.* esculentus*). First one is the primary hybridisation generating hybrid, second one is most widespread type of hybridogenesis.*[88][95]

Other examples where hybridogenesis is at least one of modes of reproduction include i.e.

- Iberian minnow *Tropidophoxinellus alburnoides* (*Squalius pyrenaicus* × hypothetical ancestor related with Anaecypris hispanica)[96]
- spined loaches *Cobitis hankugensis* × *C. longicorpus*[97]
- *Bacillus* stick insects B. rossius × Bacillus grandii benazzii[98]

5.5 See also

- Apomixis for a similar process in plants
- Arrhenotoky
- Miraculous births
- Parthenocarpy plants with seedless fruit
- Thelytoky

People

- Charles Bonnet conducted a series of experiments establishing what is now termed parthenogenesis in aphides or tree-lice
- Jacques Loeb was able to cause the eggs of sea urchins to begin embryonic development without sperm
- Gregory Goodwin Pincus experimented with parthenogenesis

5.6 References

[1] Liddell, Scott, Jones. γένεσις A.II, *A Greek-English Lexicon*, Oxford: Clarendon Press, 1940. *q.v.*.

[2] "Female Sharks Can Reproduce Alone, Researchers Find", Washington Post, Wednesday, May 23, 2007; Page A02

[3] Halliday, Tim R.; Kraig Adler (eds.) (1986). *Reptiles & Amphibians*. Torstar Books. p. 101. ISBN 0-920269-81-8.

[4] Walker, Brian (2010-11-11). "Scientists discover unknown lizard species at lunch buffet". *CNN*. Retrieved 2010-11-11.

[5] Savage, Thomas F. (September 12, 2005). "A Guide to the Recognition of Parthenogenesis in Incubated Turkey Eggs". *Oregon State University*. Retrieved 2006-10-11.

[6] Booth, W.; Johnson, D. H.; Moore, S.; Schal, C.; Vargo, E. L. (2010). "Evidence for viable, non-clonal but fatherless Boa constrictors". *Biology Letters* **7** (2): 253–256. doi:10.1098/rsbl.2010.0793. PMC 3061174. PMID 21047849.

[7] Scott, Thomas (1996). *Concise encyclopedia biology*. Walter de Gruyter. ISBN 978-3-11-010661-9.

[8] Poinar, George O, Jr; Trevor A Jackson; Nigel L Bell; Mohd B-asri Wahid (July 2002). "Elaeolenchus parthenonema n. g., n. sp. (Nematoda: Sphaerularioidea: Anandranematidae n. fam.) parasitic in the palm-pollinating weevil Elaeidobius kamerunicus Faust, with a phylogenetic synopsis of the Sphaerularioidea Lubbock, 1861". *Systematic Parasitology* **52** (3): 219–225. doi:10.1023/A:1015741820235. ISSN 0165-5752. PMID 12075153.

[9] White, Michael J.D. (1984). "Chromosomal Mechanisms in Animal Reproduction". *Bolletino di zoologia* **51** (1–2): 1–23. doi:10.1080/11250008409439455. ISSN 0373-4137.

[10] Pujade-Villar, Juli; D. Bellido; G. Segu; George Melika (2001). "Current state of knowledge of heterogony in Cynipidae (Hymenoptera, Cynipoidea)". *Sessio Conjunta dEntomologia ICHNSCL* **11** (1999): 87–107.

[11] Bernstein, H; Hopf, FA; Michod, RE (1987). "The molecular basis of the evolution of sex". *Adv Genet*. Advances in Genetics **24**: 323–370. doi:10.1016/s0065-2660(08)60012-7. ISBN 9780120176243. PMID 3324702.

[12] Harris Bernstein, Carol Bernstein and Richard E. Michod (2011). Meiosis as an Evolutionary Adaptation for DNA Repair. Chapter 19 in DNA Repair. Inna Kruman editor. InTech Open Publisher. DOI: 10.5772/25117 http://www.intechopen.com/books/dna-repair/meiosis-as-an-evolutionary-adaptation-for-dna-repair

[13] Gavrilov, I.A.; Kuznetsova, V.G. (2007). "On some terms used in the cytogenetics and reproductive biology of scale insects (Homoptera: Coccinea)" (PDF). *Comparative Cytogenetics* **1** (2): 169–174. ISSN 1993-078X.

[14] Mogie, Michael (1986). "Automixis: its distribution and status". *Biological Journal of the Linnean Society* **28** (3): 321–9. doi:10.1111/j.1095-8312.1986.tb01761.x.

[15] Zakharov, I. A. (April 2005). "Intratetrad mating and its genetic and evolutionary consequences". *Russian Journal of Genetics* **41** (4): 402–411. doi:10.1007/s11177-005-0103-z. ISSN 1022-7954. Retrieved 2011-12-19.

[16] Cosín, Darío J. Díaz, Marta Novo, and Rosa Fernández. "Reproduction of Earthworms: Sexual Selection and Parthenogenesis." In Biology of Earthworms, edited by Ayten Karaca, 24:69–86. Berlin, Heidelberg: Springer Berlin Heidelberg, 2011. http://www.springerlink.com/content/j5j72p2834355w27/.

[17] Cuellar, Orlando (1971-02-01). "Reproduction and the mechanism of meiotic restitution in the parthenogenetic lizard Cnemidophorus uniparens". *Journal of Morphology* **133** (2): 139–165. doi:10.1002/jmor.1051330203. ISSN 1097-4687. PMID 5542237.

[18] Lokki, Juhani; Esko Suomalainen; Anssi Saura; Pekka Lankinen (1975-03-01). "Genetic Polymorphism and Evolution in Parthenogenetic Animals. Ii. Diploid and Polyploid Solenobia Triquetrella (lepidoptera: Psychidae)". *Genetics* **79** (3): 513–525. PMC 1213290. PMID 1126629. Retrieved 2011-12-20.

[19] Groot, T V M; E Bruins; J A J Breeuwer (2003-02-28). "Molecular genetic evidence for parthenogenesis in the Burmese python, *Python molars bivittatus*". *Heredity* (Free full text) **90** (2): 130–135. doi:10.1038/sj.hdy.6800210. ISSN 0018-067X. PMID 12634818.

[20] Pearcy, M.; Aron, S; Doums, C; Keller, L (2004). "Conditional Use of Sex and Parthenogenesis for Worker and Queen Production in Ants". *Science* **306** (5702): 1780–3. doi:10.1126/science.1105453. PMID 15576621.

[21] Booth, Warren; Larry Million; R. Graham Reynolds; Gordon M. Burghardt; Edward L. Vargo; Coby Schal; Athanasia C. Tzika; Gordon W. Schuett (December 2011). "Consecutive Virgin Births in the New World Boid Snake, the Colombian Rainbow Boa, *Epicrates maurus*". *Journal of Heredity* **102** (6): 759–763. doi:10.1093/jhered/esr080. PMID 21868391. Retrieved 2011-12-17.

[22] Hales, Dinah F.; Alex C. C. Wilson; Mathew A. Sloane; Jean-Christophe Simon; Jean-François Legallic; Paul Sunnucks (2002). "Lack of Detectable Genetic Recombination on the X Chromosome During the Parthenogenetic Production of Female and Male Aphids". *Genetics Research* **79** (3): 203–209. doi:10.1017/S0016672302005657.

[23] Bell, G. (1982). The Masterpiece of Nature: The Evolution and Genetics of Sexuality, University of California Press, Berkeley, pp. 1- 635 (see page 295). ISBN 0520045831 ISBN 978-0520045835

[24] doi:10.1899/10-015.1
This citation will be automatically completed in the next few minutes. You can jump the queue or expand by hand

[25] van der Kooi, C.J.; Schwander, T. (2015). "Parthenogenesis: birth of a new lineage or reproductive accident?" (PDF). *Current Biology* **25**: 659–651.

[26] doi:10.1159/000195678
This citation will be automatically completed in the next few minutes. You can jump the queue or expand by hand

[27] Suomalainen E. et al. (1987). Cytology and Evolution in Parthenogenesis, Boca Raton, CRC Press

[28] Stelzer C-P, Schmidt J, Wiedlroither A, Riss S (2010) Loss of Sexual Reproduction and Dwarfing in a Small Metazoan. PLoS.

[29] Scheuerl, Thomas., et al. "Phenotypic of an Allele Causing Obligate Parthenogenesis." (2011). *Journal of Heredity* 2011:102(4):409–415. Web. 23 Oct. 2012

[30] Booth, W.; Smith, C. F.; Eskridge, P. H.; Hoss, S. K.; Mendelson, J. R.; Schuett, G. W. (2012). "Facultative parthenogenesis discovered in wild vertebrates". *Biology Letters* **8** (6): 983–985. doi:10.1098/rsbl.2012.0666. PMC 3497136. PMID 22977071.

[31] Price, A. H. (1992). Comparative behavior in lizards of the genus *Cnemidophorus* (Teiidae), with comments on the evolution of parthenogenesis in reptiles. *Copeia*, 323-331.

[32] Schut, E.; Hemmings, N.; Birkhead, T. R. (2008). "Parthenogenesis in a passerine bird, the Zebra Finch *Taeniopygia guttata*". *Ibis* **150** (1): 197–199. doi:10.1111/j.1474-919x.2007.00755.x.

[33] Chapman, Demian D.; Shivji, Mahmood S.; Louis, Ed; Sommer, Julie; Fletcher, Hugh; Prodöhl, Paulo A. (2007). "Virgin birth in a hammerhead shark". *Biology Letters* **3** (4): 425–427. doi:10.1098/rsbl.2007.0189. PMC 2390672. PMID 17519185.

[34] Vrijenhoek, R.C., R.M. Dawley, C.J. Cole, and J.P. Bogart. 1989. A list of the known unisexual vertebrates, pp. 19-23 *in*: Evolution and Ecology of Unisexual Vertebrates. R.M. Dawley and J.P. Bogart (eds.) Bulletin 466, New York State Museum, Albany, New York

[35] Hubbs, C. L.; Hubbs, L. C. (1932). "Apparent parthenogenesis in nature, in a form of fish of hybrid origin". *Science* **76** (1983): 628–630. doi:10.1126/science.76.1983.628. PMID 17730035.

[36] Kirkendall, L. R. & Normark, B. (2003) *Parthenogenesis* in Encyclopaedia of Insects (Vincent H. Resh and R. T. Carde, Eds.) Academic Press. pp. 851–856

[37] Fournier, Denis; Estoup, Arnaud; Orivel, Jérôme; Foucaud, Julien; Jourdan, Hervé; Le Breton, Julien Le; Keller, Laurent (2005). "Clonal reproduction by males and females in the little fire ant". *Nature* **435** (7046): 1230–4. doi:10.1038/nature03705. PMID 15988525.

[38] CJ van der Kooi & T Schwander (2014) Evolution of asexuality via different mechanisms in grass thrips (Thysanoptera: Aptinothrips) Evolution 86:1883-1893

[39] Eads, Brian D; Colbourne, John K; Bohuski, Elizabeth; Andrews, Justen (2007). "Profiling sex-biased gene expression during parthenogenetic reproduction in Daphnia pulex". *BMC Genomics* **8**: 464. doi:10.1186/1471-2164-8-464. PMC 2245944. PMID 18088424.

[40] Scholtz, Gerhard; Braband, Anke; Tolley, Laura; Reimann, André; Mittmann, Beate; Lukhaup, Chris; Steuerwald, Frank; Vogt, GüNter (2003). "Ecology: Parthenogenesis in an outsider crayfish". *Nature* **421** (6925): 806. doi:10.1038/421806a. PMID 12594502.

[41] Martin, Peer; Kohlmann, Klaus; Scholtz, Gerhard (2007). "The parthenogenetic Marmorkrebs (marbled crayfish) produces genetically uniform offspring". *Naturwissenschaften* **94** (10): 843–6. doi:10.1007/s00114-007-0260-0. PMID 17541537.

[42] Buřič, Miloš; Hulák, Martin; Kouba, Antonín; Petrusek, Adam; Kozák, Pavel; Etges, William J. (31 May 2011). Etges, William J, ed. "A Successful Crayfish Invader Is Capable of Facultative Parthenogenesis: A Novel Reproductive Mode in Decapod Crustaceans". *PLoS ONE* **6** (5): e20281. doi:10.1371/journal.pone.0020281. PMC 3105005. PMID 21655282.

[43] Yue GH, Wang GL, Zhu BQ, Wang CM, Zhu ZY, Lo LC (2008). "Discovery of four natural clones in a crayfish species *Procambarus clarkii*". *International Journal of Biological Sciences* **4** (5): 279–82. doi:10.7150/ijbs.4.279. PMC 2532795. PMID 18781225.

[44] "Bdelloids: No sex for over 40 million years.". TheFreeLibrary. ScienceNews. Retrieved 30 April 2011.

[45] Stelzer, C.-P.; Schmidt, J.; Wiedlroither, A.; Riss, S. (2010). "Loss of Sexual Reproduction and Dwarfing in a Small Metazoan". *PLoS ONE* **5** (9): e12854. doi:10.1371/journal.pone.0012854. PMC 2942836. PMID 20862222.

[46] Lentati, G. Benazzi (1966). "Amphimixis and pseudogamy in fresh-water triclads: Experimental reconstitution of polyploid pseudogamic biotypes". *Chromosoma* **20**: 1–14. doi:10.1007/BF00331894.

[47] Wallace, C. (1992). "arthenogenesis, sex and chromosomes in *Potamopyrgus*". *Journal of Molluscan Studies* **58** (2): 93–107. doi:10.1093/mollus/58.2.93.

[48] Ben-Ami, F.; Heller, J. (2005). "Spatial and temporal patterns of parthenogenesis and parasitism in the freshwater snail Melanoides tuberculata". *Journal of Evolutionary Biology* **18** (1): 138–146. doi:10.1111/j.1420-9101.2004.00791.x. PMID 15669970.

[49] Miranda, Nelson A. F.; Perissinotto, Renzo; Appleton, Christopher C.; Lalueza-Fox, Carles (2011). "Population Structure of an Invasive Parthenogenetic Gastropod in Coastal Lakes and Estuaries of Northern KwaZulu-Natal, South Africa". *PLoS ONE* **6** (8): e24337. doi:10.1371/journal.pone.0024337. PMC 3164166. PMID 21904629.

[50] Darevskii IS. 1967. Rock lizards of the Caucasus: systematics, ecology and phylogenesis of the polymorphic groups of Caucasian rock lizards of the subgenus *Archaeolacerta*. Nauka: Leningrad [in Russian: English translation published by the Indian National Scientific Documentation Centre, New Delhi, 1978].

[51] Tarkhnishvili DN (2012) Evolutionary History, Habitats, Diversification, and Speciation in Caucasian Rock Lizards. In: Advances in Zoology Research, Volume 2 (ed. Jenkins OP), Nova Science Publishers, Hauppauge (NY), p.79-120

[52] Self-impregnated snake in Missouri has another 'virgin birth', UPI, 21 September 2015. Retrieved 3 October 2015.

[53] Wiechmann, R. (2012). "Observations of parthenogenesis in monitor lizards". *Biawak* **6** (1): 11–21.

[54] Kinney, M.E.; Wack, R.F.; Grahn, R.A.; Lyons, L. (2013). "Parthenogenesis in a Brazilian rainbow boa (*Epicrates cenchria cenchria*)". *Zoo Biology* **32** (2): 172–176. doi:10.1002/zoo.21050. PMID 23086743.

[55] Magnusson, W.E. (1979). "Production of an embryo by an *Acrochordus javanicus* isolated for seven years". *Copeia* **1979** (4): 744–745. doi:10.2307/1443886. JSTOR 1443886.

[56] Dubach, J.; Sajewicz, A.; Pawley, R. (1997). "Parthenogenesis in the Arafura filesnake (*Acrochordus arafurae*)". *Herpetological Natural History* **5** (1): 11–18.

[57] Reynolds, R.G.; Booth, W.; Schuett, G.W.; Fitzpatrick, B.M.; Burghardt, G.M. (2012). "Successive virgin births of viable male progeny in the checkered gartersnake, *Thamnophis marcianus*". *Biological Journal of the Linnean Society* **107** (3): 566–572. doi:10.1111/j.1095-8312.2012.01954.x.

[58] Schuett, G.W.; Fernandez, P.J.; Gergits, W.F.; Casna, N.J..; Chiszar, D.; Smith, H.M.; Mitton, J.B.; Mackessy, S.P.; Odum, R.A.; Demlong, M.J. (1997). "Production of offspring in the absence of males: Evidence for facultative parthenogenesis in bisexual snakes". *Herpetological Natural History* **5** (1): 1–10.

[59] Schuett, G.W.; Fernandez, P.J.; Chiszar, D.; Smith, H.M. (1998). "Fatherless Sons: A new type of parthenogenesis in snakes". *Fauna* **1** (3): 20–25.

[60] "Virgin births discovered in wild snakes". September 12, 2012. Retrieved 2012-09-12.

[61] "No sex please, we're lizards", Roger Highfield, Daily Telegraph, 21 December 2006

[62] Virgin birth of dragons, The Hindu, 25 January 2007. Retrieved 3 February 2007.

[63] Walker, Matt (2010-11-03). "Snake has unique 'virgin birth'". *BBC News.*

[64] Lutes, Aracely A.; Diana P. Baumann; William B. Neaves; Peter Baumann (2011-06-14). "Laboratory synthesis of an independently reproducing vertebrate species". *Proceedings of the National Academy of Sciences* **108** (24): 9910–9915. doi:10.1073/pna Retrieved2011-12-23.

[65] Crews, D.; Grassman, M.; Lindzey, J. (1986). "Behavioral Facilitation of Reproduction in Sexual and Unisexual Whiptail Lizards". *Proceedings of the National Academy of Sciences* **83** (24): 9547–50. doi:10.1073/pnas.83.24.9547.

[66] Vrijenhoek RC, Parker ED. 2009. Geographical parthenogenesis: general purpose genotypes and frozen niche variation. In: Schön I, Martens K, Van Dijk P, eds. Lost sex. Berlin: Springer Publications, 99–131

[67] Murphy, RW; Darevsky, IS; MacCulloch, RD; Fu, J; Kupriyanova, LA; Upton, DE; Danielyan, F. (1997). "Old age, multiple formations or genetic plasticity? Clonal diversity in a parthenogenetic Caucasian rock lizard, Lacerta dahli". *Genetica* **101** (2): 125–130. doi:10.1023/A:1018392603062. PMID 16220367.

[68] Tarkhnishvili, D; Gavashelishvili, A; Avaliani, A; Murtskhvaladze, M; Mumladze, L (2010). "Unisexual rock lizard might be outcompeting its bisexual progenitors in the Caucasus". *Biological Journal of the Linnean Society* **101** (2): 447–460. doi:10.1111/j.1095-8312.2010.01498.x.

[69] Chapman, D. D.; Firchau, B.; Shivji, M. S. (2008). "Parthenogenesis in a large-bodied requiem shark, the blacktip". *Journal of Fish Biology* **73** (6): 1473–1477. doi:10.1111/j.1095-8649.2008.02018.x.

[70] Robinson, D. P.; Baverstock, W.; Al-Jaru, A.; Hyland, K.; Khazanehdari, K. A. (2011). "Annually recurring parthenogenesis in a zebra shark *Stegostoma fasciatum*". *Journal of Fish Biology* **79** (5): 1376–1382. doi:10.1111/j.1095-8649.2011.03110.x. PMID 22026614.

[71] "Captive shark had 'virgin birth'". *BBC News.* 2007-05-23. Retrieved 23 December 2008.

[72] "'Virgin birth' for aquarium shark". Metro.co.uk. 2008-10-10. Retrieved 2008-10-10.

[73] National Geographic, (2002). Shark gives virgin birth in Detroit. Retrieved Apr. 17, 2010, from Nationalgeographic.com Web site: http://news.nationalgeographic.com/news/2002/09/0925_020925_virginshark.html.

[74] "First Virgin Birth of Zebra Shark in Dubai". Sharkyear.com.

[75] Mittwoch, U (1978). "Parthenogenesis" (PDF). *Journal of Medical Genetics* **15** (3): 165–181. doi:10.1136/jmg.15.3.165. PMC 1013674. PMID 353283.

[76] Nestor, Karl (2009). "Parthenogenesis in turkeys". *The Tremendous Turkey.* The Ohio State University.

[77] Sarvella, P (1974). "Testes structure in normal and parthenogenetic turkeys". *The Journal of heredity* **65** (5): 287–90. PMID 4373503.

[78] Full text of "The eggs of mammals" from Internet Archive

[79] Kawahara, Manabu; Wu, Qiong; Takahashi, Nozomi; Morita, Shinnosuke; Yamada, Kaori; Ito, Mitsuteru; Ferguson-Smith, Anne C; Kono, Tomohiro (2007). "High-frequency generation of viable mice from engineered bi-maternal embryos". *Nature Biotechnology* **25** (9): 1045–50. doi:10.1038/nbt1331. PMID 17704765.

[80] Kawahara, M.; Kono, T. (2009). "Longevity in mice without a father". *Human Reproduction* **25** (2): 457–61. doi:10.1093/humre PMID 19952375. p/dep400.

[81] Bischoff, S. R.; Tsai, S.; Hardison, N.; Motsinger-Reif, A. A.; Freking, B. A.; Nonneman, D.; Rohrer, G.; Piedrahita, J. A. (2009). "Characterization of Conserved and Nonconserved Imprinted Genes in Swine". *Biology of Reproduction* **81** (5): 906–920. doi:10.1095/biolreprod.109.078139. PMC 2770020. PMID 19571260.

[82] Versieren, K; Heindryckx, B; Lierman, S; Gerris, J; De Sutter, P. (2010). "Developmental competence of parthenogenetic mouse and human embryos after chemical or electrical activation". *Reprod Biomed* **21** (6): 769–775. doi:10.1016/j.rbmo.2010.07.001. PMID 21051286.

[83] Mori, Hironori; Mizobe, Yamato; Inoue, Shin; Uenohara, Akari; Takeda, Mitsuru; Yoshida, Mitsutoshi; Miyoshi, Kazuchika (2008). "Effects of Cycloheximide on Parthenogenetic Development of Pig Oocytes Activated by Ultrasound Treatment". *Journal of Reproduction and Development* **54** (5): 364–9. doi:10.1262/jrd.20064. PMID 18635923.

[84] Revazova, E.S.; Turovets, N.A.; Kochetkova, O.D.; Kindarova, L.B.; Kuzmichev, L.N.; Janus, J.D.; Pryzhkova, M.V. (2007). "Patient-Specific Stem Cell Lines Derived from Human Parthenogenetic Blastocysts". *Cloning and Stem Cells* **9** (3): 432–49. doi:10.1089/clo.2007.0033. PMID 17594198.

[85] Revazova, E.S.; Turovets, N.A.; Kochetkova, O.D.; Agapova, L.S.; Sebastian, J.L.; Pryzhkova, M.V.; Smolnikova, V.Iu.; Kuzmichev, L.N.; Janus, J.D. (2008). "HLA Homozygous Stem Cell Lines Derived from Human Parthenogenetic Blastocysts". *Cloning and Stem Cells* **10** (1): 11–24. doi:10.1089/clo.2007.0063. PMID 18092905.

[86] Williams, Chris. "Stem cell fraudster made 'virgin birth' breakthrough: Silver lining for Korean science scandal", *The Register*, 3 August 2007.

[87] Hurtado-Gonzales, O. P.; Lamour, K. H. (2009). "Evidence for inbreeding and apomixis in close crosses ofPhytophthora capsici". *Plant Pathology* **58** (4): 715–22. doi:10.1111/j.1365-3059.2009.02059.x.

[88] Holsbeek, G.; Jooris, R. (2010). "Potential impact of genome exclusion by alien species in the hybridogenetic water frogs (*Pelophylax esculentus* complex)" (PDF). *Biol Invasions* (Springer Netherlands) **12**: 1–13. doi:10.1007/s10530-009-9427-2. ISSN 1387-3547. Retrieved 2015-06-21.

[89] Schultz, R. Jack (November–December 1969). "Hybridization, Unisexuality, and Polyploidy in the Teleost Poeciliopsis (Poeciliidae) and Other Vertebrates". *The American Naturalist* **103** (934): 605–619. doi:10.1086/282629. JSTOR 2459036.

[90] Vrijenhoek, Robert C. (1998). "Parthenogenesis and Natural Clones" (PDF). In Knobil, Ernst; Neill, Jimmy D. *Encyclopedia of Reproduction* **3**. Academic Press. pp. 695–702. ISBN 978-0-12-227020-8.

[91] Simon, J.-C.; Delmotte, F.; Rispe, C.; Crease, T. (2003). "Phylogenetic relationships between parthenogens and their sexual relatives: the possible routes to parthenogenesis in animals." (PDF). *Biological Journal of the Linnean Society* **79**: 151–163. doi:10.1046/j.1095-8312.2003.00175.x. Retrieved 2015-06-21.

[92] Vrijenhoek, J M; J C Avise; R C Vrijenhoek (1992-01-01). "An Ancient Clonal Lineage in the Fish Genus Poeciliopsis (Atheriniformes: Poeciliidae)". *Proceedings of the National Academy of Sciences* **89** (1): 348–352. Bibcode:1992PNAS...89..348Q.doi:10.1073/pnas.89.1.348. ISSN 0027-8424. Retrieved2012-03-30.

[93] "Hybridogenesis in Water Frogs". tolweb.org.

[94] Beukeboom, L. W; R. C Vrijenhoek (1998-11-01). "Evolutionary genetics and ecology of sperm-dependent parthenogenesis". *Journal of Evolutionary Biology* (free full text) **11** (6): 755–782. doi:10.1046/j.1420-9101.1998.11060755.x. ISSN 1420-9101.

[95] Vorburger, Christoph; Reyer, Heinz-Ulrich (2003). "A genetic mechanism of species replacement in European waterfrogs?" (PDF). *Conservation Genetics* (Kluwer Academic Publishers) **4** (2): 141–155. doi:10.1023/A:1023346824722. ISSN 1566-0621. Retrieved 2015-06-21.

[96] Inácio, A; Pinho, J; Pereira, PM; Comai, L; Coelho, MM (2012). "= Global Analysis of the Small RNA Transcriptome in Different Ploidies and Genomic Combinations of a Vertebrate Complex – The Squalius alburnoides". *PLoS ONE* **7** (7: e41158): 359–368. doi:10.1371/journal.pone.0041158. Retrieved 2015-06-24.

[97] Saitoh, K; Kim, I-S; Lee, E-H (2004). "= Mitochondrial Gene Introgression between Spined Loaches via Hybridogenesis". *Zool Sci* **21** (7): 795–798. doi:10.2108/zsj.21.795. PMID 15277723. Retrieved 2015-06-24.

[98] Mantovani, Barbara; Scali, Valerio (1992). "= Hybridogenesis and androgenesis in the stick-insect Bacillus rossius-Grandii benazzii (Insecta, Phasmatodea)". *Evolution* **46** (3): 783–796. doi:10.2307/2409646. JSTOR 2409646. Retrieved 2015-06-24.

5.7 Further reading

- Dawley, Robert M. & Bogart, James P. (1989). *Evolution and Ecology of Unisexual Vertebrates.* Albany, New York: New York State Museum. ISBN 1-55557-179-4.
- Fangerau, H (2005). "Can artificial parthenogenesis sidestep ethical pitfalls in human therapeutic cloning? An historical perspective". *Journal of Medical Ethics* **31** (12): 733–5. doi:10.1136/jme.2004.010199. PMC 1734065. PMID 16319240.

- Futuyma, Douglas J. & Slatkin, Montgomery. (1983). *Coevolution*. Sunderland, Mass: Sinauer Associates. ISBN 0-87893-228-3.
- Hore, T; Rapkins, R; Graves, J (2007). "Construction and evolution of imprinted loci in mammals". *Trends in Genetics* **23** (9): 440–8. doi:10.1016/j.tig.2007.07.003. PMID 17683825.
- Kono, T.; Obata, Y.; Wu, Q.; Niwa, K.; Ono, Y.; Yamamoto, Y.; Park, E.S.; Seo, J.-S.; Ogawa, H. (2004). "Birth of parthenogenetic mice that can develop to adulthood". *Nature* **428** (6985): 860–864. doi:10.1038/nature02402. PMID 15103378.
- Maynard Smith, John. (1978). *The Evolution of Sex*. Cambridge: Cambridge University Press. ISBN 0-521-29302-2.
- Michod, Richard E. & Levin, Bruce R. (1988). *The Evolution of Sex*. Sunderland, Mass: Sinauer Associates. ISBN 0-87893-459-6.
- Schlupp, Ingo (2005). "The Evolutionary Ecology of Gynogenesis". *Annual Review of Ecology, Evolution, and Systematics* **36**: 399–417. doi:10.1146/annurev.ecolsys.36.102003.152629.
- Simon, J; Rispe, Claude; Sunnucks, Paul (2002). "Ecology and evolution of sex in aphids". *Trends in Ecology & Evolution* **17**: 34–9. doi:10.1016/S0169-5347(01)02331-X.
- Stearns, Stephan C. (1988). *The Evolution of Sex and Its Consequences* (Experientia Supplementum, Vol. 55). Boston: Birkhauser. ISBN 0-8176-1807-4.

5.8 External links

- Reproductive behavior in whiptails at Crews Laboratory
- Types of asexual reproduction
- Parthenogenesis in Incubated Turkey Eggs from Oregon State University
- National Geographic NEWS: Virgin Birth Expected at Christmas – By Komodo Dragon
- BBC NEWS: 'Virgin births' for giant lizards (Komodo dragon)
- REUTERS: Komodo dragon proud mum (and dad) of five
- Female sharks capable of virgin birth
- Scientists confirm shark's 'virgin birth' Article by Steve Szkotak AP updated 1:49 a.m. ET, Fri., Oct. 10, 2008

Chapter 6

Oviparity

Oviparous animals are animals that lay eggs, with little or no other embryonic development within the mother. This is the reproductive method of most fish, amphibians, reptiles, all birds, the monotremes, and most insects, molluscs, and arachnids.

Five modes of reproduction can be differentiated [1] based on relations between zygote and parents:

- **Ovuliparity**: fertilisation is external (in arthropods and fishes, most of frogs)
- **Oviparity**: fertilisation is internal, the female lays zygotes as eggs with important vitellus (typically birds)
- **Ovo-viviparity**: or oviparity with retention of zygotes in the female's body or in the male's body, but there are no trophic interactions between zygote and parents. (*Anguis fragilis* is an example of ovo-viviparity.) In the sea horse, zygotes are retained in the male's ventral "marsupium". In the frog *Rhinoderma darwinii*, the zygotes developed in the vocal sac. In the frog *Rheobatrachus*, zygotes developed in the stomach.
- **Histotrophic viviparity**: the zygotes developed in the female's oviducts, but find their nutriments by oophagy or adelphophagy (intra-uterine cannibalism in some sharks or in the black salamander *Salamandra atra*).
- **Hemotrophic viviparity**: nutriments are provided by the female, often through placenta. In the frog *Gastrotheca ovifera*, embryos are fed by the mother through specialized gills. The lizard *Pseudomoia pagenstecheri* and most mammals exhibit a hemotrophic viviparity.

Land-dwelling animals that lay eggs, often protected by a shell, such as reptiles and insects, do so after having completed the process of internal fertilization. Water-dwelling animals, such as fish and amphibians, lay their eggs before fertilization, and the male lays its sperm on top of the newly laid eggs in a process called external fertilization.

Almost all non-oviparous fish, amphibians and reptiles are ovoviviparous, i.e. the eggs are hatched inside the mother's body (or, in case of the sea horse inside the father's). The true opposite of oviparity is placental viviparity, employed by almost all mammals (the exceptions being marsupials and monotremes).

There are only five known species of oviparous mammals (monotremes): four species of Echidna, and the Platypus.

6.1 References

[1] Thierry Lodé (2001). *Les stratégies de reproduction des animaux* (Reproduction Strategies in Animal Kingdom). Eds. Dunod Sciences. Paris.

A Southern Hawker dragonfly ovipositing

6.2 External links

- Oviparity at the US National Library of Medicine Medical Subject Headings (MeSH)

Chapter 7

Cryptobiosis

For the Doctor Who audio play, see Cryptobiosis (Doctor Who audio). For the fish disease, see Cryptobia.

Cryptobiosis is an ametabolic state of life entered by an organism in response to adverse environmental conditions such as desiccation, freezing, and oxygen deficiency. In the cryptobiotic state, all metabolic processes stop, preventing reproduction, development, and repair. An organism in a cryptobiotic state can essentially live indefinitely until environmental conditions return to being hospitable. When this occurs, the organism will return to its metabolic state of life as it was prior to the cryptobiosis.

7.1 Forms of cryptobiosis

7.1.1 Anhydrobiosis

Anhydrobiosis is the most studied form of cryptobiosis and occurs in situations of extreme desiccation. The term anhydrobiosis derives from the Greek for "life without water" and is most commonly used for the desiccation tolerance observed in certain invertebrate animals such as bdelloid rotifers, tardigrades, brine shrimp, nematodes, and at least one insect, a species of chironomid (*Polypedilum vanderplanki*). However, other life forms, including the resurrection plant *Craterostigma plantagineum*,[1] the majority of plant seeds, and many microorganisms such as bakers' yeast,[2] also exhibit desiccation tolerance. Studies have shown that some anhydrobiotic organisms can survive for decades, even centuries, in the dry state.[3]

Invertebrates undergoing anhydrobiosis often contract into a smaller shape and some proceed to form a sugar called trehalose. Desiccation tolerance in plants is associated with the production of another sugar, sucrose. These sugars are thought to protect the organism from desiccation damage.[4] In some creatures, such as bdelloid rotifers, no trehalose has been found, which has led scientists to propose other mechanisms of anhydrobiosis, possibly involving intrinsically disordered proteins.[5]

In 2011, *Caenorhabditis elegans*, a nematode that is also one of the best-studied model organisms, was shown to undergo anhydrobiosis in the dauer larva stage.[6] Further research taking advantage of genetic and biochemical tools available for this organism revealed that in addition to trehalose biosynthesis, a set of other functional pathways are involved in anhydrobiosis at the molecular level.[7] These are mainly defense mechanisms against reactive oxygen species and xenobiotics, expression of heat shock proteins and intrinsically disordered proteins as well as biosynthesis of polyunsaturated fatty acids and polyamines. Some of them are conserved among anhydrobiotic plants and animals, suggesting that anhydrobiotic ability may depend on a set of common mechanisms. Understanding these mechanisms in detail might enable modification of non-anhydrobiotic cells, tissues, organs or even organisms so that they can be preserved in a dried state of suspended animation over long time periods.

As of 2004, such an application of anhydrobiosis is being applied to vaccines. 'In vaccines, the process can produce a dry vaccine that reactivates once it is injected into the body. In theory, dry-vaccine technology could be used on any

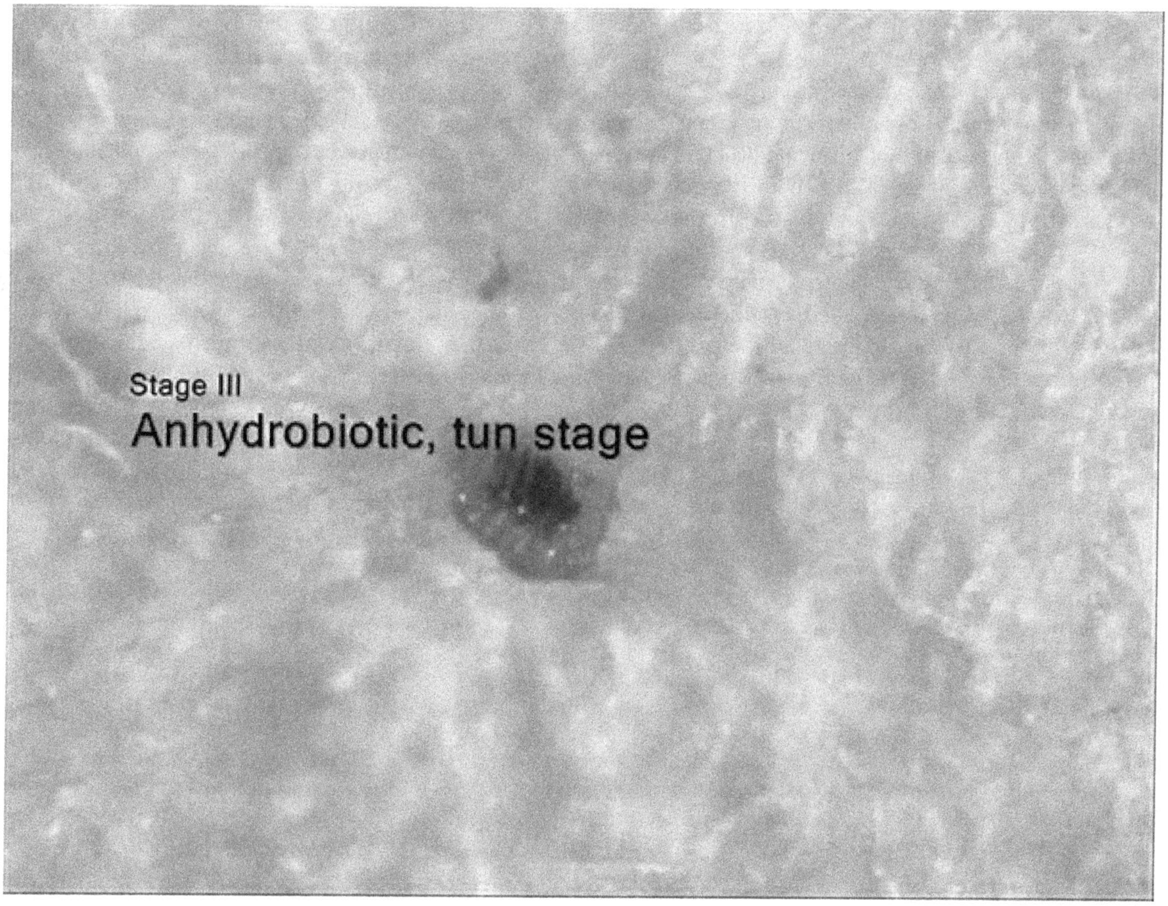

Anhydrobiosis in the tardigrade Richtersius coronifer.

vaccine, including live vaccines such as the one for measles. It could also potentially be adapted to allow a vaccine's slow release, eliminating the need for boosters. This proposes to eliminate the need for refrigerating vaccines, thus making **dry vaccines** more widely available throughout the developing world where refrigeration, electricity, and proper storage are less accessible.[8]

Based on similar principles, lyopreservation has been developed as a technique for preservation of biological samples at ambient temperatures.[9][10]

7.1.2 Anoxybiosis

In situations lacking oxygen (a.k.a., anoxia), many cryptobionts (such as M. tardigradum) take in water and become turgid and immobile, but can still survive for prolonged periods of time just as with other cryptobiological processes. While survival rate studies of organisms during supposed anoxybiosis in anoxia have historically given some conflicting results, the current consensus in the scientific community seems to be in favor of the opinion that certain ectothermic vertebrates and some invertebrates (for example, sea monkeys (Clegg et al. 1999), copepods (Marcus et al., 1994), nematodes (Crowe and Cooper, 1971), and sponge gemmules (Reiswig and Miller, 1998)) are capable of successfully surviving in a seemingly inactive state during anoxic conditions for periods of time ranging from months to decades. Studies of the metabolic activity of these idling organisms during anoxia have been mostly inconclusive, primarily due to the technical difficulty of measuring very small degrees of metabolic activity with enough reliability to conclusively prove a cryptobiotic state is occurring rather than just an extreme case of the metabolic rate depression (MRD) phenomenon exhibited by all aerobic organisms to some degree (typically 1-10% of aerobic levels) when exposed to anoxia. Thus, anoxybiosis is not considered a legitimate form of naturally occurring cryptobiosis by some comparative biologists because of these inconclusive and conflicting experimental results regarding whether a truly complete metabolic standstill actually occurs in cryptobionts during anoxic conditions, or if the metabolism is merely such a tiny fraction of the aerobic metabolic rate that it falls below the detectable limits of modern analytical methods and/or technologies. Also, many experts are skeptical of the biological feasibility of anoxybiosis, because any preservation of biological structures during anoxic conditions would imply that a seemingly impossible situation is occurring, wherein the organism is managing to prevent damage to its cellular structures from the environmental negative free energy despite being both surrounded by plenty of water and thermal energy and, most remarkably, it does so without using any free energy of its own. However (as summarized by Professor James Clegg of the UC Davis Bodega Marine laboratory in a 2001 review article), while many studies have failed to find measurable quantities of metabolic activity in potential anoxybiotic organisms, and there is also some evidence that the stress-induced protein p26 may act as a protein chaperone that requires no energy in cystic Artemia franciscana embryos, more recent data suggest that most likely an extremely specialized and slow guanine polynucleotide pathway continues to provide metabolic free energy to the *Artemia franciscana* embryos during anoxic conditions. Although it is inherently difficult to prove that not a single organism in existence anywhere is capable of entering a truly anoxybiotic state, it seems to clearly be the case that most likely at least A. franciscana (a.k.a. sea monkeys, a well-known and commonly studied cryptobiont) is only capable of approaching, but not reaching, the above-mentioned implied violation of the usual biological conversion of environmental negative free energy into the negative entropy of cellular structures via metabolic processes.

7.1.3 Chemobiosis

Chemobiosis is the cryptobiotic response to high levels of environmental toxins. It has been observed in tardigrades.[11]

7.1.4 Cryobiosis

Cryobiosis is a form of cryptobiosis that takes place in reaction to decreased temperature. Cryobiosis initiates when the water surrounding the organism's cells has been frozen, stopping molecule mobility and allowing the organism to endure the freezing temperatures until more hospitable conditions return. Organisms capable of enduring these conditions typically feature molecules that facilitate freezing of water in preferential locations while also prohibiting the growth of large ice crystals that could otherwise damage cells. One such organism is the lobster.[12]

7.1.5 Osmobiosis

Osmobiosis is the least studied of all types of cryptobiosis. Osmobiosis occurs in response to increased solute concentration in the solution the organism lives in. Little is known for certain, other than that osmobiosis appears to involve a cessation of metabolism.[13]

7.2 Examples

A commonly known organism that undergoes cryptobiosis is *Artemia salina*, also known as the brine shrimp or by its brand name Sea-monkeys. This species, which can be found in the Makgadikgadi Pans in Botswana,[14] survives over the dry season when the water of the pans evaporates, leaving a virtually desiccated lake bed.

The tardigrade, or water bear, is a widely studied and notable example,[15] partially because it can undergo all five types of cryptobiosis. While in a cryptobiotic state, the tardigrade's metabolism reduces to less than 0.01% of what is normal, and its water content can drop to 1% of normal.[16] It can withstand extreme temperature, radiation, and pressure while in a cryptobiotic state.

Some nematodes and rotifers can also undergo cryptobiosis.[17]

7.3 See also

- Biostasis
- Cryobiology
- Cryonics
- Cryptobiotic soil
- Hibernation
- Lyopreservation

7.4 References

[1] Bartels, Dorothea; Salamini, Francesco (December 2001). "Desiccation Tolerance in the Resurrection Plant Craterostigma plantagineum. A Contribution to the Study of Drought Tolerance at the Molecular Level". *Plant Physiology* **127** (4): 1346–1353. doi:10.1104/pp.010765. PMC 1540161. PMID 11743072.

[2] Calahan, Dean; Dunham, Maitreya; DeSevo, Chris; Koshland, Douglas E (October 2011). "Genetic analysis of desiccation tolerance in Sachharomyces cerevisiae". *Genetics* **189** (2): 507–519. doi:10.1534/genetics.111.130369. PMC 3189811. PMID 21840858.

[3] Shen-Miller, J; Mudgett, Mary Beth; Schopf, J William; Clarke, Steven; Berger, Rainer (November 1995). "Exceptional seed longevity and robust growth: Ancient sacred lotus from China". *American Journal of Botany* **82** (11): 1367–1380. doi:10.2307/2445863.

[4] Erkut, Cihan; Penkov, Sider; Fahmy, Karim; Kurzchalia, Teymuras V (January 2012). "How worms survive desiccation: Trehalose pro water". *Worm* **1** (1): 61–65. doi:10.4161/worm.19040. PMC 3670174. PMID 24058825.

[5] Tunnacliffe, Alan; Lapinski, Jens; McGee, Brian (September 2005). "A putative LEA protein, but no trehalose, is present in anhydrobiotic bdelloid rotifers". *Hydrobiologia* **546** (1): 315–321. doi:10.1007/s10750-005-4239-6.

[6] Erkut, Cihan; Penkov, Sider; Khesbak, Hassan; Vorkel, Daniela; Verbavatz, Jean-Marc; Fahmy, Karim; Kurzchalia, Teymuras V (August 2011). "Trehalose renders the dauer larva of Caenorhabditis elegans resistant to extreme desiccation". *Current Biology* **21** (15): 1331–1336. doi:10.1016/j.cub.2011.06.064. PMID 21782434.

[7] Erkut, Cihan; Vasilj, Andrej; Boland, Sebastian; Habermann, Bianca; Shevchenko, Andrej; Kurzchalia, Teymuras V (December 2013). "Molecular strategies of the Caenorhabditis elegans dauer larva to survive extreme desiccation". *PLoS ONE* **8** (12): e82473. doi:10.1371/journal.pone.0082473. PMC 3853187. PMID 24324795.

[8] "High hopes for fridge-free jabs". *BBC NEWS*. 2004-10-19.

[9] Yang, Geer; Gilstrap, Kyle; Zhang, Aili; Xu, Lisa X.; He, Xiaoming (1 June 2010). "Collapse temperature of solutions important for lyopreservation of living cells at ambient temperature". *Biotechnology and Bioengineering* **106** (2): 247–259. doi:10.1002/bit.22690.

[10] Chakraborty, Nilay; Chang, Anthony; Elmoazzen, Heidi; Menze, Michael A.; Hand, Steven C.; Toner, Mehmet (2011). "A Spin-Drying Technique for Lyopreservation of Mammalian Cells". *Annals of Biomedical Engineering* **39** (5): 1582–1591. doi:10.1007/s10439-011-0253-1. PMID 21293974.

[11] Møbjerg, N.; Halberg, K. A.; Jørgensen, A.; Persson, D.; Bjørn, M.; Ramløv, H.; Kristensen, R. M. (2011). "Survival in extreme environments – on the current knowledge of adaptations in tardigrades" (PDF). *Acta Physiologica* **202** (3): 409–420. doi:10.1111/j.1748-1716.2011.02252.x.

[12] "Frozen Lobsters Brought Back to Life". 18 March 2004.

[13] Møbjerg, N.; Halberg, K. A.; Jørgensen, A.; Persson, D.; Bjørn, M.; Ramløv, H.; Kristensen, R. M. (2011). "Survival in extreme environments – on the current knowledge of adaptations in tardigrades" (PDF). *Acta Physiologica* **202** (3): 409–420. doi:10.1111/j.1748-1716.2011.02252.x.

[14] C. Michael Hogan (2008) *Makgadikgadi*, The Megalithic Portal, ed. A. Burnham

[15] Illinois Wesleyan University

[16] Piper, Ross (2007), *Extraordinary Animals: An Encyclopedia of Curious and Unusual Animals*, Greenwood Press.

[17] Watanabe, Masahiko (2006). "Anhydrobiosis in invertebrates". *Appl. Entomol. Zool.* **41** (1): 15–31. doi:10.1303/aez.2006.15.

7.5 Further reading

- Clegg, J. S. (2001). "Cryptobiosis — a peculiar state of biological organization". *Comparative Biochemistry and Physiology Part B: Biochemistry and Molecular Biology* **128** (4): 613–624. doi:10.1016/S1096-4959(01)00300-1.
- David A. Wharton, *Life at the Limits: Organisms in Extreme Environments,* Cambridge University Press, 2002, hardcover, ISBN 0-521-78212-0
- Illinois Wesleyan University *Tardigrade Facts*

Chapter 8

Panspermia

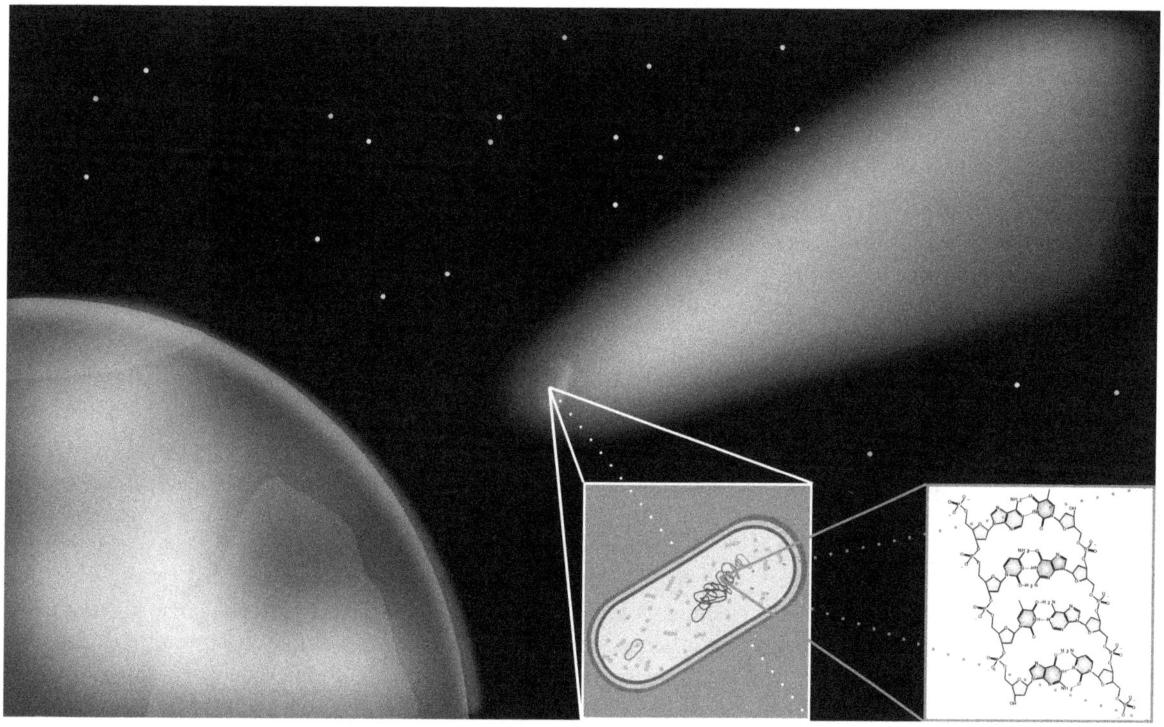

Illustration of a comet (center) transporting a bacterial life form (inset) through space to the Earth (left)

Panspermia (from Greek πᾶν *(pan)*, meaning "all", and σπέρμα *(sperma)*, meaning "seed") is the hypothesis that life exists throughout the Universe, distributed by meteoroids, asteroids, comets,[1][2] planetoids[3] and, also, by spacecraft in the form of unintended contamination by microorganisms.[4][5]

Panspermia is a hypothesis proposing that microscopic life forms that can survive the effects of space, such as extremophiles, become trapped in debris that is ejected into space after collisions between planets and small Solar System bodies that harbor life. Some organisms may travel dormant for an extended amount of time before colliding randomly with other planets or intermingling with protoplanetary disks. If met with ideal conditions on a new planet's surfaces, the organisms become active and the process of evolution begins. Panspermia is not meant to address how life began, just the method that may cause its distribution in the Universe.[6][7][8]

Pseudo-panspermia (sometimes called *"soft panspermia"* or *"molecular panspermia"*) argues that the pre-biotic organic building blocks of life originated in space and were incorporated in the solar nebula from which the planets condensed and were further —and continuously— distributed to planetary surfaces where life then emerged (abiogenesis).[9][10] From the early 1970s it was becoming evident that interstellar dust consisted of a large component of organic molecules. Interstellar molecules are formed by chemical reactions within very sparse interstellar or circumstellar clouds of dust and gas.[11] The dust plays a critical role of shielding the molecules from the ionizing effect of ultraviolet radiation emitted by stars.[12]

Several simulations in laboratories and in low Earth orbit suggest that ejection, entry and impact is survivable for some simple organisms. In 2015, "remains of biotic life" were found in 4.1 billion-year-old rocks in Western Australia, when the young Earth was about 400 million years old.[13][14] According to one of the researchers, "If life arose relatively quickly on Earth ... then it could be common in the universe."[13]

8.1 History

The first known mention of the term was in the writings of the 5th century BC Greek philosopher Anaxagoras.[15] Panspermia began to assume a more scientific form through the proposals of Jöns Jacob Berzelius (1834),[16] Hermann E. Richter (1865),[17] Kelvin (1871),[18] Hermann von Helmholtz (1879)[19][20] and finally reaching the level of a detailed hypothesis through the efforts of the Swedish chemist Svante Arrhenius (1903).[21]

Sir Fred Hoyle (1915–2001) and Chandra Wickramasinghe (born 1939) were influential proponents of panspermia.[22][23] In 1974 they proposed the hypothesis that some dust in interstellar space was largely organic (containing carbon), which Wickramasinghe later proved to be correct.[24][25][26] Hoyle and Wickramasinghe further contended that life forms continue to enter the Earth's atmosphere, and may be responsible for epidemic outbreaks, new diseases, and the genetic novelty necessary for macroevolution.[27]

In an Origins Symposium presentation on April 7, 2009, physicist Stephen Hawking stated his opinion about what humans may find when venturing into space, such as the possibility of alien life through the theory of panspermia: "Life could spread from planet to planet or from stellar system to stellar system, carried on meteors."[28]

8.2 Proposed mechanisms

Panspermia can be said to be either interstellar (between star systems) or interplanetary (between planets in the same star system);[29][30] its transport mechanisms may include comets,[31][32] radiation pressure and lithopanspermia (microorganisms embedded in rocks).[33][34][35] Interplanetary transfer of nonliving material is well documented, as evidenced by meteorites of Martian origin found on Earth.[35] Space probes may also be a viable transport mechanism for interplanetary cross-pollination in our Solar System or even beyond. However, space agencies have implemented planetary protection procedures to reduce the risk of planetary contamination,[36][37] although, as recently discovered, some microorganisms, such as Tersicoccus phoenicis, may be resistant to procedures used in spacecraft assembly clean room facilities.[4][5] In 2012, mathematician Edward Belbruno and astronomers Amaya Moro-Martín and Renu Malhotra proposed that gravitational low energy transfer of rocks among the young planets of stars in their birth cluster is commonplace, and not rare in the general galactic stellar population.[38][39] Deliberate directed panspermia from space to seed Earth[40] or sent from Earth to seed other solar systems have also been proposed.[41][42][43][44] One twist to the hypothesis by engineer Thomas Dehel (2006), proposes that plasmoid magnetic fields ejected from the magnetosphere may move the few spores lifted from the Earth's atmosphere with sufficient speed to cross interstellar space to other systems before the spores can be destroyed.[45][46]

8.2.1 Radiopanspermia

In 1903, Svante Arrhenius published in his article *The Distribution of Life in Space*,[47] the hypothesis now called radiopanspermia, that microscopic forms of life can be propagated in space, driven by the radiation pressure from stars.[48] Arrhenius argued that particles at a critical size below 1.5 μm would be propagated at high speed by radiation pressure of the Sun. However, because its effectiveness decreases with increasing size of the particle, this mechanism holds for very tiny particles only, such as single bacterial spores.[49] The main criticism of radiopanspermia hypothesis came from Shklovskii and Sagan, who pointed out the proofs of the lethal action of space radiations (UV and X-rays) in the cosmos.[50] Regardless of the evidence, Wallis and Wickramasinghe argued in 2004 that the transport of individual bacteria or clumps of bacteria, is overwhelmingly more important than lithopanspermia in terms of numbers of microbes transferred, even accounting for the death rate of unprotected bacteria in transit.[51]

Then, data gathered by the orbital experiments ERA, BIOPAN, EXOSTACK and EXPOSE, determined that isolated spores, including those of *B. subtilis*, were killed by several orders of magnitude if exposed to the full space environment for a mere few seconds, but if shielded against solar UV, the spores were capable of surviving in space for up to 6 years while embedded in clay or meteorite powder (artificial meteorites).[49][52] Though minimal protection is required to shelter a spore against UV radiation, exposure to solar UV and cosmic ionizing radiation of unprotected DNA, break it up into its bases.[53][54][55] Also, exposing DNA to the ultrahigh vacuum of space alone is sufficient to cause DNA damage, so the transport of unprotected DNA or RNA during interplanetary flights powered solely by light pressure is extremely unlikely.[55] The feasibility of other means of transport for the more massive shielded spores into the outer solar system - for example, thru gravitational capture by comets - is at this time unknown.

Based on experimental data on radiation effects and DNA stability, it has been concluded that for such long travel times, boulder sized rocks which are greater than or equal to 1 meter in diameter are required to effectively shield resistant microorganisms, such as bacterial spores against galactic cosmic radiation.[56][57] These results clearly negate the radiopanspermia hypothesis, which requires single spores accelerated by the radiation pressure of the Sun, requiring many years to travel between the planets, and support the likelihood of interplanetary transfer of microorganisms within asteroids or comets, the so-called **lithopanspermia** hypothesis.[49][52]

8.2.2 Lithopanspermia

Lithopanspermia, the transfer of organisms in rocks from one planet to another either through interplanetary or interstellar space, remains speculative. Although there is no evidence that lithopanspermia has occurred in our own Solar System, the various stages have become amenable to experimental testing.[58]

- **Planetary ejection** — For lithopanspermia to occur, microorganisms must survive ejection from a planetary surface which involves extreme forces of acceleration and shock with associated temperature excursions. Hypothetical values of shock pressures experienced by ejected rocks are obtained with Martian meteorites, which suggest the shock pressures of approximately 5 to 55 GPa, acceleration of 3×10^6 m/s^2 and jerk of 6×10^9 m/s^3 and post-shock temperature increases of about 1 K to 1000 K.[59][60] To determine the effect of acceleration during ejection on microorganisms, rifle and ultracentrifuge methods were successfully used under simulated outer space conditions.[58]
- **Survival in transit** — The survival of microorganisms has been studied extensively using both simulated facilities and in low Earth orbit. A large number of microorganisms have been selected for exposure experiments. It is possible to separate these microorganisms into two groups, the human-borne, and the extremophiles. Studying the human-borne microorganisms is significant for human welfare and future manned missions; whilst the extremophiles are vital for studying the physiological requirements of survival in space.[58]
- **Atmospheric entry** — An important aspect of the lithopanspermia hypothesis to test is that microbes situated on or within rocks could survive hypervelocity entry from space through Earth's atmosphere (Cockell, 2008). As with planetary ejection, this is experimentally tractable, with sounding rockets and orbital vehicles being used for microbiological experiments.[58][59] *B. subtilis* spores inoculated onto granite domes were subjected to hypervelocity atmospheric transit (twice) by launch to a ~120 km altitude on an Orion two-stage rocket. The spores were shown to have survived on the sides of the rock, but they did not survive on the forward-facing surface that was subjected to a maximum temperature of 145 °C.[61] In separate experiments, as part of the ESA STONE experiment, numerous organisms were embedded in different types or rocks and were mounted in the heat shield of six Foton re-entry capsules. During reentry, the rock samples were subjected to temperatures and pressure loads comparable to those experienced in meteorites.[62] The exogenous arrival of photosynthetic microorganisms could have quite profound consequences for the course of biological evolution on the inoculated planet. As photosynthetic organisms must be close to the surface of a rock to obtain sufficient light energy, atmospheric transit might act as a filter against them by ablating the surface layers of the rock. Although cyanobacteria have been shown to survive the desiccating, freezing conditions of space in orbital experiments, this would be of no benefit as the STONE experiment showed that they cannot survive atmospheric entry.[63] Thus, non-photosynthetic organisms deep within rocks have a chance

to survive the exit and entry process. (See also: Impact survival.) Research presented at the European Planetary Science Congress in 2015 suggests that ejection, entry and impact is survivable for some simple organisms.[64]

8.2.3 Accidental panspermia

Thomas Gold, a professor of astronomy, suggested in 1960 the hypothesis of "Cosmic Garbage", that life on Earth might have originated accidentally from a pile of waste products dumped on Earth long ago by extraterrestrial beings.[65]

8.2.4 Directed panspermia

Main article: Directed panspermia

Directed panspermia concerns the deliberate transport of microorganisms in space, sent to Earth to start life here, or sent from Earth to seed new solar systems with life by introduced species of microorganisms on lifeless planets. The Nobel prize winner Francis Crick, along with Leslie Orgel proposed that life may have been purposely spread by an advanced extraterrestrial civilization,[40] but considering an early "RNA world" Crick noted later that life may have originated on Earth.[66] It has been suggested that 'directed' panspermia was proposed in order to counteract various objections, including the argument that microbes would be inactivated by the space environment and cosmic radiation before they could make a chance encounter with Earth.[67]

Conversely, active directed panspermia has been proposed to secure and expand life in space.[43] This may be motivated by biotic ethics that values, and seeks to propagate, the basic patterns of our organic gene/protein life-form.[68] The panbiotic program would seed new solar systems nearby, and clusters of new stars in interstellar clouds. These young targets, where local life would not have formed yet, avoid any interference with local life.

For example, microbial payloads launched by solar sails at speeds up to 0.0001 *c* (30,000 m/s) would reach targets at 10 to 100 light-years in 0.1 million to 1 million years. Fleets of microbial capsules can be aimed at clusters of new stars in star-forming clouds, where they may land on planets or captured by asteroids and comets and later delivered to planets. Payloads may contain extremophiles for diverse environments and cyanobacteria similar to early microorganisms. Hardy multicellular organisms (rotifer cysts) may be included to induce higher evolution.[69]

The probability of hitting the target zone can be calculated from $P(target) = \frac{A(target)}{\pi(dy)^2} = \frac{ar(target)^2v^2}{(tp)^2d^4}$ where *A*(target) is the cross-section of the target area, *dy* is the positional uncertainty at arrival; *a* - constant (depending on units), *r*(target) is the radius of the target area; *v* the velocity of the probe; (tp) the targeting precision (arcsec/yr); and *d* the distance to the target, guided by high-resolution astrometry of 1×10^{-5} arcsec/yr (all units in SIU). These calculations show that relatively near target stars(Alpha PsA, Beta Pictoris) can be seeded by milligrams of launched microbes; while seeding the Rho Ophiochus star-forming cloud requires hundreds of kilograms of dispersed capsules.[43]

Directed panspermia to secure and expand life in space is becoming possible because of developments in solar sails, precise astrometry, extrasolar planets, extremophiles and microbial genetic engineering. After determining the composition of chosen meteorites, astroecologists performed laboratory experiments that suggest that many colonizing microorganisms and some plants could obtain many of their chemical nutrients from asteroid and cometary materials.[70] However, the scientists noted that phosphate (PO_4) and nitrate (NO_3–N) critically limit nutrition to many terrestrial lifeforms.[70] With such materials, and energy from long-lived stars, microscopic life planted by directed panspermia could find an immense future in the galaxy.[71]

A number of publications since 1979 have proposed the idea that directed panspermia could be demonstrated to be the origin of all life on Earth if a distinctive 'signature' message were found, deliberately implanted into either the genome or the genetic code of the first microorganisms by our hypothetical progenitor.[72][73][74][75] In 2013 a team of physicists claimed that they had found mathematical and semiotic patterns in the genetic code which, they believe, is evidence for such a signature.[76][77][78] Further investigations are needed.

A microscopic ball made of titanium and vanadium was found in Earth's upper atmosphere in early 2015. Milton Wainwright, a UK researcher and astrobiologist at the University of Buckingham claimed in a tabloid that the metal ball "could contain DNA." He speculates that it could be an alien device sent to Earth by extraterrestrials in order to continue seeding the planet with life.[79]

8.2.5 Pseudo-panspermia

Further information: List of interstellar and circumstellar molecules and Abiogenesis § Extraterrestrial organic molecules

Pseudo-panspermia (sometimes called soft panspermia, molecular panspermia or quasi-panspermia) proposes that the organic molecules used for life originated in space and were incorporated in the solar nebula, from which the planets condensed and were further —and continuously— distributed to planetary surfaces where life then emerged (abiogenesis).[9][10] From the early 1970s it was becoming evident that interstellar dust consisted of a large component of organic molecules. The first suggestion came from Chandra Wickramasinghe, who proposed a polymeric composition based on the molecule formaldehyde (CH_2O).[80] Interstellar molecules are formed by chemical reactions within very sparse interstellar or circumstellar clouds of dust and gas. Usually this occurs when a molecule becomes ionized, often as the result of an interaction with cosmic rays. This positively charged molecule then draws in a nearby reactant by electrostatic attraction of the neutral molecule's electrons. Molecules can also be generated by reactions between neutral atoms and molecules, although this process is generally slower.[11] The dust plays a critical role of shielding the molecules from the ionizing effect of ultraviolet radiation emitted by stars.[12]

A 2008 analysis of $^{12}C/^{13}C$ isotopic ratios of organic compounds found in the Murchison meteorite indicates a non-terrestrial origin for these molecules rather than terrestrial contamination. Biologically relevant molecules identified so far include uracil, an RNA nucleobase, and xanthine.[81][82] These results demonstrate that many organic compounds which are components of life on Earth were already present in the early Solar System and may have played a key role in life's origin.[83]

In August 2009, NASA scientists identified one of the fundamental chemical building-blocks of life (the amino acid glycine) in a comet for the first time.[84]

On August 2011, a report, based on NASA studies with meteorites found on Earth, was published suggesting building blocks of DNA (adenine, guanine and related organic molecules) may have been formed extraterrestrially in outer space.[85][86][87] In October 2011, scientists reported that cosmic dust contains complex organic matter ("amorphous organic solids with a mixed aromatic-aliphatic structure") that could be created naturally, and rapidly, by stars.[88][89][90] One of the scientists suggested that these complex organic compounds may have been related to the development of life on Earth and said that, "If this is the case, life on Earth may have had an easier time getting started as these organics can serve as basic ingredients for life."[88]

On August 2012, and in a world first, astronomers at Copenhagen University reported the detection of a specific sugar molecule, glycolaldehyde, in a distant star system. The molecule was found around the protostellar binary *IRAS 16293-2422*, which is located 400 light years from Earth.[91][92] Glycolaldehyde is needed to form ribonucleic acid, or RNA, which is similar in function to DNA. This finding suggests that complex organic molecules may form in stellar systems prior to the formation of planets, eventually arriving on young planets early in their formation.[93]

In September 2012, NASA scientists reported that polycyclic aromatic hydrocarbons (PAHs), subjected to interstellar medium (ISM) conditions, are transformed, through hydrogenation, oxygenation and hydroxylation, to more complex organics- "a step along the path towardamino acidsandnucleotides,the raw materials ofproteinsandDNA,respectively".[Further,as a result of these transformations,the PAHs lose theirspectroscopic signaturewhich could be one of the reasons" for the lack of PAH detection in interstellar ice grains, particularly the outer regions of cold, dense clouds or the upper molecular layers ofprotoplanetary disks."[94][95]

In 2013, the Atacama Large Millimeter Array (ALMA Project) confirmed that researchers have discovered an important pair of prebiotic molecules in the icy particles in interstellar space (ISM). The chemicals, found in a giant cloud of gas about 25,000 light-years from Earth in ISM, may be a precursor to a key component of DNA and the other may have a role in the formation of an important amino acid. Researchers found a molecule called cyanomethanimine, which produces adenine, one of the four nucleobases that form the "rungs" in the ladder-like structure of DNA. The other molecule, called ethanamine, is thought to play a role in forming alanine, one of the twenty amino acids in the genetic code. Previously, scientists thought such processes took place in the very tenuous gas between the stars. The new discoveries, however, suggest that the chemical formation sequences for these molecules occurred not in gas, but on the surfaces of ice grains in interstellar space.[96] NASA ALMA scientist Anthony Remijan stated that finding these molecules in an interstellar gas cloud means that important building blocks for DNA and amino acids can 'seed' newly formed planets with the chemical precursors for life.[97]

In March 2013, a simulation experiment indicate that dipeptides (pairs of amino acids) that can be building blocks of proteins, can be created in interstellar dust.[98]

In February 2014, NASA announced a greatly upgraded database for tracking polycyclic aromatic hydrocarbons (PAHs) in the universe. According to scientists, more than 20% of the carbon in the universe may be associated with PAHs, possible starting materials for the formation of life. PAHs seem to have been formed shortly after the Big Bang, are widespread throughout the universe, and are associated with new stars and exoplanets.[99]

In March 2015, NASA scientists reported that, for the first time, complex DNA and RNA organic compounds of life, including uracil, cytosine and thymine, have been formed in the laboratory under outer space conditions, using starting chemicals, such as pyrimidine, found in meteorites. Pyrimidine, like polycyclic aromatic hydrocarbons (PAHs), the most carbon-rich chemical found in the Universe, may have been formed in red giants or in interstellar dust and gas clouds, according to the scientists.[100]

8.3 Extraterrestrial life

Main article: Extraterrestrial life

The chemistry of life may have begun shortly after the Big Bang, 13.8 billion years ago, during a habitable epoch when the Universe was only 10–17 million years old.[101][102][103] According to the panspermia hypothesis, microscopic life—distributed by meteoroids, asteroids and other small Solar System bodies—may exist throughout the universe.[104] Nonetheless, Earth is the only place in the universe known to harbor life.[105][106] The sheer number of planets in the Milky Way galaxy, however, may make it probable that life has arisen somewhere else in the galaxy and the universe. It is generally agreed that the conditions required for the evolution of intelligent life as we know it are probably exceedingly rare in the universe, while simultaneously noting that simple single-celled microorganisms may be more likely.[107]

The extrasolar planet results from the Kepler mission estimate 100–400 billion exoplanets, with over 3,500 as candidates or confirmed exoplanets.[108] On 4 November 2013, astronomers reported, based on Kepler space mission data, that there could be as many as 40 billion Earth-sized planets orbiting in the habitable zones of sun-like stars and red dwarf stars within the Milky Way Galaxy.[109][110] 11 billion of these estimated planets may be orbiting sun-like stars.[111] The nearest such planet may be 12 light-years away, according to the scientists.[109][110]

It is estimated that space travel over cosmic distances would take an incredibly long time to an outside observer, and with vast amounts of energy required. However, there are reasons to hypothesize that faster-than-light interstellar space travel might be feasible. This has been explored by NASA scientists since at least 1995.[112]

8.3.1 Hypotheses on extraterrestrial sources of illnesses

Hoyle and Wickramasinghe have speculated that several outbreaks of illnesses on Earth are of extraterrestrial origins, including the 1918 flu pandemic, and certain outbreaks of polio and mad cow disease. For the 1918 flu pandemic they hypothesized that cometary dust brought the virus to Earth simultaneously at multiple locations—a view almost universally dismissed by experts on this pandemic. Hoyle also speculated that HIV came from outer space.[113] After Hoyle's death, *The Lancet* published a letter to the editor from Wickramasinghe and two of his colleagues,[114] in which they hypothesized that the virus that causes severe acute respiratory syndrome (SARS) could be extraterrestrial in origin and not originated from chickens. *The Lancet* subsequently published three responses to this letter, showing that the hypothesis was not evidence-based, and casting doubts on the quality of the experiments referenced by Wickramasinghe in his letter.[115][116][117] A 2008 encyclopedia notes that "Like other claims linking terrestrial disease to extraterrestrial pathogens, this proposal was rejected by the greater research community."[113]

8.3.2 Case studies

- A meteorite originating from Mars known as ALH84001 was shown in 1996 to contain microscopic structures resembling small terrestrial nanobacteria. When the discovery was announced, many immediately conjectured that these were fossils and were the first evidence of extraterrestrial life — making headlines around the world. Public interest soon started to dwindle as most experts started to agree that these structures were not indicative of life, but could instead be formed abiotically from organic molecules. However, in November 2009, a team of scientists at Johnson Space Center, including David McKay, reasserted that there was "strong evidence that life may have existed on ancient Mars", after having reexamined the meteorite and finding magnetite crystals.[118][119]

- On May 11, 2001, two researchers from the University of Naples claimed to have found live extraterrestrial bacteria inside a meteorite. Geologist Bruno D'Argenio and molecular biologist Giuseppe Geraci claim the bacteria were wedged inside the crystal structure of minerals, but were resurrected when a sample of the rock was placed in a culture medium. They believe that the bacteria were not terrestrial because they survived when the sample was sterilized at very high temperature and washed with alcohol. They also claim that the bacteria's DNA is unlike any on Earth.[120][121] They presented a report on May 11, 2001, concluding that this is the first evidence of extraterrestrial life, documented in its genetic and morphological properties. Some of the bacteria they discovered were found inside meteorites that have been estimated to be over 4.5 billion years old, and were determined to be related to modern day *Bacillus subtilis* and *Bacillus pumilis* bacteria on Earth but appears to be a different strain.[122]

- An Indian and British team of researchers led by Chandra Wickramasinghe reported on 2001 that air samples over Hyderabad, India, gathered from the stratosphere by the Indian Space Research Organization, contained clumps of living cells. Wickramasinghe calls this "unambiguous evidence for the presence of clumps of living cells in air samples from as high as 41 km, above which no air from lower down would normally be transported".[123][124] Two bacterial and one fungal species were later independently isolated from these filters which were identified as *Bacillus simplex*, *Staphylococcus pasteuri* and *Engyodontium album* respectively.[125][126] The experimental procedure suggested that these were not the result of laboratory contamination, although similar isolation experiments at separate laboratories were unsuccessful.

  A reaction report at NASA Ames indicated skepticism towards the premise that Earth life cannot travel to and reside at such altitudes.[127]

  Pushkar Ganesh Vaidya from the Indian Astrobiology Research Centre reported in 2009 that "the three microorganisms captured during the balloon experiment do not exhibit any distinct adaptations expected to be seen in microorganisms occupying a cometary niche".[128][129]

- In 2005 an improved experiment was conducted by ISRO. On April 10, 2005 air samples were collected from the upper atmosphere at altitudes ranging from 20 km to more than 40 km. The samples were tested at two labs in India. The labs found 12 bacterial and 6 different fungal species in these samples. The fungi were *Penicillium decumbens*, *Cladosporium cladosporioides*, *Alternaria sp.* and *Tilletiopsis albescens*. Out of the 12 bacterial samples, three were identified as new species and named *Janibacter hoyeli.sp.nov* (after Fred Hoyle), *Bacillus isronensis.sp.nov* (named after ISRO) and *Bacillus aryabhati* (named after the ancient Indian mathematician, Aryabhata). These three new species showed that they were more resistant to UV radiation than similar bacteria.[130][131]

  Atmospheric sampling by NASA in 2010 before and after hurricanes, collected 314 different types of bacteria; the study suggests that large-scale convection during tropical storms and hurricanes can then carry this material from the surface higher up into the atmosphere.[132][133]

- On January 10, 2013, Chandra Wickramasinghe found fossil diatom frustules in what he thinks is a new kind of carbonaceous meteorite called Polonnaruwa that landed in the North Central Province of Sri Lanka on 29 December 2012.[134] Early on, there was criticism that that Wickramasinghe's report was not an examination of an actual meteorite but of some terrestrial rock passed off as a meteorite.[135]

  Wickramasinghe's team remark that they are aware that a large number of unrelated stones have been submitted for analysis, and have no knowledge regarding the nature, source or origin of the stones their critics have examined, so Wickramasinghe clarifies that he is using the stones submitted by the Medical Research Institute in Sri Lanka.[136] In response to the criticism from other scientists, Wickramasinghe performed X-ray diffraction[137] and isotope[136] analyses to verify its meteoritic origin. His analysis revealed a 95% silica and 3% quartz content,[137] and interpreted this result as a "carbonaceous meteorite of unknown type".[137] In addition, Wickramasinghe's team remarked that the temperature at which sand must be heated by lightning to melt and form a fulgurite (1770 °C) would have vaporized and burned all carbon-rich organisms and melted and thus destroyed the delicately marked silica frustules of the diatoms,[136] and that the oxygen isotope data confirms its meteoric origin.[136] Wickramasinghe's team also argues that since living diatoms require nitrogen fixation to synthetize amino acids, proteins, DNA, RNA and other life-critical biomolecules, a population of extraterrestrial cyanobacteria must have been a required component of the comet (Polonnaruwa meteorite) "ecosystem".[136]

- In 2013, Dale Warren Griffin, a microbiologist working at the United States Geological Survey noted that viruses are the most numerous entities on Earth. Griffin speculates that viruses evolved in comets and on other planets and moons may be pathogenic to humans, so he proposed to also look for viruses on moons and planets of the Solar System.[138]

8.3.3 Hoaxes

A separate fragment of the Orgueil meteorite (kept in a sealed glass jar since its discovery) was found in 1965 to have a seed capsule embedded in it, whilst the original glassy layer on the outside remained undisturbed. Despite great initial excitement, the seed was found to be that of a European Juncaceae or Rush plant that had been glued into the fragment and camouflaged using coal dust. The outer "fusion layer" was in fact glue. Whilst the perpetrator of this hoax is unknown, it is thought he sought to influence the 19th century debate on spontaneous generation — rather than panspermia — by demonstrating the transformation of inorganic to biological matter.[139]

8.4 Extremophiles

See also: Extremophile

Until the 1970s, life was believed to depend on its access to sunlight. Even life in the ocean depths, where sunlight cannot reach, was believed to obtain its nourishment either from consuming organic detritus rained down from the surface waters or from eating animals that did.[140] However, in 1977, during an exploratory dive to the Galapagos Rift in the deep-sea exploration submersible *Alvin*, scientists discovered colonies of assorted creatures clustered around undersea volcanic features known as black smokers.[140] It was soon determined that the basis for this food chain is a form of bacterium that derives its energy from oxidation of reactive chemicals, such as hydrogen or hydrogen sulfide, that bubble up from the Earth's interior. This chemosynthesis revolutionized the study of biology by revealing that terrestrial life need not be Sun-dependent; it only requires water and an energy gradient in order to exist.

It is now known that extremophiles, microorganisms with extraordinary capability to thrive in the harshest environments on Earth, can specialize to thrive in the deep-sea,[141][142][143] ice, boiling water, acid, the water core of nuclear reactors, salt crystals, toxic waste and in a range of other extreme habitats that were previously thought to be inhospitable for life.[144][145][146][147] Living bacteria found in ice core samples retrieved from 3,700 metres (12,100 ft) deep at Lake Vostok in Antarctica, have provided data for extrapolations to the likelihood of microorganisms surviving frozen in extraterrestrial habitats or during interplanetary transport.[148] Also, bacteria have been discovered living within warm rock deep in the Earth's crust.[149]

In order to test some these organism's potential resilience in outer space, plant seeds and spores of bacteria, fungi and ferns have been exposed to the harsh space environment.[146][147][150] Spores are produced as part of the normal life cycle of many plants, algae, fungi and some protozoans, and some bacteria produce endospores or cysts during times of stress. These structures may be highly resilient to ultraviolet and gamma radiation, desiccation, lysozyme, temperature, starvation and chemical disinfectants, while metabolically inactive. Spores germinate when favourable conditions are restored after exposure to conditions fatal to the parent organism.

Although computer models suggest that a captured meteoroid would typically take some tens of millions of years before collision with a neighboring solar system planet,[38] there are documented viable Earthly bacterial spores that are 40 million years old that are very resistant to radiation,[38][44] and others able to resume life after being dormant for 25 million years,[151] suggesting that lithopanspermia life-transfers are possible via meteorites exceeding 1m in size.[38]

The discovery of deep-sea ecosystems, along with advancements in the fields of astrobiology, observational astronomy and discovery of large varieties of extremophiles, opened up a new avenue in astrobiology by massively expanding the number of possible extraterrestrial habitats and possible transport of hardy microbial life through vast distances.[58]

8.4.1 Research in outer space

The question of whether certain microorganisms can survive in the harsh environment of outer space has intrigued biologists since the beginning of spaceflight, and opportunities were provided to expose samples to space. The first American tests were made in 1966, during the Gemini IX and XII missions, when samples of bacteriophage T1 and spores of *Penicillium roqueforti* were exposed to outer space for 16.8 h and 6.5 h, respectively.[49][58] Other basic life sciences research in low Earth orbit started in 1966 with the Soviet biosatellite program Bion and the U.S. Biosatellite program. Thus, the plausibility of panspermia can be evaluated by examining life forms on Earth for their capacity to survive in space.[152] The following experiments carried on low Earth orbit specifically tested some aspects of panspermia or lithopanspermia:

ERA

The Exobiology Radiation Assembly (ERA) was a 1992 experiment on board the European Retrievable Carrier (EURECA) on the biological effects of space radiation. EURECA was an unmanned 4.5 tonne satellite with a payload of 15 experiments.[153] It was an astrobiology mission developed by the European Space Agency (ESA). Spores of different strains of *Bacillus subtilis* and the *Escherichia coli* plasmid pUC19 were exposed to selected conditions of space (space vacuum and/or defined wavebands and intensities of solar ultraviolet radiation). After the approximately 11-month mission, their responses were studied in terms of survival, mutagenesis in the *his* (*B. subtilis*) or *lac* locus (pUC19), induction of DNA strand breaks, efficiency of DNA repair systems, and the role of external protective agents. The data were compared with those of a simultaneously running ground control experiment:[154][155]

- The survival of spores treated with the vacuum of space, however shielded against solar radiation, is substantially increased, if they are exposed in multilayers and/or in the presence of glucose as protective.
- All spores in "artificial meteorites", i.e. embedded in clays or simulated Martian soil, are killed.
- Vacuum treatment leads to an increase of mutation frequency in spores, but not in plasmid DNA.
- Extraterrestrial solar ultraviolet radiation is mutagenic, induces strand breaks in the DNA and reduces survival substantially.
- Action spectroscopy confirms results of previous space experiments of a synergistic action of space vacuum and solar UV radiation with DNA being the critical target.
- The decrease in viability of the microorganisms could be correlated with the increase in DNA damage.
- The purple membranes, amino acids and urea were not measurably affected by the dehydrating condition of open space, if sheltered from solar radiation. Plasmid DNA, however, suffered a significant amount of strand breaks under these conditions.[154]

BIOPAN

BIOPAN is a multi-user experimental facility installed on the external surface of the Russian Foton descent capsule. Experiments developed for BIOPAN are designed to investigate the effect of the space environment on biological material after exposure between 13 to 17 days.[156] The experiments in BIOPAN are exposed to solar and cosmic radiation, the space vacuum and weightlessness, or a selection thereof. Of the 6 missions flown so far on BIOPAN between 1992 and 2007, dozens of experiments were conducted, and some analyzed the likelihood of panspermia. Some bacteria, lichens (*Xanthoria elegans*, *Rhizocarpon geographicum* and their mycobiont cultures, the black Antarctic microfungi *Cryomyces minteri* and *Cryomyces antarcticus*), spores, and even one animal (tardigrades) were found to have survived the harsh outer space environment and cosmic radiation.[157][158][159][160]

EXOSTACK

The German EXOSTACK experiment was deployed in 7 April 1984 on board the Long Duration Exposure Facility statellite. 30% of *Bacillus subtilis* spores survived the nearly 6 years exposure when embedded in salt crystals, whereas 80% survived in the presence of glucose, which stabilize the structure of the cellular macromolecules, especially during vacuum-induced dehydration.[49][161]

If shielded against solar UV, spores of *B. subtilis* were capable of surviving in space for up to 6 years, especially if embedded in clay or meteorite powder (artificial meteorites). The data support the likelihood of interplanetary transfer of microorganisms within meteorites, the so-called lithopanspermia hypothesis.[49]

EXPOSE

EXPOSE is a multi-user facility mounted outside the International Space Station dedicated to astrobiology experiments.[150] Results from the orbital mission, especially the experiments *SEEDS*[162] and *LiFE*,[163] concluded that after an 18-month exposure, some seeds and lichens (*Stichococcus sp.* and *Acarospora sp.*, a lichenized fungal genus) may be capable to survive interplanetary travel if sheltered inside comets or rocks from cosmic radiation and UV radiation.[150][164] The survival of some lichen species in space has also been characterized in simulated laboratory experiments.[165][166]

A separate experiment on EXPOSE called Beer was designed to find microbes that could be used in life-support recycling equipment and future "bio-mining" projects on Mars. It carried group of microbes called OU-20 resembling cyanobacteria genus *Gloeocapsa*, and it survived 553 days exposure outside the ISS.[167]

Rosetta

In 2014, the *Rosetta* spacecraft arrived at COMET 67P/Churyumov–Gerasimenko. A few months after arriving at the comet, *Rosetta* released a small lander, named *Philae*, onto its surface. The plan was to investigate Churyumov-Gerasimenko up close for two years. *Philae's* battery has since died; however scientists hope that as the comet travels toward the sun greater solar energy will recharge *Philae* (via its solar panels) and *Philae* will resume operation. Rosetta's Project Scientist, Gerhard Schwehm, stated that sterilization is generally not crucial since comets are usually regarded as objects where prebiotic molecules can be found, but not living microorganisms.[168] Notwithstanding, other scientists think it will be an opportunity to gather evidence for one of panspermia's hypotheses: the possibility of both active and dormant microbes inside comets.[7][8]

In July 2015, scientists reported that upon the first touchdown of the *Philae* lander on comet 67/P 's surface, measurements by the COSAC and Ptolemy instruments revealed sixteen organic compounds, four of which were seen for the first time on a comet, including acetamide, acetone, methyl isocyanate and propionaldehyde.[169][170][171]

Phobos LIFE

The *Phobos LIFE* or *Living Interplanetary Flight Experiment*, was developed by the Planetary Society and intended to send selected microorganisms on a three-year interplanetary round-trip in a small capsule aboard the Russian Fobos-Grunt spacecraft in 2011. Unfortunately, the spacecraft suffered technical difficulties soon after launch and fell back to Earth, so the experiment was never carried out. The experiment would have tested one aspect of panspermia: lithopanspermia, the hypothesis that life could survive space travel, if protected inside rocks blasted by impact off one planet to land on another.[172][173][174][175]

8.5 Criticism

Panspermia is criticized because it does not answer the question of the origin of life but merely places it on another celestial body. It was also criticized because it could not be tested experimentally. Furthermore, it was suggested that single spores will not survive the physical forces and environment of outer space.[176]

The concept of panspermia was revived when technology provided the opportunity to study the survival of bacterial spores ín the harsh environment of space.[58] Wallis and Wickramasinghe argued in 2004 that the transport of individual bacteria or clumps of bacteria, is overwhelmingly more important than lithopanspermia in terms of numbers of microbes transferred, even accounting for the death rate of unprotected bacteria in transit.[177] Then it was found that isolated spores of *B. subtilis* were killed by several orders of magnitude if exposed to the full space environment for a mere few seconds. These results clearly negate the original panspermia hypothesis, which requires single spores as space travelers accelerated by the radiation pressure of the Sun, requiring many years to travel between the planets. However, if shielded against solar UV, spores of *Bacillus subtilis* were capable of surviving in space for up to 6 years, especially if embedded in clay or meteorite powder (artificial meteorites). The data support the likelihood of interplanetary transfer of microorganisms within meteorites, the so-called **lithopanspermia** hypothesis.[49]

8.6 Science fiction

- Jack Finney's novel *The Body Snatchers* (1955) and the subsequent film adaptations describe spores drifting through space to arrive on the surface of Earth, though the premise is most fully discussed in the second version *Invasion of the Body Snatchers (1978 film).*
- In Ursula K. Le Guin's series the Hainish Cycle (1964–2014), Earth and other planets are seeded by the Hain using genetic engineering.
- Michael Crichton's 1969 novel, *The Andromeda Strain*, is based on the panspermiatic premise of a meteor bringing an crystalline alien bacterium to Earth. The phrase “Andromeda Strain” has become a shorthand for mysterious infectious diseases.
- Stephen King's short story "Weeds" (1976), later adapted into the Creepshow vignette “The Lonesome Death of Jordy Verrill” (1982; starring King,) involves a meteor crashing to Earth which carries with it a virulent plant/fungus which spreads rapidly.
- In the *Star Trek: The Next Generation* episode, "The Chase" (season 6, episode 20, April 26, 1993), the common humanoid form and genetic compatibility of alien species throughout the Alpha Quadrant is revealed to have resulted from directed panspermia by an earlier species of intelligent humanoid progenitors who seeded the many planets with their own DNA.
- In the 1990s TV series *Space: Above and Beyond*, about a war between humanity and an alien species known as the Chigs, it is eventually revealed that life on the Chig's planet and the Chigs themselves evolved from Earth bacteria carried there by an asteroid billions of years ago.
- Tess Gerritsen's novel, *Gravity* (1999), involves the exposure of astronauts aboard the Space Shuttle and International Space Station, to a chimera based on Archaeons, that were recovered from the Galapagos Rift.
- The plot of 2001 American science fiction comedy *Evolution* follows college professor Ira Kane (David Duchovny) and geologist Harry Block (Orlando Jones) who investigate a meteor crash in Arizona. They discover that the meteor is harboring extraterrestrial life which is evolving very quickly into large, diverse and outlandish creatures.
- The plot of the 2001 short film *Horses on Mars* centers on microbes as characters spreading into the inner solar system from Mars four billion years ago, with the main character making it to Venus while his friends land on Earth. His friends on Earth successfully evolve and send him a message via the Venera 13 lander, and later eventually make the trip back home to Mars as space-faring creatures, but without the main character, whose unsophisticated attempt to make it back to Mars ends in failure.

- In the reimagined *Battlestar Galactica*, season 3, episodes 6 and 7 ("Torn", November 3, 2006; "A Measure of Salvation", November 10, 2006), a Cylon basestar discovers an ancient beacon and takes it on board, whereupon a deadly virus from the beacon infects the Cylons. Doctor Cottle determines the Cylon infection to be a three-thousand-year-old strain of Lymphocytic choriomeningitis. Admiral Adama and President Roslin speculate that the beacon was accidentally infected prior to placement by ancient human colonists on their way from Kobol to Earth. Adama remarks, “An entire race almost wiped out because someone forgot to wipe their nose.”
- The premise of Gareth Edwards’s 2010 film *Monsters* is that a NASA deep space probe crashes, bringing back with it an alien species requiring the U.S. and Mexican military to quarantine a large district of the border region.
- The opening sequence of Ridley Scott’s 2012 *Alien* prequel, *Prometheus* depicts a humanoid species, referred to as 'the Engineers’, seeding what is presumably the early Earth by disintegrating the body of one of their members and spilling his DNA into the water of the planet. At the climax of the film it is revealed that for unknown reasons the Engineers deemed their experiment to have been a failure and intended to end it by eradicating all life on Earth.
- The novels, "The Ice Limit" (2000) and “The Lost Island” (2014), by Douglas Preston and Lincoln Child, make references to panspermia.
- The novel *Titan* by Stephen Baxter ends with human astronauts seeding the moon Titan with bacteria. The bacteria eventually evolve into creatures that intentionally spread primitive lifeforms to other star systems.

8.7 See also

- Abiogenesis
- Anthropic principle
- Astrobiology
- Cryptobiosis
- Drake equation
- Fermi paradox
- Fine-tuned Universe
- Interplanetary contamination
- Last universal ancestor
- List of microorganisms tested in outer space
- Planetary protection
- Rare Earth hypothesis
- Red rain in Kerala

8.8 References

[1] Wickramasinghe, Chandra (2011). “Bacterial morphologies supporting cometary panspermia: a reappraisal”. *International Journal of Astrobiology* **10** (1): 25–30. Bibcode:2011IJAsB..10...25W. doi:10.1017/S1473550410000157.

[2] Napier, William (October 2011). “Exchange of Biomaterial Between Planetary Systems” (PDF) **16**. pp. 6616–6642.

[3] Rampelotto, P. H. (2010). Panspermia: A promising field of research. In: Astrobiology Science Conference. Abs 5224.

[4] Forward planetary contamination like *Tersicoccus phoenicis*, that has shown resistance to methods usually used in spacecraft assembly clean rooms: Madhusoodanan, Jyoti (May 19, 2014). “Microbial stowaways to Mars identified”. *Nature (journal)*. doi:10.1038/nature.2014.15249. Retrieved May 23, 2014.

[5] Webster, Guy (November 6, 2013). “Rare New Microbe Found in Two Distant Clean Rooms”. *NASA.gov*. Retrieved November 6, 2013.

[6] A variation of the panspermia hypothesis is **necropanspermia** which is described by astronomer Paul Wesson as follows: “The vast majority of organisms reach a new home in the Milky Way in a technically dead state ... Resurrection may, however, be possible.” Grossman, Lisa (2010-11-10). “All Life on Earth Could Have Come From Alien Zombies”. *Wired*. Retrieved 10 November 2010.

[7] Hoyle, F. and Wickramasinghe, N.C., 1981. Evolution from Space (Simon & Schuster Inc., NY, 1981 and J.M. Dent and Son, Lond, 1981), ch3 pp. 35-49.

[8] Wickramasinghe, J., Wickramasinghe, C. and Napier, W., 2010. Comets and the Origin of Life (World Scientific, Singapore. 1981), ch6 pp. 137-154.

[9] Klyce, Brig (2001). “Panspermia Asks New Questions”. Retrieved 25 July 2013.

[10] Klyce, Brig (2001). Kingsley, Stuart A; Bhathal, Ragbir, eds. “The Search for Extraterrestrial Intelligence (SETI) in the Optical Spectrum III”. *Proc. SPIE Vol. 4273*. The Search for Extraterrestrial Intelligence (SETI) in the Optical Spectrum III **4273**: 11. Bibcode:2001SPIE.4273...11K. doi:10.1117/12.435366. |chapter= ignored (help)

[11] Dalgarno, A. (2006). “The galactic cosmic ray ionization rate”. *Proceedings of the National Academy of Sciences* **103** (33): 12269–73. Bibcode:2006PNAS..10312269D. doi:10.1073/pnas.0602117103. PMC 1567869. PMID 16894166.

[12] Brown, Laurie M.; Pais, Abraham; Pippard, A. B. (1995). “The physics of the interstellar medium”. *Twentieth Century Physics* (2nd ed.). CRC Press. p. 1765. ISBN 0-7503-0310-7.

[13] Borenstein, Seth (19 October 2015). “Hints of life on what was thought to be desolate early Earth”. *Excite* (Yonkers, NY: Mindspark Interactive Network). Associated Press. Retrieved 2015-10-20.

[14] Bell, Elizabeth A.; Boehnike, Patrick; Harrison, T. Mark; et al. (19 October 2015). “Potentially biogenic carbon preserved in a 4.1 billion-year-old zircon” (PDF). *Proc. Natl. Acad. Sci. U.S.A.* (Washington, D.C.: National Academy of Sciences). doi:10.1073/pnas.1517557112. ISSN 1091-6490. Retrieved 2015-10-20. Early edition, published online before print.

[15] Margaret O'Leary (2008) Anaxagoras and the Origin of Panspermia Theory, iUniverse publishing Group, # ISBN 978-0-595-49596-2

[16] Berzelius (1799-1848), J. J. “Analysis of the Alais meteorite and implications about life in other worlds”.

[17] Lynn J. Rothschild; Adrian M. Lister (June 2003). *Evolution on Planet Earth - The Impact of the Physical Environment*. Academic Press. pp. 109–127. ISBN 978-0-12-598655-7.

[18] Thomson (Lord Kelvin), W. (1871). “Inaugural Address to the British Association Edinburgh. “We must regard it as probably to the highest degree that there are countless seed-bearing meteoritic stones moving through space."". *Nature* **4** (92): 261–278 [262]. Bibcode:1871Natur...4..261.. doi:10.1038/004261a0.

[19] “The word: Panspermia”. *New Scientist* (2541). 7 March 2006. Retrieved 25 July 2013.

[20] “History of Panspermia”. Retrieved 25 July 2013.

[21] Arrhenius, S., *Worlds in the Making: The Evolution of the Universe*. New York, Harper & Row, 1908.

[22] Napier, W.M. (2007). “Pollination of exoplanets by nebulae”. *Int.J.Astrobiol* **6** (3): 223–228. Bibcode:2007IJAsB...6..223N. doi:10.1017/S1473550407003710.

[23] Line, M.A. (2007). “Panspermia in the context of the timing of the origin of life and microbial phylogeny”. *Int. J. Astrobiol.* 3 **6** (3): 249–254. Bibcode:2007IJAsB...6..249L. doi:10.1017/S1473550407003813.

[24] Wickramasinghe, D. T.; Allen, D. A. (1980). “The 3.4-µm interstellar absorption feature”. *Nature* **287** (5782): 518–519. Bibcode:1980Natur.287..518W. doi:10.1038/287518a0.

[25] Allen, D. A.; Wickramasinghe, D. T. (1981). "Diffuse interstellar absorption bands between 2.9 and 4.0 μm". *Nature* **294** (5838): 239–240. Bibcode:1981Natur.294..239A. doi:10.1038/294239a0.

[26] Wickramasinghe, D. T.; Allen, D. A. (1983). "Three components of 3?4 ?m absorption bands". *Astrophysics and Space Science* **97** (2): 369–378. Bibcode:1983Ap&SS..97..369W. doi:10.1007/BF00653492.

[27] Fred Hoyle; Chandra Wickramasinghe & John Watson (1986). *Viruses from Space and Related Matters*. University College Cardiff Press.

[28] Weaver, Rheyanne (April 7, 2009). "Ruminations on other worlds". *statepress.com*. Retrieved 25 July 2013.

[29] Khan, Amina (7 March 2014). "Did two planets around nearby star collide? Toxic gas holds hints". *LA Times*. Retrieved 9 March 2014.

[30] Dent, W. R. F.; Wyatt, M. C.; Roberge, A.; Augereau, J.- C.; et al. (6 March 2014). "Molecular Gas Clumps from the Destruction of Icy Bodies in the β Pictoris Debris Disk". *Science* **343** (6178): 1490–1492. Bibcode:2014Sci...343.1490D. doi:10.1126/science.1248726. Retrieved 9 March 2014.

[31] Wickramasinghe, Chandra; Wickramasinghe, Chandra; Napier, William (2009). *Comets and the Origin of Life*. World Scientific Press. doi:10.1142/6008. ISBN 978-981-256-635-5.

[32] Wall, Mike. "Comet Impacts May Have Jump-Started Life on Earth". space.com. Retrieved 1 August 2013.

[33] Weber, P; Greenberg, J. M. (1985). "Can spores survive in interstellar space?".*Nature***316**(6027): 403–407. Bibcode:1985Natu doi:10.1038/316403a0.

[34] Melosh, H. J. (1988). "The rocky road to panspermia". *Nature* **332** (6166): 687–688. Bibcode:1988Natur.332..687M. doi:10.1038/332687a0. PMID 11536601.

[35] C. Mileikowsky; F. A. Cucinotta; J. W. Wilson; B. Gladman; et al. (2000). "Risks threatening viable transfer of microbes between bodies in our solar system". *Planetary and Space Science* **48** (11): 1107–1115. Bibcode:2000P&SS...48.1107M. doi:10.1016/S0032-0633(00)00085-4.

[36] Studies Focus On Spacecraft Sterilization

[37] European Space Agency: Dry heat sterilisation process to high temperatures

[38] Edward Belbruno; Amaya Moro-Martı́n; Malhotra, Renu & Savransky, Dmitry (2012). "Chaotic Exchange of Solid Material between Planetary". *Astrobiology* **12** (8): 754–74. arXiv:1205.1059. Bibcode:2012AsBio..12..754B. doi:10.1089/ast.2012.0825. PMC 3440031. PMID 22897115.

[39] Slow-moving rocks better odds that life crashed to Earth from space News at Princeton, September 24, 2012.

[40] Crick, F. H.; Orgel, L. E. (1973). "Directed Panspermia".*Icarus***19**(3): 341–348. Bibcode:1979JBIS...32..419M.doi:10.1016/0 1035(73)90110-3.

[41] Mautner, Michael N. (2000). *Seeding the Universe with Life: Securing Our Cosmological Future* (PDF). Washington D. C.: Legacy Books (www.amazon.com). ISBN 0-476-00330-X.

[42] Mautner, M; Matloff, G. (1979). "Directed panspermia: A technical evaluation of seeding nearby solar systems" (PDF). *J. British Interplanetary Soc.* **32**: 419.

[43] Mautner, M. N. (1997). "Directed panspermia. 3. Strategies and motivation for seeding star-forming clouds" (PDF). *J. British Interplanetary Soc.* **50**: 93–102. Bibcode:1997JBIS...50...93M.

[44] BBC Staff (23 August 2011). "Impacts 'more likely' to have spread life from Earth". BBC. Retrieved 24 August 2011.

[45] "Electromagnetic space travel for bugs? - space - 21 July 2006 - New Scientist Space". Space.newscientist.com. Archived from the original on January 11, 2009. Retrieved December 8, 2014.

[46] Dehel, T. (2006-07-23). "Uplift and Outflow of Bacterial Spores via Electric Field". *36th COSPAR Scientific Assembly. Held 16–23 July 2006* (Adsabs.harvard.edu) **36**: 1. arXiv:hep-ph/0612311. Bibcode:2006cosp...36....1D.

[47] "Die Verbreitung des Lebens im Weltenraum" (the "Distribution of Life in Space"). Published in Die Umschau. 1903.

[48] *Ancient micronauts: interplanetary transport of microbes by cosmic impacts.* Wayne L. Nicholson. Trends in Microbiology, Vol. 17, No. 6. (June 2009), pp. 243-250, doi:10.1016/j.tim.2009.03.004

[49] Horneck, G.; Klaus, D. M.; Mancinelli, R. L. (2010). "Space Microbiology". *Microbiology and Molecular Biology Reviews* **74** (1): 121–56. doi:10.1128/MMBR.00016-09. PMC 2832349. PMID 20197502.

[50] I. S. Shklovskii; Carl Sagan (1966). *Intelligent Life in the Universe.* Emerson-Adams Press, Incorporated. ISBN 9781892803023.

[51] Wickramasinghe, M.K.; Wickramasinghe, C. (2004). "Interstellar transfer of planetary microbiota". *Mon. Not.R. Astr. Soc.* **348**: 52–57. Bibcode:2004MNRAS.348...52W. doi:10.1111/j.1365-2966.2004.07355.x.

[52] *Protection of Bacterial Spores in Space, a Contribution to the Discussion on Panspermia.* Gerda Horneck, Petra Rettberg, Günther Reitz, Jörg Wehner, Ute Eschweiler, Karsten Strauch, Corinna Panitz, Verena Starke, Christa Baumstark-Khan. Origins of life and evolution of the biosphere. December 2001, Volume 31, Issue 6, pp. 527-547.

[53] R.O. Rahn, J.L. Hosszu, Influence of relative humidity on the photochemistry of DNA films, Biochim. Biophys Acta 190 (1969) 126–131.

[54] M.H. Patrick, D.M. Gray, Independence of photproduct formation on DNA conformation, Photochem. Photobiol. 24 (1976) 507–513.

[55] Wayne L. Nicholson; Andrew C. Schuerger; Peter Setlow (21 January 2005). "The solar UV environment and bacterial spore UV resistance: considerations for Earth-to-Mars transport by natural processes and human spaceflight" (PDF). *Mutation Research* **571** (1–2): 249–264. doi:10.1016/j.mrfmmm.2004.10.012. PMID 15748651. Retrieved 2 August 2013.

[56] Clark BC., Planetary interchange of bioactive material: probability factors and implications Origins Life Evol Biosphere 2001; 31: 185-97

[57] Mileikowsky C. et al. Natural Transfer of Microbes in space, part I: from Mars to Earth and Earth to Mars Icarus 2000; 145; 391-427

[58] Olsson-Francis, Karen; Cockell, Charles S. (2010). "Experimental methods for studying microbial survival in extraterrestrial environments". *Journal of Microbiological Methods* **80** (1): 1–13. doi:10.1016/j.mimet.2009.10.004. PMID 19854226.

[59] Cockell, Charles S. (2007). "The Interplanetary Exchange of Photosynthesis". *Origins of Life and Evolution of Biospheres* **38**: 87–104. Bibcode:2008OLEB...38...87C. doi:10.1007/s11084-007-9112-3.

[60] Horneck, Gerda; Stöffler, Dieter; Ott, Sieglinde; Hornemann, Ulrich; et al. (2008). "Microbial Rock Inhabitants Survive Hypervelocity Impacts on Mars-Like Host Planets: First Phase of Lithopanspermia Experimentally Tested". *Astrobiology* **8** (1): 17–44. Bibcode:2008AsBio...8...17H. doi:10.1089/ast.2007.0134. PMID 18237257.

[61] Fajardo-Cavazos, Patricia; Link, Lindsey; Melosh, H. Jay; Nicholson, Wayne L. (2005). "Bacillus subtilisSpores on Artificial Meteorites Survive Hypervelocity Atmospheric Entry: Implications for Lithopanspermia". *Astrobiology* **5** (6): 726–36. Bibcode:2005AsBio...5..726F. doi:10.1089/ast.2005.5.726. PMID 16379527.

[62] Brack, A.; Baglioni, P.; Borruat, G.; Brandstätter, F.; et al. (2002). "Do meteoroids of sedimentary origin survive terrestrial atmospheric entry? The ESA artificial meteorite experiment STONE". *Planetary and Space Science* **50** (7–8): 763–772. Bibcode:2002P&SS...50..763B. doi:10.1016/S0032-0633(02)00018-1.

[63] Cockell, Charles S.; Brack, André; Wynn-Williams, David D.; Baglioni, Pietro; et al. (2007). "Interplanetary Transfer of Photosynthesis: An Experimental Demonstration of a Selective Dispersal Filter in Planetary Island Biogeography". *Astrobiology* **7** (1): 1–9. Bibcode:2007AsBio...7....1C. doi:10.1089/ast.2006.0038. PMID 17407400.

[64] "Could Life Have Survived a Fall to Earth?". *EPSC.* 12 September 2013. Retrieved 2015-04-21.

[65] Gold, T. "Cosmic Garbage", Air Force and Space Digest, 65 (May 1960).

[66] "Anticipating an RNA world. Some past speculations on the origin of life: where are they today?" by L. E. Orgel and F. H. C. Crick in *FASEB J.* (1993) Volume 7 pages 238-239.

[67] Clark, Benton C. Clark (February 2001). "Planetary Interchange of Bioactive Material: Probability Factors and Implications". *Origins of life and evolution of the biosphere* **31** (1–2): 185–197. Bibcode:2001OLEB...31..185C. doi:10.1023/A:1006757011007. PMID 11296521.

[68] Mautner, Michael N. (2009). "Life-centered ethics, and the human future in space" (PDF). *Bioethics* **23** (8): 433–440. doi:10.1111/j.1467-8519.2008.00688.x. PMID 19077128.

[69] Mautner, Michael Noah Ph.D. (2000). *Seeding the Universe with Life: Securing our Cosmological Future* (PDF). Legacy Books (www.amazon.com). ISBN 0-476-00330-X.

[70] Mautner, Michael N. (2002). "Planetary bioresources and astroecology. 1. Planetary microcosm bioessays of Martian and meteorite materials: soluble electrolytes, nutrients, and algal and plant responses" (PDF). *Icarus* **158** (1): 72–86. Bibcode:2002Icar..1 doi:10.1006/icar.2002.6841. PMID 12449855.

[71] Mautner, Michael N. (2005). "Life in the cosmological future: Resources, biomass and populations" (PDF). *Journal of the British Interplanetary Society* **58**: 167–180. Bibcode:2005JBIS...58..167M.

[72] G. Marx (1979). "Message through time". *Acta Astronautica* **6** (1–2): 221–225. Bibcode:1979AcAau...6..221M. doi:10.1016/0094-5765(79)90158-9.

[73] H. Yokoo; T. Oshima (1979). "Is bacteriophage φX174 DNA a message from an extraterrestrial intelligence?". *Icarus* **38** (1): 148–153. Bibcode:1979Icar...38..148Y. doi:10.1016/0019-1035(79)90094-0.

[74] Overbye, Dennis (26 June 2007). "Human DNA, the Ultimate Spot for Secret Messages (Are Some There Now?)". Retrieved 2014-10-09.

[75] Davies, Paul C.W. (2010). *The Eerie Silence: Renewing Our Search for Alien Intelligence*. Boston, Massachusetts: Houghton Mifflin Harcourt. ISBN 978-0-547-13324-9.

[76] V. I. shCherbak; M. A. Makukov (2013). "The "Wow! signal" of the terrestrial genetic code". *Icarus* **224** (1): 228–242. Bibcode:2013Icar..224..228S. doi:10.1016/j.icarus.2013.02.017.

[77] Makukov, Maxim (4 October 2014). "Claim to have identified extraterrestrial signal in the universal genetic code thereby confirming directed panspermia.". *Maxim Makukov*. The New Reddit Journal of Science. Retrieved 2014-10-09.

[78] M. A. Makukov; V. I. shCherbak (2014). "Space ethics to test directed panspermia". *Life Sciences in Space Research* **3**: 10–17. doi:10.1016/j.lssr.2014.07.003.

[79] "'Seed of Life' From Outer Space Suggests Aliens Created Life On Earth, U.K. Scientists Say". *The Inquisitr*. February 13, 2015. Retrieved 2015-03-11.

[80] N.C. Wickramasinghe, Formaldehyde Polymers in Interstellar Space, Nature, 252, 462, 1974.

[81] Martins, Zita; Botta, Oliver; Fogel, Marilyn L.; Sephton, Mark A.; Glavin, Daniel P.; Watson, Jonathan S.; Dworkin, Jason P.; Schwartz, Alan W.; Ehrenfreund, Pascale (2008). "Extraterrestrial nucleobases in the Murchison meteorite". *Earth and Planetary Science Letters* **270**: 130–136. Bibcode:2008E&PSL.270..130M. doi:10.1016/j.epsl.2008.03.026.

[82] AFP Staff (20 August 2009). "We may all be space aliens: study". AFP. Archived from the original on June 17, 2008. Retrieved 8 November 2014.

[83] Martins, Zita; Botta, Oliver; Fogel, Marilyn L.; Sephton, Mark A.; Glavin, Daniel P.; Watson, Jonathan S.; Dworkin, Jason P.; Schwartz, Alan W.; Ehrenfreund, Pascale (2008). "Extraterrestrial nucleobases in the Murchison meteorite". *Earth and Planetary Science Letters* **270**: 130–136. Bibcode:2008E&PSL.270..130M. doi:10.1016/j.epsl.2008.03.026.

[84] "'Life chemical' detected in comet". *NASA* (BBC News). 18 August 2009. Retrieved 6 March 2010.

[85] Callahan, M. P.; Smith, K. E.; Cleaves, H. J.; Ruzicka, J.; et al. (2011). "Carbonaceous meteorites contain a wide range of extraterrestrial nucleobases". *Proceedings of the National Academy of Sciences* **108** (34): 13995–8. Bibcode:2011PNAS..10813995C. doi:10.1073/pnas.1106493108. PMC 3161613. PMID 21836052.

[86] Steigerwald, John (8 August 2011). "NASA Researchers: DNA Building Blocks Can Be Made in Space". NASA. Retrieved 10 August 2011.

[87] ScienceDaily Staff (9 August 2011). "DNA Building Blocks Can Be Made in Space, NASA Evidence Suggests". ScienceDaily. Retrieved 9 August 2011.

[88] Chow, Denise (26 October 2011). "Discovery: Cosmic Dust Contains Organic Matter from Stars". Space.com. Retrieved 26 October 2011.

[89] ScienceDaily Staff (26 October 2011). "Astronomers Discover Complex Organic Matter Exists Throughout the Universe". ScienceDaily. Retrieved 27 October 2011.

[90] Kwok, Sun; Zhang, Yong (2011). "Mixed aromatic–aliphatic organic nanoparticles as carriers of unidentified infrared emission features". *Nature* **479** (7371): 80–3. Bibcode:2011Natur.479...80K. doi:10.1038/nature10542. PMID 22031328.

[91] Than, Ker (August 29, 2012). "Sugar Found In Space". *National Geographic*. Retrieved August 31, 2012.

[92] Staff (August 29, 2012). "Sweet! Astronomers spot sugar molecule near star". AP News. Retrieved August 31, 2012.

[93] Jørgensen, Jes K.; Favre, Cécile; Bisschop, Suzanne E.; Bourke, Tyler L.; et al. (2012). "Detection of the Simplest Sugar, Glycolaldehyde, in a Solar-Type Protostar with Alma". *The Astrophysical Journal* **757**: L4. Bibcode:2012ApJ...757L...4J. doi:10.1088/2041-8205/757/1/L4.

[94] Staff (September 20, 2012). "NASA Cooks Up Icy Organics to Mimic Life's Origins". Space.com. Retrieved September 22, 2012.

[95] Gudipati, Murthy S.; Yang, Rui (2012). "In-Situ Probing of Radiation-Induced Processing of Organics in Astrophysical Ice Analogs—Novel Laser Desorption Laser Ionization Time-Of-Flight Mass Spectroscopic Studies". *The Astrophysical Journal* **756**: L24. Bibcode:2012ApJ...756L..24G. doi:10.1088/2041-8205/756/1/L24.

[96] Loomis, Ryan A.; Zaleski, Daniel P.; Steber, Amanda L.; Neill, Justin L.; et al. (2013). "The Detection of Interstellar Ethanimine (Ch3Chnh) from Observations Taken During the Gbt Primos Survey". *The Astrophysical Journal* **765**: L9. Bibcode:2013ApJ...765L...9L. doi:10.1088/2041-8205/765/1/L9.

[97] Finley, Dave,*Discoveries Suggest Icy Cosmic Start for Amino Acids and DNA Ingredients,* The National Radio Astronomy Observatory, Feb. 28, 2013

[98] Kaiser, R. I.; Stockton, A. M.; Kim, Y. S.; Jensen, E. C.; et al. (March 5, 2013). "On the Formation of Dipeptides in Interstellar Model Ices". *The Astrophysical Journal* **765** (2): 111. Bibcode:2013ApJ...765..111K. doi:10.1088/0004-637X/765/2/111. Lay summary – *Phys.org*.

[99] Hoover, Rachel (February 21, 2014). "Need to Track Organic Nano-Particles Across the Universe? NASA's Got an App for That". *NASA*. Retrieved 22 February 2014.

[100] Marlaire, Ruth (3 March 2015). "NASA Ames Reproduces the Building Blocks of Life in Laboratory". *NASA*. Retrieved 5 March 2015.

[101] Loeb, Abraham (October 2014). "The Habitable Epoch of the Early Universe". *International Journal of Astrobiology* (Cambridge) **13** (4): 337–39. Bibcode:2014IJAsB..13..337L. doi:10.1017/S1473550414000196. Retrieved 15 December 2014.

[102] Loeb, Abraham(2 December 2013). "The Habitable Epoch of the Early Universe"(PDF).*Arxiv***13**(4): 337–339. arXiv:1312. Bibcode:2014IJAsB..13..337L. doi:10.1017/S1473550414000196. Retrieved 15 December 2014.

[103] Dreifus, Claudia (2 December 2014). "Much-Discussed Views That Go Way Back – Avi Loeb Ponders the Early Universe, Nature and Life". *The New York Times*. Retrieved 3 December 2014.

[104] Rampelotto, P.H. (2010). "Panspermia: A Promising Field Of Research" (PDF). *Astrobiology Science Conference*. Harvard: USRA. Retrieved 3 December 2014. External link in |work= (help)

[105] Graham, Robert W (February 1990). "Extraterrestrial Life in the Universe" (PDF). *Technical Memorandum* (Lewis Research Center, OH: NASA). 102363. Retrieved 7 July 2014.

[106] Altermann, Wladyslaw (2008). "From Fossils to Astrobiology – A Roadmap to Fata Morgana?". In Seckbach, Joseph; Walsh, Maud. *From Fossils to Astrobiology: Records of Life on Earth and the Search for Extraterrestrial Biosignatures* **12**. p. xvii. ISBN 1-4020-8836-1.

[107] Webb, Stephen (2002), *If the universe is teeming with aliens, where is everybody? Fifty solutions to the Fermi paradox and the problem of extraterrestrial life*, Copernicus, Springer.

[108] Steffen, Jason H.; Batalha, Natalie M.; Borucki, William J; Buchhave, Lars A.; et al. (9 November 2010). "Five Kepler target stars that show multiple transiting exoplanet candidates". *Astrophysical Journal* **725**: 1226–41. arXiv:1006.2763. Bibcode:2010ApJ...725.1226S. doi:10.1088/0004-637X/725/1/1226.

[109] Overbye, Dennis (November 4, 2013). "Far-Off Planets Like the Earth Dot the Galaxy". *The New York Times*. Retrieved 5 November 2013.

[110] Petigura, Eric A.; Howard, Andrew W.; Marcy, Geoffrey W (October 31, 2013). "Prevalence of Earth-size planets orbiting Sun-like stars". *Proceedings of the National Academy of Sciences of the United States of America* **110** (48): 19273–78. Bibcode:2013PNAS..11019273P. doi:10.1073/pnas.1319909110. Retrieved 5 November 2013.

[111] Khan, Amina (November 4, 2013). "Milky Way may host billions of Earth-size planets". *The Los Angeles Times*. Retrieved 5 November 2013.

[112] Crawford, I.A. (Sep 1995). "Some Thoughts on the Implications of Faster-Than-Light Interstellar Space Travel". *Quarterly Journal of the Royal Astronomical Society* **36** (3): 205. Bibcode:1995QJRAS..36..205C.

[113] Byrne, Joseph Patrick (2008). "Panspermia". *Encyclopedia of Pestilence, Pandemics, and Plagues* (entry). ABC-CLIO. pp. 454–55. ISBN 978-0-313-34102-1.

[114] Wickramasinghe, C; Wainwright, M; Narlikar, J (May 24, 2003). "SARS—a clue to its origins?". *Lancet* **361** (9371): 1832. doi:10.1016/S0140-6736(03)13440-X. PMID 12781581.

[115] Willerslev, E; Hansen, AJ; Rønn, R; Nielsen, OJ (Aug 2, 2003). "Panspermia – true or false?". *Lancet* **362** (9381): 406; author reply 407–8. doi:10.1016/S0140-6736(03)14039-1. PMID 12907025.

[116] Bhargava, PM (Aug 2, 2003). "Panspermia – true or false?". *Lancet* **362** (9381): 407; author reply 407–8. doi:10.1016/S0140-6736(03)14041-X. PMID 12907028.

[117] Ponce de Leon, S; Lazcano, A (Aug 2, 2003). "Panspermia – true or false?". *Lancet* **362** (9381): 406–7; author reply 407–8. doi:10.1016/s0140-6736(03)14040-8. PMID 12907026.

[118] "New Study Adds to Finding of Ancient Life Signs in Mars Meteorite". NASA. 2009-11-30. Retrieved 1 December 2009.

[119] Thomas-Keprta, K.; Clemett, S; McKay, D; Gibson, E & Wentworth, S (2009). "Origin of Magnetite Nanocrystals in Martian Meteorite ALH84001".*Geochimica et Cosmochimica Acta***73**(73): 6631–6677. Bibcode:2009GeCoA..73.6631T.doi:10.1016/

[120] "Alien visitors". *New Scientist Space*. 11 May 2001. Retrieved 20 August 2009.

[121] D'Argenio, Bruno; Geraci, Giuseppe & del Gaudio, Rosanna (March 2001). "Microbes in rocks and meteorites: a new form of life unaffected by time, temperature, pressure". *Rendiconti Lincei* **12** (1): 51–68. doi:10.1007/BF02904521. Retrieved 13 October 2009.

[122] Geraci, Giuseppe; del Gaudio, Rosanna; D'Argenio, Bruno (2001), "Microbes in rocks and meteorites: a new form of life unaffected by time, temperature, pressure" (PDF), *Rend. Fis. Acc. Linceis* **12**: 51–68 .

[123] "Scientists Say They Have Found Extraterrestrial Life in the Stratosphere But Peers Are Skeptical: Scientific American". Sciam. 2001-07-31. Retrieved 20 August 2009.

[124] Narlikar, JV; Lloyd, D; Wickramasinghe, NC; Turner; Al-Mufti; Wallis; Wainwright; Rajaratnam; Shivaji; Reddy; Ramadurai; Hoyle (2003). "Balloon experiment to detect micro-organisms in the outer space". *Astrophys Space Science* **285** (2): 555–62. Bibcode:2003Ap&SS.285..555N. doi:10.1023/A:1025442021619. Missing |last4= in Authors list (help)

[125] Wainwright, M; Wickramasinghe, N.C; Narlikar, J.V; Rajaratnam, P. "Microorganisms cultured from stratospheric air samples obtained at 41km". Retrieved 11 May 2007.

[126] Wainwright, M (2003). "A microbiologist looks at panspermia".*Astrophys Space Science***285**(2): 563–70. Bibcode:2003Ap&S doi:10.1023/A:1025494005689.

[127] Stenger, Richard (2000-11-24). "Space: Scientists discover possible microbe from space". *CNN*. Retrieved 20 August 2009.

[128] Vaidya, Pushkar Ganesh (July 2009). "Critique on Vindication of Panspermia" (PDF). *Apeiron* **16** (3). Retrieved 28 November 2009.

[129] *Mumbai scientist challenges theory that bacteria came from space*, IN: AOL.

[130] *Janibacter hoylei sp. nov.*, *Bacillus isronensis sp. nov.* and *Bacillus aryabhattai sp. nov.*, isolated from cryotubes used for collecting air from upper atmosphere. *International Journal of Systematic and Evolutionary Microbiology* 2009. http://ijs.sgmjournals.org/cgi/content/abstract/ijs.0.002527-0v1

[131] Discovery of New Microorganisms in the Stratosphere.

[132] Timothy Oleson (May 5, 2013). "Lofted by hurricanes, bacteria live the high life". *NASA* (Earth Magazine). Retrieved 21 September 2013.

[133] Helen Shen (28 January 2013). "High-flying bacteria spark interest in possible climate effects".*Nature News*. doi:10.1038/nature Retrieved 21 September 2013.

[134] Wickramasinghe, N. C.; Wallis, J.; Wallis, D. H.; Samaranayake, Anil (January 10, 2013). "Fossil Diatoms in a New Carbonaceous Meteorite" (PDF). *Journal of Cosmology* (Journal of Cosmology) **21** (37): 1–14. arXiv:1303.2398. Bibcode:2013JCos... 21.9560W.Retrieved January14, 2013.

[135] Phil Plait (15 January 2013). "No, Diatoms Have Not Been Found in a Meteorite". *Slate.com - Astronomy*. Retrieved 16 January 2013.

[136] Wallis, Jamie; Miyake, Nori; Hoover, Richard B.; Oldroyd, Andrew; et al. (5 March 2013). "The Polonnaruwa meteorite: oxygen isotope, crystalline and biological composition" (PDF). *Journal of Cosmology* **22** (2): 1845. arXiv:1303.1845. Bibcode:2013JCos...2210004W. Retrieved 7 March 2013.

[137] Wickramasinghe, N.C.; J. Wallis; N. Miyake; Anthony Oldroyd; et al. (4 February 2013). "Authenticity of the life-bearing Polonnaruwa meteorite" (PDF). *Journal of Cosmology*. Retrieved 4 February 2013.

[138] Griffin, Dale Warren (14 August 2013). "The Quest for Extraterrestrial Life: What About the Viruses?". *Astrobiology* **13** (8): 774–783. Bibcode:2013AsBio..13..774G. doi:10.1089/ast.2012.0959.

[139] Edward Anders, Eugene R. DuFresne,Ryoichi Hayatsu, Albert Cavaille, Ann DuFresne, and Frank W. Fitch. "Contaminated Meteorite", *Science, New Series*, Volume 146, Issue 3648 (Nov.27, 1964), 1157-1161.

[140] Chamberlin, Sean (1999). "Black Smokers and Giant Worms". *Fullerton College*. Retrieved 11 February 2011.

[141] Choi, Charles Q. (17 March 2013). "Microbes Thrive in Deepest Spot on Earth". LiveScience. Retrieved 17 March 2013.

[142] Oskin, Becky (14 March 2013). "Intraterrestrials: Life Thrives in Ocean Floor". LiveScience. Retrieved 17 March 2013.

[143] Glud, Ronnie; Wenzhöfer, Frank; Middelboe, Mathias; Oguri, Kazumasa; et al. (17 March 2013). "High rates of microbial carbon turnover in sediments in the deepest oceanic trench on Earth". *Nature Geoscience* **6** (4): 284–288. Bibcode:2013NatGe...6.. 284G.doi:10.1038/ngeo1773. Retrieved17March2013.

[144] Carey, Bjorn (7 February 2005). "Wild Things: The Most Extreme Creatures". *Live Science*. Retrieved 20 October 2008.

[145] Cavicchioli, R. (Fall 2002). "Extremophiles and the search for extraterrestrial life".*Astrobiology***2**(3): 281–92. Bibcode:2002. doi:10.1089/153110702762027862. PMID 12530238.

[146] The BIOPAN experiment MARSTOX II of the FOTON M-3 mission July 2008.

[147] Surviving the Final Frontier. 25 November 2002.

[148] Christner, Brent C. (2002). "Detection, recovery, isolation, and characterization of bacteria in glacial ice and Lake Vostok accretion ice". *Ohio State University*. Retrieved 4 February 2011.

[149] Nanjundiah, V. (2000). "The smallest form of life yet?" (PDF). *Journal of Biosciences* **25** (1): 9–10. doi:10.1007/BF02985175. PMID 10824192.

[150] Rabbow, Elke Rabbow; Gerda Horneck; Petra Rettberg; Jobst-Ulrich Schott; et al. (9 July 2009). "EXPOSE, an Astrobiological Exposure Facility on the International Space Station - from Proposal to Flight" (PDF). *Orig Life Evol Biosph* **39** (6): 581–98. doi:10.1007/s11084-009-9173-6. PMID 19629743. Retrieved 8 July 2013.

[151] Bacterium revived from 25 million year sleep Digital Center for Microbial Ecology

[152] Tepfer, David Tepfer (December 2008). "The origin of life, panspermia and a proposal to seed the Universe". *Plant Science* **175** (6): 756–760. doi:10.1016/j.plantsci.2008.08.007.

[153] "Exobiology and Radiation Assembly (ERA)". *ESA*. NASA. 1992. Retrieved 22 July 2013.

[154] Zhang, K. Dose; A. Bieger-Dose; R. Dillmann; M. Gill; et al. (1995). "ERA-experiment "space biochemistry"". *Advances in Space Research* **16** (8): 119–129. doi:10.1016/0273-1177(95)00280-R. PMID 11542696.

[155] Vaisberg, Horneck G; Eschweiler U; Reitz G; Wehner J; et al. (1995). "Biological responses to space: results of the experiment "Exobiological Unit" of ERA on EURECA I". *Adv Space Res.* **16** (8): 105–18. Bibcode:1995AdSpR..16..105V. doi:10.1016/0273-1177(95)00279-N. PMID 11542695.

[156] "BIOPAN Pan for exposure to space environment". *Kayser Italia.* 2013. Retrieved 17 July 2013.

[157] De La Torre Noetzel, Rosa (2008). "Experiment lithopanspermia: Test of interplanetary transfer and re-entry process of epi- and endolithic microbial communities in the FOTON-M3 Mission". *37th COSPAR Scientific Assembly. Held 13–20 July 2008* **37**: 660. Bibcode:2008cosp...37..660D.

[158] "Life in Space for Life ion Earth - Biosatelite Foton M3". June 26, 2008. Retrieved 13 October 2009.

[159] Jönsson, K. Ingemar Jönsson; Elke Rabbow; Ralph O. Schill; Mats Harms-Ringdahl; et al. (9 September 2008). "Tardigrades survive exposure to space in low Earth orbit". *Current Biology* **18** (17): R729–R731. doi:10.1016/j.cub.2008.06.048. PMID 18786368.

[160] de Vera; J.P.P.; et al. (2010). "COSPAR 2010 Conference". Research Gate. Retrieved 17 July 2013

[161] Paul Clancy (Jun 23, 2005). *Looking for Life, Searching the Solar System.* Cambridge University Press. Retrieved 26 March 2014.

[162] Tepfer, David Tepfer; Andreja Zalar & Sydney Leach. (May 2012). "Survival of Plant Seeds, Their UV Screens, and nptII DNA for 18 Months Outside the International Space Station". *Astrobiology* **12** (5): 517–528. Bibcode:2012AsBio..12..517T. doi:10.1089/ast.2011.0744. PMID 22680697.

[163] Scalzi, Giuliano Scalzi; Laura Selbmann; Laura Zucconi; Elke Rabbow; et al. (1 June 2012). "LIFE Experiment: Isolation of Cryptoendolithic Organisms from Antarctic Colonized Sandstone Exposed to Space and Simulated Mars Conditions on the International Space Station". *Origins of Life and Evolution of Biospheres* **42** (2 – 3): 253–262. doi:10.1007/s11084-012-9282-5.

[164] Onofri, Silvano Onofri; Rosa de la Torre; Jean-Pierre de Vera; Sieglinde Ott; et al. (May 2012). "Survival of Rock-Colonizing Organisms After 1.5 Years in Outer Space". *Astrobiology* **12** (5): 508–516. Bibcode:2012AsBio..12..508O. doi:10.1089/ast.201 1.0736.PMID22680696.

[165] Baldwin, Emily (26 April 2012). "Lichen survives harsh Mars environment". Skymania News. Retrieved 27 April 2012.

[166] de Vera, J.-P.; Kohler, Ulrich (26 April 2012). "The adaptation potential of extremophiles to Martian surface conditions and its implication for the habitability of Mars" (PDF). European Geosciences Union. Retrieved 27 April 2012.

[167] Amos, Jonathan (23 August 2010). "Beer microbes live 553 days outside ISS". *BBC News - Science and Technology.* Retrieved 31 July 2013.

[168] "Nol bugs please, this is a clean planet!". European Space Agency (ESA). 30 July 2002. Retrieved 16 July 2013.

[169] Jordans, Frank (30 July 2015). "Philae probe finds evidence that comets can be cosmic labs". *The Washington Post.* Associated Press. Retrieved 30 July 2015.

[170] "Science on the Surface of a Comet". European Space Agency. 30 July 2015. Retrieved 30 July 2015.

[171] Bibring, J.-P.; Taylor, M.G.G.T.; Alexander, C.; Auster, U.; Biele, J.; Finzi, A. Ercoli; Goesmann, F.; Klingehoefer, G.; Kofman, W.; Mottola, S.; Seidenstiker, K.J.; Spohn, T.; Wright, I. (31 July 2015). "Philae's First Days on the Comet - Introduction to Special Issue". *Science* **349** (6247): 493. doi:10.1126/science.aac5116. Retrieved 30 July 2015.

[172] "LIFE Experiment". Planetary.org. Retrieved 20 August 2009.

[173] "Living interplanetary flight experiment: an experiment on survivability of microorganisms during interplanetary transfer" (PDF). Retrieved 20 August 2009.

[174] "Projects: LIFE Experiment: Phobos". The Planetary Society. Retrieved 2 April 2011.

[175] Zak, Anatoly (1 September 2008). "Mission Possible". *Air & Space Magazine.* Smithsonian Institution. Retrieved 26 May 2009.

[176] Nussinov, M. D; Lysenko, S. V (1983), "Cosmic vacuum prevents radiopanspermia", *Orig. Life* **13**: 153–64.

[177] Wickramasinghe, M.K.; Wickramasinghe, C. (2004). "Interstellar transfer of planetary microbiota,". *Mon. Not.R. Astr. Soc.* **348**: 52–57. Bibcode:2004MNRAS.348...52W. doi:10.1111/j.1365-2966.2004.07355.x.

8.9 Further reading

- Crick, F (1981), *Life, Its Origin and Nature*, Simon & Schuster, ISBN 0-7088-2235-5.
- Hoyle, F (1983), *The Intelligent Universe*, London: Michael Joseph, ISBN 0-7181-2298-4.

8.10 External links

- A.E. Zlobin, 2013, Tunguska similar impacts and origin of life (mathematical theory of origin of life; incoming of pattern recognition algorithm due to comets)
- Francis Crick's notes for a lecture on directed panspermia, dated 5 November 1976.
- "Earth sows its seeds in space". *Nature News*. 23 February 2004. doi:10.1038/news040216-20 (inactive 2015-02-01).
- Warmflash, D.; Weiss, B. (24 October 2005). "Did Life Come from Another World?". *Scientific American* **293**: 64–71. doi:10.1038/scientificamerican1105-64.

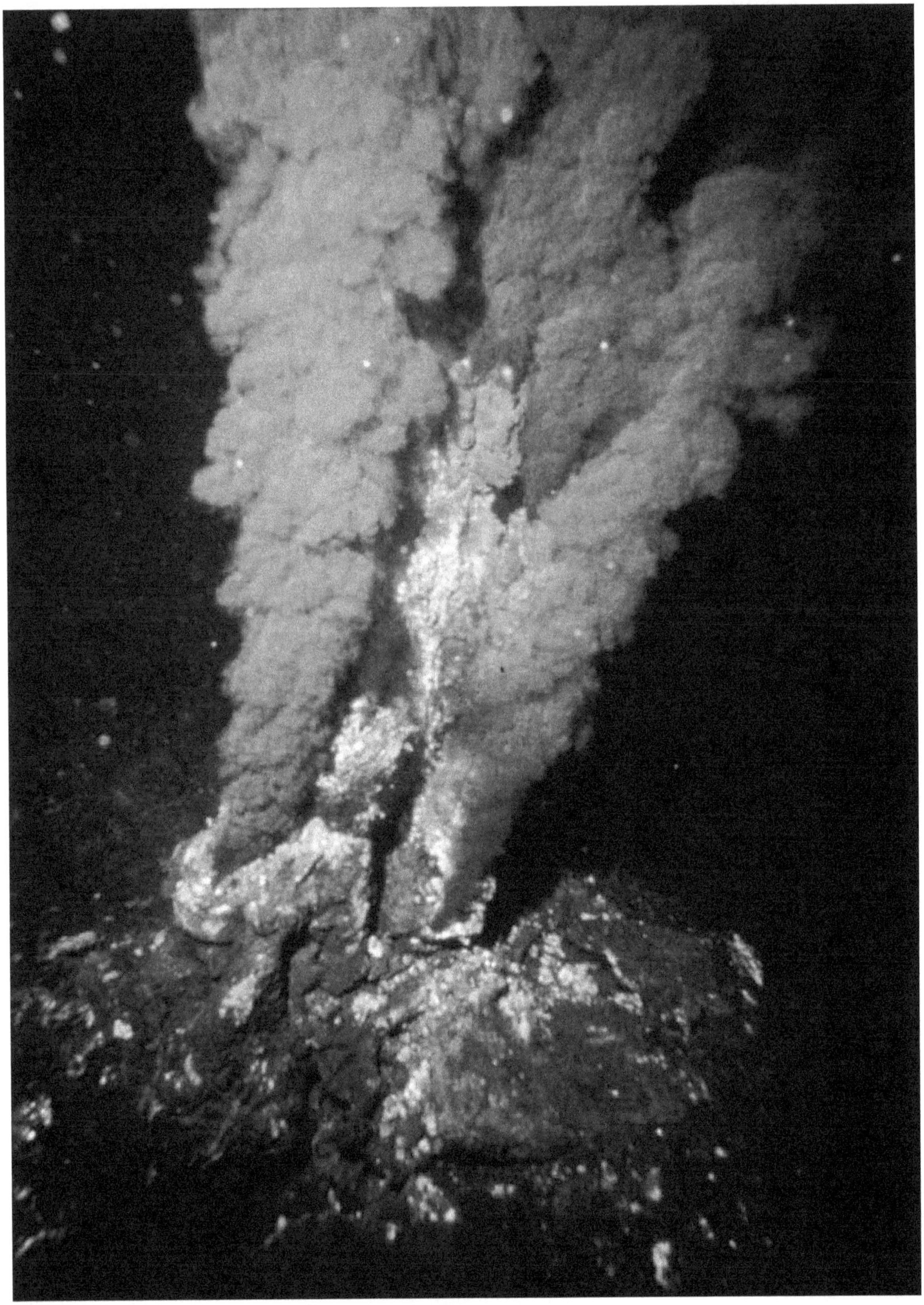

Hydrothermal vents are able to support extremophile bacteria on Earth and may also support life in other parts of the cosmos.

EURECA facility deployment in 1992

EXOSTACK on the Long Duration Exposure Facility satellite.

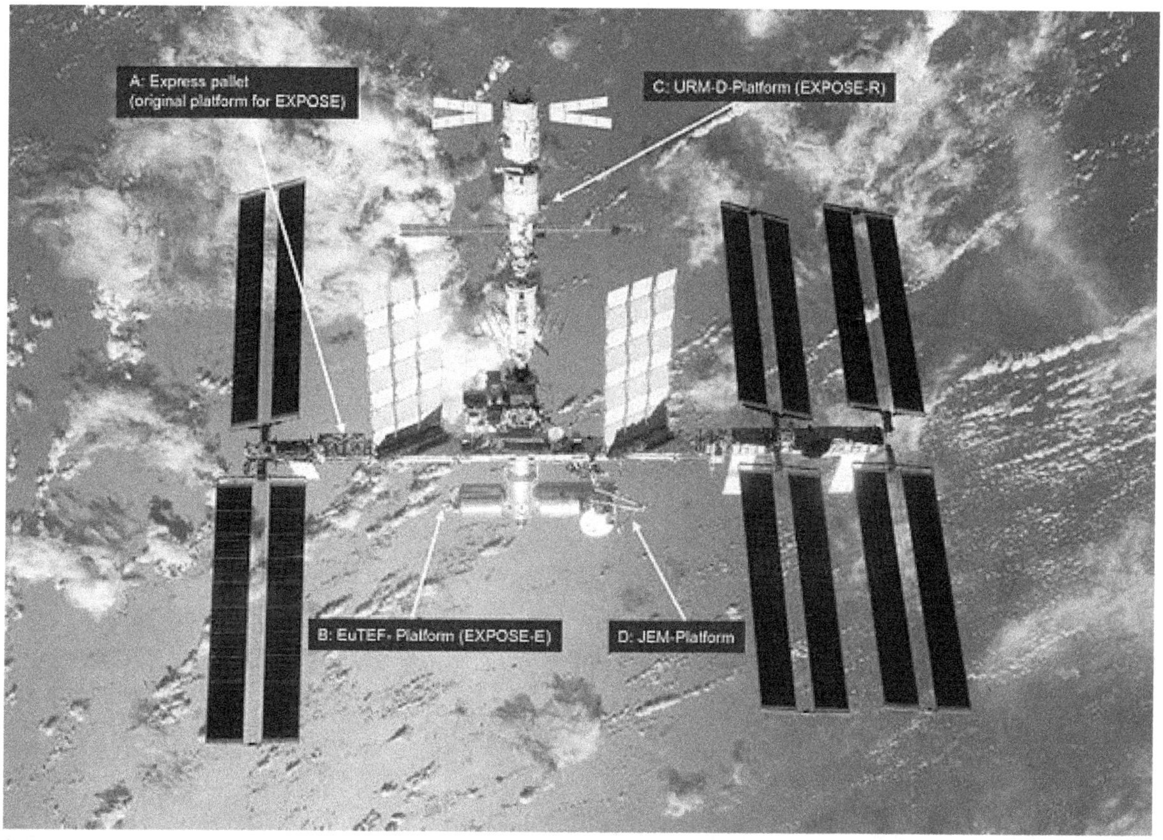

Location of the astrobiology EXPOSE-E and EXPOSE-R facilities on the International Space Station

Chapter 9

Abiogenesis

“Origin of life” redirects here. For non-scientific views on the origins of life, see Creation myth.

Abiogenesis (Brit.: /ˌeɪbaɪ.ɵˈdʒɛnɨsɪs/ ***AY**-by-oh-**JEN**-ə-siss*[1] U.S. English pronunciation: /ˌeɪˌbaɪoʊˈdʒɛnɪsɪs/),[2] or **biopoiesis**,[3]

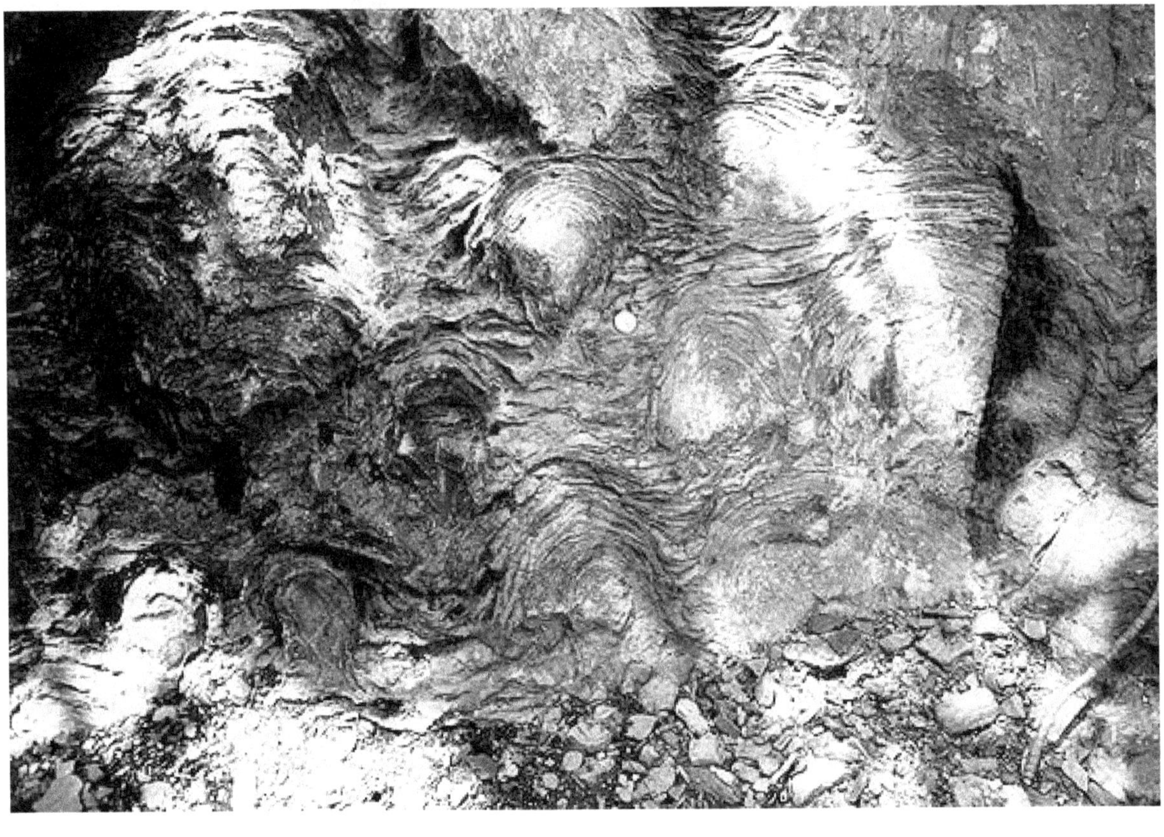

Precambrian stromatolites in the Siyeh Formation, Glacier National Park. In 2002, a paper in the scientific journal Nature *suggested that these 3.5 Ga (billion years) old geological formations contain fossilized cyanobacteria microbes. This suggests they are evidence of one of the earliest known life forms on Earth.*

is the natural process of life arising from non-living matter, such as simple organic compounds.[4][5][6][7] It is thought to have occurred on Earth between 3.8 and 4 billion years ago, and is studied through a combination of laboratory experiments and extrapolation from the genetic information of modern organisms in order to make reasonable conjectures about what pre-life chemical reactions may have given rise to a living system.[8]

The study of abiogenesis involves three main types of considerations: the geophysical, the chemical, and the biological,[9] with more recent approaches attempting a synthesis of all three. Many approaches investigate how self-replicating molecules, or their components, came into existence. It is generally accepted that current life on Earth descended from an RNA world,[10] although RNA-based life may not have been the first life to have existed.[11][12] The Miller–Urey experiment and similar experiments demonstrated that most amino acids, basic chemicals of life, can be synthesized from inorganic compounds in conditions intended to be similar to early Earth. Several mechanisms of organic molecule synthesis have been investigated, including lightning and radiation. Other approaches ("metabolism first" hypotheses) focus on understanding how catalysis in chemical systems on the early Earth might have provided the precursor molecules necessary for self-replication.[13] Complex organic molecules have been found in the Solar System and in interstellar space, and these molecules may have provided starting material for the development of life on Earth.[14][15][16][17]

The panspermia hypothesis suggests that microscopic life was distributed by meteoroids, asteroids and other small Solar System bodies and that life may exist throughout the Universe.[18] It is speculated that the biochemistry of life may have begun shortly after the Big Bang, 13.8 billion years ago, during a habitable epoch when the age of the universe was only 10–17 million years.[19][20] Panspermia hypothesis answers the question as from whence life, not how life came to be; it only relocates the origin of life to a locale outside the Earth.

Nonetheless, Earth is the only place in the Universe known to harbor life.[21][22] The age of the Earth is about 4.54 billion years.[23][24][25] The earliest undisputed evidence of life on Earth dates at least from 3.5 billion years ago,[26][27][28] during the Eoarchean Era after a geological crust started to solidify following the earlier molten Hadean Eon. There are microbial mat fossils found in 3.48 billion-year-old sandstone discovered in Western Australia.[29][30][31] Other early physical evidence of a biogenic substance is graphite in 3.7 billion-year-old metasedimentary rocks discovered in southwestern Greenland[32] as well as "remains of biotic life" found in 4.1 billion-year-old rocks in Western Australia.[33][34] According to one of the researchers, "If life arose relatively quickly on Earth ... then it could be common in the universe."[33]

9.1 Early geophysical conditions

Main article: Timeline of the evolutionary history of life

Based on recent computer model studies, the complex organic molecules necessary for life may have formed in the protoplanetary disk of dust grains surrounding the Sun before the formation of the Earth.[35] According to the computer studies, this same process may also occur around other stars that acquire planets.[35] (Also see Extraterrestrial organic molecules).

The Hadean Earth is thought to have had a secondary atmosphere, formed through degassing of the rocks that accumulated from planetesimal impactors. At first, it was thought that the Earth's atmosphere consisted of hydrides—methane, ammonia and water vapour—and that life began under such reducing conditions, which are conducive to the formation of organic molecules. During its formation, the Earth lost a significant part of its initial mass, with a nucleus of the heavier rocky elements of the protoplanetary disk remaining.[36] According to later models, suggested by study of ancient minerals, the atmosphere in the late Hadean period consisted largely of nitrogen and carbon dioxide, with smaller amounts of carbon monoxide, hydrogen, and sulfur compounds.[37] As Earth lacked the gravity to hold any molecular hydrogen, this component of the atmosphere would have been rapidly lost during the Hadean period, along with the bulk of the original inert gases. The solution of carbon dioxide in water is thought to have made the seas slightly acidic, giving it a pH of about 5.5.[38] The atmosphere at the time has been characterized as a "gigantic, productive outdoor chemical laboratory."[39] It may have been similar to the mixture of gases released today by volcanoes, which still support some abiotic chemistry.[39]

Oceans may have appeared first in the Hadean Eon, as soon as two hundred million years (200 Ma) after the Earth was formed, in a hot 100 °C (212 °F) reducing environment, and the pH of about 5.8 rose rapidly towards neutral.[40] This has been supported by the dating of 4.404 Ga-old zircon crystals from metamorphosed quartzite of Mount Narryer in Western Australia, which are evidence that oceans and continental crust existed within 150 Ma of Earth's formation.[41] Despite the likely increased vulcanism and existence of many smaller tectonic "platelets," it has been suggested that between 4.4 and 4.3 Ga (billion year), the Earth was a water world, with little if any continental crust, an extremely turbulent atmosphere and a hydrosphere subject to intense ultraviolet (UV) light, from a T Tauri stage Sun, cosmic radiation and continued bolide impacts.[42]

The Hadean environment would have been highly hazardous to modern life. Frequent collisions with large objects, up to

500 kilometres (310 mi) in diameter, would have been sufficient to sterilise the planet and vaporise the ocean within a few months of impact, with hot steam mixed with rock vapour becoming high altitude clouds that would completely cover the planet. After a few months, the height of these clouds would have begun to decrease but the cloud base would still have been elevated for about the next thousand years. After that, it would have begun to rain at low altitude. For another two thousand years, rains would slowly have drawn down the height of the clouds, returning the oceans to their original depth only 3,000 years after the impact event.[43]

9.1.1 The earliest biological evidence for life on Earth

The earliest life on Earth existed before 3.5 billion years ago,[26][27][28] during the Eoarchean Era when sufficient crust had solidified following the molten Hadean Eon. Physical evidence has been found in biogenic graphite in 3.7 billion-year-old metasedimentary rocks from southwestern Greenland[32] and microbial mat fossils found in 3.48 billion-year-old sandstone from Western Australia.[29][31] Evidence of early life in rocks from Akilia Island, near the Isua supracrustal belt in southwestern Greenland, dating to 3.7 billion years ago have shown biogenic carbon isotopes.[44] At Strelley Pool, in the Pilbarra region of Western Australia, compelling evidence of early life has been found in pyrite-bearing sandstone in a fossilized beach, that showed rounded tubular cells that oxidised sulfur by photosynthesis in the absence of oxygen.[45] More recently, UCLA geochemists have found evidence that life likely existed on Earth at least 4.1 billion years ago — 300 million years earlier than previous research suggested.[33][34][46]

In the earlier period between 3.8 and 4.1 Ga, changes in the orbits of the giant planets may have caused a heavy bombardment by asteroids and comets[47] that pockmarked the Moon and the other inner planets (Mercury, Mars, and presumably Earth and Venus). This would likely have repeatedly sterilized the planet, had life appeared before that time.[39] Geologically, the Hadean Earth would have been far more active than at any other time in its history. Studies of meteorites suggests that radioactive isotopes such as aluminium-26 with a half-life of 7.17×10^5 years, and potassium-40 with a half-life of 1.250×10^9 years, isotopes mainly produced in supernovae, were much more common.[48] Coupled with internal heating as a result of gravitational sorting between the core and the mantle, there would have been a great deal of mantle convection, with the probable result of many more smaller and much more active tectonic plates than now exist.

The time periods between such devastating environmental events give time windows for the possible origin of life in the early environments. A study by Kevin A. Maher and David J. Stevenson shows that if the deep marine hydrothermal setting provides a suitable site for the origin of life, then abiogenesis could have happened as early as 4.0 to 4.2 Ga, whereas if it occurred at the surface of the Earth, abiogenesis could only have occurred between 3.7 and 4.0 Ga.[49]

9.2 Conceptual history

9.2.1 Spontaneous generation

Main article: Spontaneous generation

Belief in spontaneous generation of certain forms of life from non-living matter goes back to Aristotle and ancient Greek philosophy and continued to have support in Western scholarship until the 19th century.[50] This belief was paired with a belief in heterogenesis, i.e., that one form of life derived from a different form (e.g., bees from flowers).[51] Classical notions of spontaneous generation held that certain complex, living organisms are generated by decaying organic substances. According to Aristotle, it was a readily observable truth that aphids arise from the dew that falls on plants, flies from putrid matter, mice from dirty hay, crocodiles from rotting logs at the bottom of bodies of water, and so on.[52] In the 17th century, people began to question such assumptions. In 1646, Sir Thomas Browne published his *Pseudodoxia Epidemica* (subtitled *Enquiries into Very many Received Tenets, and commonly Presumed Truths*), which was an attack on false beliefs and "vulgar errors." His contemporary, Alexander Ross, erroneously refuted him, stating: "To question this [Ed.: i.e., spontaneous generation], is to question Reason, Sense, and Experience: If he doubts of this, let him go to *Ægypt*, and there he will finde the fields swarming with mice begot of the mud of *Nylus*, to the great calamity of the Inhabitants."[53][54]

In 1665, Robert Hooke published the first drawings of a microorganism. Hooke was followed in 1676 by Antonie van

Leeuwenhoek, who drew and described microorganisms that are now thought to have been protozoa and bacteria.[55] Many felt the existence of microorganisms was evidence in support of spontaneous generation, since microorganisms seemed too simplistic for sexual reproduction, and asexual reproduction through cell division had not yet been observed. Van Leeuwenhoek took issue with the ideas common at the time that fleas and lice could spontaneously result from putrefaction, and that frogs could likewise arise from slime. Using a broad range of experiments ranging from sealed and open meat incubation and the close study of insect reproduction he became, by the 1680s, convinced that spontaneous generation was incorrect.[56]

The first experimental evidence against spontaneous generation came in 1668 when Francesco Redi showed that no maggots appeared in meat when flies were prevented from laying eggs. It was gradually shown that, at least in the case of all the higher and readily visible organisms, the previous sentiment regarding spontaneous generation was false. The alternative seemed to be biogenesis: that every living thing came from a pre-existing living thing (*omne vivum ex ovo*, Latin for "every living thing from an egg").

In 1768, Lazzaro Spallanzani demonstrated that microbes were present in the air, and could be killed by boiling. In 1861, Louis Pasteur performed a series of experiments that demonstrated that organisms such as bacteria and fungi do not spontaneously appear in sterile, nutrient-rich media, but only could only appear by invasion from without.

The belief that self-ordering by spontaneous generation was impossible begged for an alternative. By the middle of the 19th century, the theory of biogenesis had accumulated so much evidential support, due to the work of Pasteur and others, that the alternative theory of spontaneous generation had been effectively disproven. John Desmond Bernal, a pioneer in X-ray crystallography, suggested that earlier theories such as spontaneous generation were based upon an explanation that life was continuously created as a result of chance events.[57]

9.2.2 The origin of the terms *biogenesis* and *abiogenesis*

Main article: Biogenesis

The term biogenesis is usually credited to either Henry Charlton Bastian or to Thomas Henry Huxley.[58] Bastian used the term (around 1869) in an unpublished exchange with John Tyndall to mean *life-origination or commencement.* In 1870, Huxley, as new president of the British Association for the Advancement of Science, delivered an address entitled *Biogenesis and Abiogenesis*.[59] In it he introduced the term *biogenesis* (with an opposite meaning to Bastian) and also introduced the term *abiogenesis*:

> And thus the hypothesis that living matter always arises by the agency of pre-existing living matter, took definite shape; and had, henceforward, a right to be considered and a claim to be refuted, in each particular case, before the production of living matter in any other way could be admitted by careful reasoners. It will be necessary for me to refer to this hypothesis so frequently, that, to save circumlocution, I shall call it the hypothesis of *Biogenesis*; and I shall term the contrary doctrine–that living matter may be produced by not living matter–the hypothesis of *Abiogenesis*.[59]

Subsequently, in the preface to Bastian's 1871 book, *The Modes of Origin of Lowest Organisms*,[60] the author refers to the possible confusion with Huxley's usage and he explicitly renounced his own meaning:

> A word of explanation seems necessary with regard to the introduction of the new term *Archebiosis*. I had originally, in unpublished writings, adopted the word *Biogenesis* to express the same meaning—viz., life-origination or commencement. But in the mean time the word *Biogenesis* has been made use of, quite independently, by a distinguished biologist [Huxley], who wished to make it bear a totally different meaning. He also introduced the word *Abiogenesis*. I have been informed, however, on the best authority, that neither of these words can—with any regard to the language from which they are derived—be supposed to bear the meanings which have of late been publicly assigned to them. Wishing to avoid all needless confusion, I therefore renounced the use of the word *Biogenesis*, and being, for the reason just given, unable to adopt the other term, I was compelled to introduce a new word, in order to designate the process by which living matter is supposed to come into being, independently of pre-existing living matter.[61]

9.2.3 Pasteur and Darwin

Pasteur remarked, about a finding of his in 1864 which he considered definitive, "Never will the doctrine of spontaneous generation recover from the mortal blow struck by this simple experiment."[62][63] One alternative was that life's origins on Earth had come from somewhere else in the Universe. Periodically resurrected (see Panspermia, above) Bernal said that this approach "is equivalent in the last resort to asserting the operation of metaphysical, spiritual entities... it turns on the argument of creation by design by a creator or demiurge."[64] Such a theory, Bernal said was unscientific and a number of scientists defined life as a result of an inner *life force*, which in the late 19th century was championed by Henri Bergson.

The concept of evolution proposed by Charles Darwin put an end to these metaphysical theologies. In a letter to Joseph Dalton Hooker on 1 February 1871,[65] Darwin discussed the suggestion that the original spark of life may have begun in a "warm little pond, with all sorts of ammonia and phosphoric salts, light, heat, electricity, &c., present, that a proteine compound was chemically formed ready to undergo still more complex changes." He went on to explain that "at the present day such matter would be instantly devoured or absorbed, which would not have been the case before living creatures were formed." He had written to Hooker in 1863 stating that "It is mere rubbish, thinking at present of the origin of life; one might as well think of the origin of matter.". In *On the Origin of Species* he had referred to life having been "created", by which he "really meant 'appeared' by some wholly unknown process", but had soon regretted using the old-testament term "creation".[66]

9.2.4 "Primordial soup" hypothesis

Further information: Miller–Urey experiment

No new notable research or theory on the subject appeared until 1924, when Alexander Oparin reasoned that atmospheric oxygen prevents the synthesis of certain organic compounds that are necessary building blocks for the evolution of life. In his book *The Origin of Life*,[67][68] Oparin proposed that the "spontaneous generation of life" that had been attacked by Louis Pasteur did in fact occur once, but was now impossible because the conditions found on the early Earth had changed, and preexisting organisms would immediately consume any spontaneously generated organism. Oparin argued that a "primeval soup" of organic molecules could be created in an oxygenless atmosphere through the action of sunlight. These would combine in ever more complex ways until they formed coacervate droplets. These droplets would "grow" by fusion with other droplets, and "reproduce" through fission into daughter droplets, and so have a primitive metabolism in which factors that promote "cell integrity" survive, and those that do not become extinct. Many modern theories of the origin of life still take Oparin's ideas as a starting point.

Robert Shapiro has summarized the "primordial soup" theory of Oparin and J. B. S. Haldane in its "mature form" as follows:[69]

1. The early Earth had a chemically reducing atmosphere.
2. This atmosphere, exposed to energy in various forms, produced simple organic compounds ("monomers").
3. These compounds accumulated in a "soup" that may have concentrated at various locations (shorelines, oceanic vents etc.).
4. By further transformation, more complex organic polymers—and ultimately life—developed in the soup.

About this time, Haldane suggested that the Earth's prebiotic oceans—different from their modern counterparts—would have formed a "hot dilute soup" in which organic compounds could have formed. Bernal called this idea *biopoiesis* or *biopoesis*, the process of living matter evolving from self-replicating but nonliving molecules,[57][70] and proposed that biopoiesis passes through a number of intermediate stages.

One of the most important pieces of experimental support for the "soup" theory came in 1952. Stanley L. Miller and Harold C. Urey, performed an experiment that demonstrated how organic molecules could have spontaneously formed from inorganic precursors, under conditions like those posited by the Oparin-Haldane Hypothesis. The now-famous

Charles Darwin in 1879

Miller–Urey experiment used a highly reducing mixture of gases—methane, ammonia and hydrogen—to form basic organic monomers, such as amino acids.[71] In the Miller–Urey experiment, a mixture of water, hydrogen, methane, and ammonia was cycled through an apparatus that delivered electrical sparks to the mixture. After one week, it was found that about 10% to 15% of the carbon in the system was now in the form of a racemic mixture of organic compounds, including amino acids, which are the building blocks of proteins. This provided direct experimental support for the second point of the "soup" theory, and it is around the remaining two points of the theory that much of the debate now centers.

Bernal shows that based upon this and subsequent work there is no difficulty in principle in forming most of the molecules we recognise as the basic molecules of life from their inorganic precursors. The underlying hypothesis held by Oparin, Haldane, Bernal, Miller and Urey, for instance, was that multiple conditions on the primeval Earth favored chemical reactions that synthesized the same set of complex organic compounds from such simple precursors. A 2011 reanalysis of the saved vials containing the original extracts that resulted from the Miller and Urey experiments, using current and more advanced analytical equipment and technology, has uncovered more biochemicals than originally discovered in the 1950s. One of the more important findings was 23 amino acids, far more than the five originally found.[72] However, Bernal said that "it is not enough to explain the formation of such molecules, what is necessary," he says, "is a physical-chemical explanation of the origins of these molecules that suggests the presence of suitable sources and sinks for free energy."[73]

9.2.5 Proteinoid microspheres

Main article: Proteinoid

In trying to uncover the intermediate stages of abiogenesis mentioned by Bernal, Sidney W. Fox in the 1950s and 1960s studied the spontaneous formation of peptide structures under conditions that might plausibly have existed early in Earth's history. He demonstrated that amino acids could spontaneously form small chains called peptides. In one of his experiments, he allowed amino acids to dry out as if puddled in a warm, dry spot in prebiotic conditions. He found that, as they dried, the amino acids formed long, often cross-linked, thread-like, submicroscopic polypeptide molecules now named "proteinoid microspheres."[74]

In another experiment using a similar method to set suitable conditions for life to form, Fox collected volcanic material from a cinder cone in Hawaii. He discovered that the temperature was over 100 °C (212 °F) just 4 inches (100 mm) beneath the surface of the cinder cone, and suggested that this might have been the environment in which life was created—molecules could have formed and then been washed through the loose volcanic ash and into the sea. He placed lumps of lava over amino acids derived from methane, ammonia and water, sterilized all materials, and baked the lava over the amino acids for a few hours in a glass oven. A brown, sticky substance formed over the surface and when the lava was drenched in sterilized water a thick, brown liquid leached out. It turned out that the amino acids had combined to form proteinoids, and the proteinoids had combined to form small globules that Fox called "microspheres." His proteinoids were not cells, although they formed clumps and chains reminiscent of cyanobacteria, but they contained no functional nucleic acids or any encoded information. Based upon such experiments, Colin S. Pittendrigh stated in December 1967 that "laboratories will be creating a living cell within ten years," a remark that reflected the typical contemporary levels of innocence of the complexity of cell structures.[75]

9.3 Current models

There is still no "standard model" of the origin of life. Most currently accepted models draw at least some elements from the framework laid out by Alexander Oparin (in 1924) and J. B. S. Haldane (in 1925), who postulated the molecular or chemical evolution theory of life.[76] According to them, the first molecules constituting the earliest cells "were synthesized under natural conditions by a slow process of molecular evolution, and these molecules then organized into the first molecular system with properties with biological order."[76] Oparin and Haldane suggested that the atmosphere of the early Earth may have been chemically reducing in nature, composed primarily of methane (CH_4), ammonia (NH_3), water (H_2O), hydrogen sulfide (H_2S), carbon dioxide (CO_2) or carbon monoxide (CO), and phosphate (PO_4^{3-}), with molecular oxygen (O_2) and ozone (O_3) either rare or absent. According to later models, the atmosphere in the late Hadean period consisted largely of nitrogen (N_2) and carbon dioxide, with smaller amounts of carbon monoxide, hydrogen (H_2), and sulfur compounds;[77] while it did lack molecular oxygen and ozone,[78] it wasn't as chemically reducing as Oparin and Haldane supposed. In the atmosphere proposed by Oparin and Haldane, electrical activity can catalyze the creation of certain basic small molecules (monomers) of life, such as amino acids. This was demonstrated in the Miller–Urey experiment reported in 1953.

Bernal coined the term *biopoiesis* in 1949 to refer to the origin of life.[79] In 1967, he suggested that it occurred in three "stages": 1) the origin of biological monomers; 2) the origin of biological polymers; and 3) the evolution from molecules to cells. He suggested that evolution commenced between stage 1 and 2. The first stage is now fairly well understood, and the discovery of alkaline vents and the similarity with the "proton pump" found as the basis of biological life has begun to provide evidence about the second stage. Bernal considered the third, the discovery of methods by which biological reactions were incorporated behind cell walls, to be the most difficult. Modern work on the self organising capacities by which cell membranes self-assemble, and the work on micropores in various substrates is seen as a halfway house towards the development of independent free-living cells, and research into this is an ongoing effort.[38][80][81]

The chemical processes that took place on the early Earth are called *chemical evolution*. Both Manfred Eigen and Sol Spiegelman demonstrated that evolution, including replication, variation, and natural selection, can occur in populations of molecules as well as in organisms.[39] Spiegelman took advantage of natural selection to synthesize the Spiegelman Monster, which had a genome with just 218 nucleotide bases. Eigen built on Spiegelman's work and produced a similar system with just 48 or 54 nucleotides.[82]

Chemical evolution was followed by the initiation of biological evolution, which led to the first cells.[39] No one has yet synthesized a "protocell" using basic components with the necessary properties of life (the so-called "bottom-up-approach"). Without such a proof-of-principle, explanations have tended to focus on chemosynthesis.[83] However, some researchers are working in this field, notably Steen Rasmussen and Jack W. Szostak. Others have argued that a "top-down approach" is more feasible. One such approach, successfully attempted by Craig Venter and others at The Institute for Genomic Research, involves engineering existing prokaryotic cells with progressively fewer genes, attempting to discern at which point the most minimal requirements for life were reached.[84][85]

9.4 Chemical origin of organic molecules

The elements, except for hydrogen, ultimately derive from stellar nucleosynthesis. Complex molecules, including organic molecules, form naturally both in space and on planets.[14] There are two possible sources of organic molecules on the early Earth:

1. Terrestrial origins – organic molecule synthesis driven by impact shocks or by other energy sources (such as UV light, redox coupling, or electrical discharges) (e.g., Miller's experiments)

2. Extraterrestrial origins – formation of organic molecules in interstellar dust clouds and rain down on planets.[86][87] (See pseudo-panspermia)

Estimates of the production of organics from these sources suggest that the Late Heavy Bombardment before 3.5 Ga within the early atmosphere made available quantities of organics comparable to those produced by terrestrial sources.[88][89]

It has been estimated that the Late Heavy Bombardment may also have effectively sterilised the Earth's surface to a depth of tens of metres. If life evolved deeper than this, it would have also been shielded from the early high levels of ultraviolet radiation from the T Tauri stage of the Sun's evolution. Simulations of geothermically heated oceanic crust yield far more organics than those found in the Miller-Urey experiments (see below). In the deep hydrothermal vents, Everett Shock has found "there is an enormous thermodynamic drive to form organic compounds, as seawater and hydrothermal fluids, which are far from equilibrium, mix and move towards a more stable state."[90] Shock has found that the available energy is maximised at around 100 – 150 degrees Celsius, precisely the temperatures at which the hyperthermophilic bacteria and thermoacidophilic archaea have been found, at the base of the phylogenetic tree of life closest to the Last Universal Common Ancestor (LUCA).[91]

9.4.1 Chemical synthesis

While features of self-organization and self-replication are often considered the hallmark of living systems, there are many instances of abiotic molecules exhibiting such characteristics under proper conditions. Stan Palasek showed that self-assembly of ribonucleic acid (RNA) molecules can occur spontaneously due to physical factors in hydrothermal vents.[92] Virus self-assembly within host cells has implications for the study of the origin of life,[93] as it lends further credence to the hypothesis that life could have started as self-assembling organic molecules.[94][95]

Multiple sources of energy were available for chemical reactions on the early Earth. For example, heat (such as from geothermal processes) is a standard energy source for chemistry. Other examples include sunlight and electrical discharges (lightning), among others.[39] Unfavorable reactions can also be driven by highly favorable ones, as in the case of iron-sulfur chemistry. For example, this was probably important for carbon fixation (the conversion of carbon from its inorganic form to an organic one).[note 1] Carbon fixation via iron-sulfur chemistry is highly favorable, and occurs at neutral pH and 100 °C (212 °F). Iron-sulfur surfaces, which are abundant near hydrothermal vents, are also capable of producing small amounts of amino acids and other biological metabolites.[39]

Formamide produces all four ribonucleotides and other biological molecules when warmed in the presence of various terrestrial minerals. Formamide is ubiquitous in the Universe, produced by the reaction of water and hydrogen cyanide (HCN). It has several advantages as a biotic precursor, including the ability to easily become concentrated through the evaporation of water.[96][97] Although HCN is poisonous, it only affects aerobic organisms (eukaryotes and aerobic bacteria), which did not yet exist. It can play roles in other chemical processes as well, such as the synthesis of the amino acid glycine.[39]

In 1961, it was shown that the nucleic acid purine base adenine can be formed by heating aqueous ammonium cyanide solutions.[98] Other pathways for synthesizing bases from inorganic materials were also reported.[99] Leslie E. Orgel and colleagues have shown that freezing temperatures are advantageous for the synthesis of purines, due to the concentrating effect for key precursors such as hydrogen cyanide.[100] Research by Stanley L. Miller and colleagues suggested that while adenine and guanine require freezing conditions for synthesis, cytosine and uracil may require boiling temperatures.[101] Research by the Miller group notes the formation of seven different amino acids and 11 types of nucleobases in ice when ammonia and cyanide were left in a freezer from 1972 to 1997.[102][103] Other work demonstrated the formation of s-triazines (alternative nucleobases), pyrimidines (including cytosine and uracil), and adenine from urea solutions subjected to freeze-thaw cycles under a reductive atmosphere (with spark discharges as an energy source).[104] The explanation given for the unusual speed of these reactions at such a low temperature is eutectic freezing. As an ice crystal forms, it stays pure: only molecules of water join the growing crystal, while impurities like salt or cyanide are excluded. These impurities become crowded in microscopic pockets of liquid within the ice, and this crowding causes the molecules to collide more often. Mechanistic exploration using quantum chemical methods provide a more detailed understanding of some of the chemical processes involved in chemical evolution, and a partial answer to the fundamental question of molecular biogenesis.[105]

At the time of the Miller–Urey experiment, scientific consensus was that the early Earth had a reducing atmosphere with compounds relatively rich in hydrogen and poor in oxygen (e.g., CH_4 and NH_3 as opposed to CO_2 and nitrogen dioxide (NO_2)). However, current scientific consensus describes the primitive atmosphere as either weakly reducing or neutral[106][107] (see also Oxygen Catastrophe). Such an atmosphere would diminish both the amount and variety of amino acids that could be produced, although studies that include iron and carbonate minerals (thought present in early oceans) in the experimental conditions have again produced a diverse array of amino acids.[106] Other scientific research has focused on two other potential reducing environments: outer space and deep-sea thermal vents.[108][109][110]

The spontaneous formation of complex polymers from abiotically generated monomers under the conditions posited by the "soup" theory is not at all a straightforward process. Besides the necessary basic organic monomers, compounds that would have prohibited the formation of polymers were also formed in high concentration during the Miller–Urey and Joan Oró experiments.[111] The Miller–Urey experiment, for example, produces many substances that would react with the amino acids or terminate their coupling into peptide chains.[112]

A research project completed in March 2015 by John D. Sutherland and others found that a network of reactions beginning with hydrogen cyanide and hydrogen sulfide, in streams of water irradiated by UV light, could produce the chemical components of proteins and lipids, as well as those of RNA,[113][114] while not producing a wide range of other compounds.[115] The researchers used the term "cyanosulfidic" to describe this network of reactions.[114]

9.4.2 Autocatalysis

Main article: Autocatalysis

Autocatalysts are substances that catalyze the production of themselves and therefore are "molecular replicators." The simplest self-replicating chemical systems are autocatalytic, and typically contain three components: a product molecule and two precursor molecules. The product molecule joins together the precursor molecules, which in turn produce more product molecules from more precursor molecules. The product molecule catalyzes the reaction by providing a complementary template that binds to the precursors, thus bringing them together. Such systems have been demonstrated both in biological macromolecules and in small organic molecules.[116][117] Systems that do not proceed by template mechanisms, such as the self-reproduction of micelles and vesicles, have also been observed.[117]

It has been proposed that life initially arose as autocatalytic chemical networks.[118] British ethologist Richard Dawkins wrote about autocatalysis as a potential explanation for the origin of life in his 2004 book *The Ancestor's Tale*.[119] In his book, Dawkins cites experiments performed by Julius Rebek, Jr. and his colleagues in which they combined amino adenosine and pentafluorophenyl esters with the autocatalyst amino adenosine triacid ester (AATE). One product was a variant of AATE, which catalysed the synthesis of themselves. This experiment demonstrated the possibility that autocatalysts could exhibit competition within a population of entities with heredity, which could be interpreted as a rudimentary form of natural selection.[120][121]

In the early 1970s, Manfred Eigen and Peter Schuster examined the transient stages between the molecular chaos and a self-replicating hypercycle in a prebiotic soup.[122] In a hypercycle, the information storing system (possibly RNA) produces an enzyme, which catalyzes the formation of another information system, in sequence until the product of the last aids in the formation of the first information system. Mathematically treated, hypercycles could create quasispecies, which through natural selection entered into a form of Darwinian evolution. A boost to hypercycle theory was the discovery of ribozymes capable of catalyzing their own chemical reactions. The hypercycle theory requires the existence of complex biochemicals, such as nucleotides, which do not form under the conditions proposed by the Miller–Urey experiment.

It has been shown that early error prone translation machinery can be stable against an error catastrophe of the type that had been envisaged as problematical known as "Orgel's paradox" caused caused by catalytic activities that would be disruptive.[123][124][125]

9.4.3 Homochirality

Main article: Homochirality

Homochirality refers to the geometric property of some materials that are composed of chiral units. Chiral refers to nonsuperimposable 3D forms that are mirror images of one another, as are left and right hands. Living organisms use molecules that have the same chirality ("handedness"): with almost no exceptions,[126] amino acids are left-handed while nucleotides and sugars are right-handed. Chiral molecules can be synthesized, but in the absence of a chiral source or a chiral catalyst, they are formed in a 50/50 mixture of both enantiomers (called a racemic mixture). Known mechanisms for the production of non-racemic mixtures from racemic starting materials include: asymmetric physical laws, such as the electroweak interaction; asymmetric environments, such as those caused by circularly polarized light, quartz crystals, or the Earth's rotation; and statistical fluctuations during racemic synthesis.[127]

Once established, chirality would be selected for.[128] A small bias (enantiomeric excess) in the population can be amplified into a large one by asymmetric autocatalysis, such as in the Soai reaction.[129] In asymmetric autocatalysis, the catalyst is a chiral molecule, which means that a chiral molecule is catalysing its own production. An initial enantiomeric excess, such as can be produced by polarized light, then allows the more abundant enantiomer to outcompete the other.[130]

Clark has suggested that homochirality may have started in outer space, as the studies of the amino acids on the Murchison meteorite showed that L-alanine is more than twice as frequent as its D form, and L-glutamic acid was more than three times prevalent than its D counterpart. Various chiral crystal surfaces can also act as sites for possible concentration and assembly of chiral monomer units into macromolecules.[131] Compounds found on meteorites suggest that the chirality of life derives from abiogenic synthesis, since amino acids from meteorites show a left-handed bias, whereas sugars show a predominantly right-handed bias, the same as found in living organisms.[132]

9.5 Self-enclosement, reproduction, duplication and the RNA world

9.5.1 Protocells

Main article: Protocell

A protocell is a self-organized, self-ordered, spherical collection of lipids proposed as a stepping-stone to the origin of life.[133] A central question in evolution is how simple protocells first arose and differed in reproductive contribution to the following generation driving the evolution of life. Although a functional protocell has not yet been achieved in a laboratory setting, there are scientists who think the goal is well within reach.[134][135][136]

Self-assembled vesicles are essential components of primitive cells.[133] The second law of thermodynamics requires that the Universe move in a direction in which disorder (or entropy) increases, yet life is distinguished by its great degree of organization. Therefore, a boundary is needed to separate life processes from non-living matter.[137] Researchers Irene A. Chen and Jack W. Szostak amongst others, suggest that simple physicochemical properties of elementary protocells can give rise to essential cellular behaviors, including primitive forms of differential reproduction competition and energy storage. Such cooperative interactions between the membrane and its encapsulated contents could greatly simplify the transition from simple replicating molecules to true cells.[135] Furthermore, competition for membrane molecules would favor stabilized membranes, suggesting a selective advantage for the evolution of cross-linked fatty acids and even the phospholipids of today.[135] Such micro-encapsulation would allow for metabolism within the membrane, the exchange of small molecules but the prevention of passage of large substances across it.[138] The main advantages of encapsulation include the increased solubility of the contained cargo within the capsule and the storage of energy in the form of a electrochemical gradient.

A 2012 study led by Armen Y. Mulkidjanian of Germany's University of Osnabrück, suggests that inland pools of condensed and cooled geothermal vapour have the ideal characteristics for the origin of life.[139] Scientists confirmed in 2002 that by adding a montmorillonite clay to a solution of fatty acid micelles (lipid spheres), the clay sped up the rate of vesicles formation 100-fold.[136]

Another protocell model is the Jeewanu. First synthesized in 1963 from simple minerals and basic organics while exposed to sunlight, it is still reported to have some metabolic capabilities, the presence of semipermeable membrane, amino acids, phospholipids, carbohydrates and RNA-like molecules.[140][141] However, the nature and properties of the Jeewanu remains to be clarified.

Electrostatic interactions induced by short, positively charged, hydrophobic peptides containing 7 amino acids in length or fewer, can attach RNA to a vesicle membrane, the basic cell membrane.[142]

9.5.2 RNA world

Main article: RNA world

The RNA world hypothesis describes an early Earth with self-replicating and catalytic RNA but no DNA or proteins.[144] It is generally accepted that current life on Earth descends from an RNA world,[10][145] although RNA-based life may not have been the first life to exist.[11][12] This conclusion is drawn from many independent lines of evidence, such as the observations that RNA is central to the translation process and that small RNAs can catalyze all of the chemical groups and information transfers required for life.[12][146] The structure of the ribosome has been called the "smoking gun," as it showed that the ribosome is a ribozyme, with a central core of RNA and no amino acid side chains within 18 angstroms of the active site where peptide bond formation is catalyzed.[11] The concept of the RNA world was first proposed in 1962 by Alexander Rich,[147] and the term was coined by Walter Gilbert in 1986.[12][148]

Possible precursors for the evolution of protein synthesis include a mechanism to synthesize short peptide cofactors or from a mechanism for the duplication of RNA. It is likely that the ancestral ribosome was composed entirely of RNA, although some roles have since been taken over by proteins. Major remaining questions on this topic include identifying the selective force for the evolution of the ribosome and determining how the genetic code arose.[149]

Eugene Koonin said, "Despite considerable experimental and theoretical effort, no compelling scenarios currently exist for the origin of replication and translation, the key processes that together comprise the core of biological systems and the apparent pre-requisite of biological evolution. The RNA World concept might offer the best chance for the resolution of this conundrum but so far cannot adequately account for the emergence of an efficient RNA replicase or the translation system. The MWO [Ed.: "many worlds in one"] version of the cosmological model of eternal inflation could suggest a way out of this conundrum because, in an infinite multiverse with a finite number of distinct macroscopic histories (each repeated an infinite number of times), emergence of even highly complex systems by chance is not just possible but inevitable."[150]

Viral origins and the RNA World

Recent evidence for a "virus first" hypothesis, which may support theories of the RNA world have been suggested in new research.[151] One of the problems for those studying viral origins and evolution is their high rate of mutation, particularly the case in RNA retroviruses like HIV/AIDS.[152] A 2015 study compared protein fold structures across different branches of the tree of life, where researchers can reconstruct the evolutionary histories of the folds and of the organisms whose genomes code for them. They argue that protein folds are better markers of ancient events because their three-dimensional structures can be maintained even as the sequences that code for them begin to change.[151] Thus, viral protein repertoire retain traces of ancient evolutionary history that can be recovered using advanced bioinformatics approaches. According to their data, "the prolonged pressure of genome and particle size reduction eventually reduced virocells into modern viruses (identified by the complete loss of cellular makeup), whereas other coexisting cellular lineages diversified into modern cells.[153] The data suggest that viruses originated from multiple ancient cells and co-existed with the ancestors of modern cells.[151] These ancient cells likely contained segmented RNA genomes.[151][154]

9.5.3 RNA synthesis and replication

The RNA world has spurred scientists to try to determine if RNA molecules could have spontaneously formed that were capable of catalyzing their own replication.[155][156][157] Evidence suggests that the chemical conditions, including the presence of boron, molybdenum and oxygen needed for the initial production of RNA molecules, may have been better on the planet Mars than on the planet Earth.[155][156] If so, life-suitable molecules originating on Mars, may have later migrated to Earth via meteor ejections.[155][156]

A number of hypotheses of modes of formation have been put forward. As of 1994, there were difficulties in the abiotic synthesis of the nucleotides cytosine and uracil.[158] Subsequent research has shown possible routes of synthesis; for example, formamide produces all four ribonucleotides and other biological molecules when warmed in the presence of various terrestrial minerals.[96][97] Early cell membranes could have formed spontaneously from proteinoids, which are protein-like molecules produced when amino acid solutions are heated while in the correct concentration of aqueous solution. These are seen to form micro-spheres which are observed to behave similarly to membrane-enclosed compartments. Other possibilities include systems of chemical reactions that take place on clay substrates or on the surface of pyrite rocks.

Factors supportive of an important role for RNA in early life include its ability to act both to store information and to catalyze chemical reactions (as a ribozyme); its many important roles as an intermediate in the expression and maintenance of the genetic information (in the form of DNA) in modern organisms; and the ease of chemical synthesis of at least the components of the RNA molecule under the conditions that approximated the early Earth. Relatively short RNA molecules have been artificially produced in labs, which are capable of replication.[159] Such replicase RNA, which functions as both code and catalyst provides its own template upon which copying can occur. Jack W. Szostak has shown that certain catalytic RNAs can join smaller RNA sequences together, creating the potential for self-replication. If these conditions were present, Darwinian natural selection would favour the proliferation of such autocatalytic sets, to which further functionalities could be added.[160] Tracey A. Lincoln and Gerald F. Joyce have identified such autocatalytic systems of RNA capable of self-sustained replication.[161] The RNA replication systems, which include two ribozymes that catalyze each other's synthesis, showed a doubling time of the product of about one hour, and were subject to natural selection.[162] In evolutionary competition experiments, this led to the emergence of new systems which replicated more efficiently.[11] This was the first demonstration of evolutionary adaptation occurring in a molecular genetic system.[162]

Depending on the specific definition for life being used, life can be considered to have emerged when RNA chains began to express the basic conditions necessary for natural selection to operate as conceived by Darwin: heritability, variation of type, and differential reproductive output. Fitness of an RNA replicator (its per capita rate of increase) would likely be a function of adaptive capacities that were intrinsic (in the sense that they were determined by the nucleotide sequence) and the availability of resources.[163][164] The three primary adaptive capacities may have been (1) the capacity to replicate with moderate fidelity (giving rise to both heritability and variation of type), (2) the capacity to avoid decay, and (3) the capacity to acquire and process resources.[163][164] These capacities would have been determined initially by the folded configurations of the RNA replicators that, in turn, would be encoded in their individual nucleotide sequences. Relative reproductive success among different replicators would have depended on the relative values of these adaptive capacities.

9.5.4 Pre-RNA world

It is possible that a different type of nucleic acid, such as PNA, TNA or GNA, was the first to emerge as a self-reproducing molecule, only later replaced by RNA.[165][166] Larralde *et al.*, say that "the generally accepted prebiotic synthesis of ribose, the formose reaction, yields numerous sugars without any selectivity."[167] and they conclude that their "results suggest that the backbone of the first genetic material could not have contained ribose or other sugars because of their instability." The ester linkage of ribose and phosphoric acid in RNA is known to be prone to hydrolysis.[168]

Pyrimidine ribonucleosides and their respective nucleotides have been prebiotically synthesised by a sequence of reactions which by-pass the free sugars, and are assembled in a stepwise fashion by using nitrogenous or oxygenous chemistries. Sutherland has demonstrated high yielding routes to cytidine and uridine ribonucleotides built from small 2 and 3 carbon fragments such as glycolaldehyde, glyceraldehyde or glyceraldehyde-3-phosphate, cyanamide and cyanoacetylene. One of the steps in this sequence allows the isolation of enantiopure ribose aminooxazoline if the enantiomeric excess of glyceraldehyde is 60% or greater.[169] This can be viewed as a prebiotic purification step, where the said compound spontaneously crystallised out from a mixture of the other pentose aminooxazolines. Ribose aminooxazoline can then react with cyanoacetylene in a mild and highly efficient manner to give the alpha cytidine ribonucleotide. Photoanomerization with UV light allows for inversion about the 1' anomeric centre to give the correct beta stereochemistry.[170] In 2009 they showed that the same simple building blocks allow access, via phosphate controlled nucleobase elaboration, to 2',3'-cyclic pyrimidine nucleotides directly, which are known to be able to polymerise into RNA. This paper also highlights the possibility for the photo-sanitization of the pyrimidine-2',3'-cyclic phosphates.[171]

9.6 Origin of biological metabolism

Laboratory research suggests that metabolism-like reactions could have occurred naturally in early oceans, before the first organisms evolved.[13][172] The findings suggests that metabolism predates the origin of life and evolved through the chemical conditions that prevailed in the world's earliest oceans. Reconstructions in laboratories show that some of these reactions can produce RNA, and some others resemble two essential reaction cascades of metabolism: glycolysis and the pentose phosphate pathway, that provide essential precursors for nucleic acids, amino acids and lipids.[172] Following are some observed discoveries and related hypotheses.

9.6.1 Iron–sulfur world

Main article: Iron–sulfur world theory

Proposed in the 1980s by Günter Wächtershäuser, encouraged and supported by Karl R. Popper,[173][174][175] in his iron–sulfur world theory, this hypothesis postulates the evolution of pre-biotic chemical pathways as the start toward the evolution of life. It presents a consistent system of tracing today's biochemistry back to ancestral reactions that provide alternative pathways to the synthesis of organic building blocks from simple gaseous compounds.

In contrast to the classical Miller experiments, which depend on external sources of energy (such as simulated lightning or ultraviolet irradiation), "Wächtershäuser systems" come with a built-in source of energy, sulfides of iron and other minerals (e.g., pyrite). The energy released from redox reactions of these metal sulfides is not only available for the synthesis of organic molecules. It is therefore hypothesized that such systems may be able to evolve into autocatalytic sets of self-replicating, metabolically active entities that would predate the life forms known today.[13][172] The experiment produced a relatively small yield of dipeptides (0.4% to 12.4%) and a smaller yield of tripeptides (0.10%) although under the same conditions, dipeptides were quickly broken down.[176]

Several models reject the idea of the self-replication of a "naked-gene" but postulate the emergence of a primitive metabolism which could provide a safe environment for the later emergence of RNA replication. The centrality of the Krebs cycle (citric acid cycle) to energy production in aerobic organisms, and in drawing in carbon dioxide and hydrogen ions in biosynthesis of complex organic chemicals, suggests that it was one of the first parts of the metabolism to evolve.[177] Somewhat in agreement with these notions, geochemist Michael Russell has proposed that "the purpose of life is to hydrogenate carbon dioxide" (as part of a "metabolism-first," rather than a "genetics-first," scenario).[178][179] Physicist Jeremy England of MIT has proposed that thermodynamically, life was bound to eventually arrive, as based on established physics, he mathematically indicates "...that when a group of atoms is driven by an external source of energy (like the sun or chemical fuel) and surrounded by a heat bath (like the ocean or atmosphere), it will often gradually restructure itself in order to dissipate increasingly more energy. This could mean that under certain conditions, matter inexorably acquires the key physical attribute associated with life."[180][181]

One of the earliest incarnations of this idea was put forward in 1924 with Oparin's notion of primitive self-replicating vesicles which predated the discovery of the structure of DNA. Variants in the 1980s and 1990s include Wächtershäuser's iron–sulfur world theory and models introduced by Christian de Duve based on the chemistry of thioesters. More abstract and theoretical arguments for the plausibility of the emergence of metabolism without the presence of genes include a mathematical model introduced by Freeman Dyson in the early 1980s and Stuart Kauffman's notion of collectively autocatalytic sets, discussed later in that decade.

Orgel summarized his analysis of the proposal by stating, "There is at present no reason to expect that multistep cycles such as the reductive citric acid cycle will self-organize on the surface of FeS/FeS_2 or some other mineral."[182] It is possible that another type of metabolic pathway was used at the beginning of life. For example, instead of the reductive citric acid cycle, the "open" acetyl-CoA pathway (another one of the five recognised ways of carbon dioxide fixation in nature today) would be compatible with the idea of self-organisation on a metal sulfide surface. The key enzyme of this pathway, carbon monoxide dehydrogenase/acetyl-CoA synthase harbours mixed nickel-iron-sulfur clusters in its reaction centers and catalyses the formation of acetyl-CoA (which may be regarded as a modern form of acetyl-thiol) in a single step.

9.6.2 Zn-World hypothesis

The Zn-World (zinc world) theory of Armen Y. Mulkidjanian[183] is an extension of Wächtershäuser's pyrite hypothesis. Wächtershäuser based his theory of the initial chemical processes leading to informational molecules (i.e., RNA, peptides) on a regular mesh of electric charges at the surface of pyrite that may have made the primeval polymerization thermodynamically more favourable by attracting reactants and arranging them appropriately relative to each other.[184] The Zn-World theory specifies and differentiates further.[183][185] Hydrothermal fluids rich in H_2S interacting with cold primordial ocean (or Darwin's "warm little pond") water leads to the precipitation of metal sulfide particles. Oceanic vent systems and other hydrothermal systems have a zonal structure reflected in ancient volcanogenic massive sulfide deposits (VMS) of hydrothermal origin. They reach many kilometers in diameter and date back to the Archean Eon. Most abundant are pyrite (FeS_2), chalcopyrite ($CuFeS_2$), and sphalerite (ZnS), with additions of galena (PbS) and alabandite (MnS). ZnS and MnS have a unique ability to store radiation energy, e.g., provided by UV light. Since during the relevant time window of the origins of replicating molecules the primordial atmospheric pressure was high enough (>100 bar, about 100 atmospheres) to precipitate near the Earth's surface and UV irradiation was 10 to 100 times more intense than now, the unique photosynthetic properties mediated by ZnS provided just the right energy conditions to energize the synthesis of informational and metabolic molecules and the selection of photostable nucleobases.

The Zn-World theory has been further filled out with experimental and theoretical evidence for the ionic constitution of the interior of the first proto-cells before archaea, bacteria and proto-eukaryotes evolved. Archibald Macallum noted the resemblance of organism fluids such as blood, and lymph to seawater;[186] however, the inorganic composition of all cells differ from that of modern seawater, which led Mulkidjanian and colleagues to reconstruct the "hatcheries" of the first cells combining geochemical analysis with phylogenomic scrutiny of the inorganic ion requirements of universal components of modern cells. The authors conclude that ubiquitous, and by inference primordial, proteins and functional systems show affinity to and functional requirement for K^+, Zn^{2+}, Mn^{2+}, and phosphate. Geochemical reconstruction shows that the ionic composition conducive to the origin of cells could not have existed in what we today call marine settings but is compatible with emissions of vapor-dominated zones of what we today call inland geothermal systems. Under the oxygen depleted, CO_2-dominated primordial atmosphere, the chemistry of water condensates and exhalations near geothermal fields would resemble the internal milieu of modern cells. Therefore, the precellular stages of evolution may have taken place in shallow "Darwin ponds" lined with porous silicate minerals mixed with metal sulfides and enriched in K^+, Zn^{2+}, and phosphorus compounds.[187][188]

9.6.3 Deep sea vent hypothesis

The deep sea vent, or alkaline hydrothermal vent, theory for the origin of life on Earth posits that life may have begun at submarine hydrothermal vents,[189] William Martin and Michael Russell have suggested "that life evolved in structured iron monosulphide precipitates in a seepage site hydrothermal mound at a redox, pH and temperature gradient between sulphide-rich hydrothermal fluid and iron(II)-containing waters of the Hadean ocean floor. The naturally arising, three-dimensional compartmentation observed within fossilized seepage-site metal sulphide precipitates indicates that these inorganic compartments were the precursors of cell walls and membranes found in free-living prokaryotes. The known capability of FeS and NiS to catalyse the synthesis of the acetyl-methylsulphide from carbon monoxide and methyl-sulphide, constituents of hydrothermal fluid, indicates that pre-biotic syntheses occurred at the inner surfaces of these metal-sulphide-walled compartments,..."[190] These form where hydrogen-rich fluids emerge from below the sea floor, as a result of serpentinization of ultra-mafic olivine with seawater and a pH interface with carbon dioxide-rich ocean water. The vents form a sustained chemical energy source derived from redox reactions, in which electron donors, such as molecular hydrogen, react with electron acceptors, such as carbon dioxide (see Iron–sulfur world theory). These are highly exothermic reactions.[note 2]

Michael Russell demonstrated that alkaline vents created an abiogenicproton motive force(PMF)chemiosmoticgradient,[190]in which conditions are ideal for an abiogenic hatchery for life. Their microscopic compartments " provide a natural meansof concentrating organic molecules," composed of iron-sulfur minerals such asmackinawite, endowed these mineral cellswith the catalytic properties envisaged by Wächtershäuser.[177]This movement of ions across the membrane depends ona combination of two factors:

1. Diffusion force caused by concentration gradient—all particles including ions tend to diffuse from higher concentration to lower.
2. Electrostatic force caused by electrical potential gradient—cations like protons H^+ tend to diffuse down the electrical potential, anions in the opposite direction.

These two gradients taken together can be expressed as an electrochemical gradient, providing energy for abiogenic synthesis. The proton motive force can be described as the measure of the potential energy stored as a combination of proton and voltage gradients across a membrane (differences in proton concentration and electrical potential).

Jack W. Szostak suggested that geothermal activity provides greater opportunities for the origination of life in open lakes where there is a buildup of minerals. In 2010, based on spectral analysis of sea and hot mineral water, Ignat Ignatov and Oleg Mosin demonstrated that life may have predominantly originated in hot mineral water. The hot mineral water that contains bicarbonate and calcium ions has the most optimal range.[191] This is similar case as the origin of life in hydrothermal vents, but with bicarbonate and calcium ions in hot water. This water has a pH of 9–11 and is possible to have the reactions in seawater. According to Melvin Calvin, certain reactions of condensation-dehydration of amino acids and nucleotides in individual blocks of peptides and nucleic acids can take place in the primary hydrosphere with pH 9-11 at a later evolutionary stage.[192] Some of these compounds like hydrocyanic acid (HCN) have been proven in the experiments of Miller. This is the environment in which the stromatolites have been created. David Ward of Montana State University described the formation of stromatolites in hot mineral water at the Yellowstone National Park. Stromatolites survive in hot mineral water and in proximity to areas with volcanic activity.[193] Processes have evolved in the sea near geysers of hot mineral water. In 2011, Tadashi Sugawara from the University of Tokyo created a protocell in hot water.[194]

Experimental research and computer modeling suggest that the surfaces of mineral particles inside hydrothermal vents have catalytic properties to enzymes and are able to create simple organic molecules, such as methanol (CH_3OH) and formic, acetic and pyruvic acid out of the dissolved CO_2 in the water.[195][196]

9.6.4 Thermosynthesis

Today's bioenergetic process of fermentation is carried out by either the aforementioned citric acid cycle or the Acetyl-CoA pathway, both of which have been connected to the primordial Iron–sulfur world. In a different approach, the thermosynthesis hypothesis considers the bioenergetic process of chemiosmosis, which plays an essential role in cellular respiration and photosynthesis, more basal than fermentation: the ATP synthase enzyme, which sustains chemiosmosis, is proposed as the currently extant enzyme most closely related to the first metabolic process.[197][198]

First, life needed an energy source to bring about the condensation reaction that yielded the peptide bonds of proteins and the phosphodiester bonds of RNA. In a generalization and thermal variation of the binding change mechanism of today's ATP synthase, the "first protein" would have bound substrates (peptides, phosphate, nucleosides, RNA 'monomers') and condensed them to a reaction product that remained bound until after a temperature change it was released by thermal unfolding.

The energy source under the thermosynthesis hypothesis was thermal cycling, the result of suspension of protocells in a convection current, as is plausible in a volcanic hot spring; the convection accounts for the self-organization and dissipative structure required in any origin of life model. The still ubiquitous role of thermal cycling in germination and cell division is considered a relic of primordial thermosynthesis.

By phosphorylating cell membrane lipids, this "first protein" gave a selective advantage to the lipid protocell that contained the protein. This protein also synthesized a library of many proteins, of which only a minute fraction had thermosynthesis capabilities. As proposed by Dyson,[9] it propagated functionally: it made daughters with similar capabilities, but it did not copy itself. Functioning daughters consisted of different amino acid sequences.

Whereas the Iron–sulfur world identifies a circular pathway as the most simple, the thermosynthesis hypothesis does not even invoke a pathway: ATP synthase's binding change mechanism resembles a physical adsorption process that yields free energy,[199] rather than a regular enzyme's mechanism, which decreases the free energy. It has been claimed that the emergence of cyclic systems of protein catalysts is implausible.[200]

9.7 Other models of abiogenesis

9.7.1 Clay hypothesis

Montmorillonite, an abundant clay, is a catalyst for the polymerization of RNA and for the formation of membranes from lipids.[201] A model for the origin of life using clay was forwarded by Alexander Graham Cairns-Smith in 1985 and explored as a plausible mechanism by several scientists.[202] The clay hypothesis postulates that complex organic molecules arose gradually on a pre-existing, non-organic replication surfaces of silicate crystals in solution.

At the Rensselaer Polytechnic Institute, James P. Ferris' studies have also confirmed that clay minerals of montmorillonite catalyze the formation of RNA in aqueous solution, by joining nucleotides to form longer chains.[203]

In 2007, Bart Kahr from the University of Washington and colleagues reported their experiments that tested the idea that crystals can act as a source of transferable information, using crystals of potassium hydrogen phthalate. "Mother" crystals with imperfections were cleaved and used as seeds to grow "daughter" crystals from solution. They then examined the distribution of imperfections in the new crystals and found that the imperfections in the mother crystals were reproduced in the daughters, but the daughter crystals also had many additional imperfections. For gene-like behavior to be observed, the quantity of inheritance of these imperfections should have exceeded that of the mutations in the successive generations, but it did not. Thus Kahr concluded that the crystals "were not faithful enough to store and transfer information from one generation to the next."[204]

9.7.2 Gold's "deep-hot biosphere" model

In the 1970s, Thomas Gold proposed the theory that life first developed not on the surface of the Earth, but several kilometers below the surface. It is claimed that discovery of microbial life below the surface of another body in our Solar System would lend significant credence to this theory. Thomas Gold also asserted that a trickle of food from a deep, unreachable, source is needed for survival because life arising in a puddle of organic material is likely to consume all of its food and become extinct. Gold's theory is that the flow of such food is due to out-gassing of primordial methane from the Earth's mantle; more conventional explanations of the food supply of deep microbes (away from sedimentary carbon compounds) is that the organisms subsist on hydrogen released by an interaction between water and (reduced) iron compounds in rocks.

9.7.3 Panspermia

Main article: Panspermia

Exogenesis is related to, but not the same as, the notion of panspermia. Neither hypothesis actually answers the question of how life first originated, but merely shifts it to another planet or a comet. However, the advantage of an extraterrestrial origin of primitive life is that life is not required to have evolved on each planet it occurs on, but rather in a single location, and then spread about the galaxy to other star systems via cometary and/or meteorite impact. Evidence to support the hypothesis is scant, but it finds support in studies of Martian meteorites found in Antarctica and in studies of extremophile microbes' survival in outer space.[205][206][207][208][209][210][211]

9.7.4 Extraterrestrial organic molecules

See also: List of interstellar and circumstellar molecules and Panspermia § Pseudo-panspermia

An organic compound is any member of a large class of gaseous, liquid, or solid chemicals whose molecules contain carbon. Carbon is the fourth most abundant element in the Universe by mass after hydrogen, helium, and oxygen.[212] Carbon is abundant in the Sun, stars, comets, and in the atmospheres of most planets.[213] Organic compounds are relatively common in space, formed by "factories of complex molecular synthesis" which occur in molecular clouds and circumstellar envelopes, and chemically evolve after reactions are initiated mostly by ionizing radiation.[14][214][215][216] Based on computer model studies, the complex organic molecules necessary for life may have formed on dust grains in the protoplanetary disk surrounding the Sun before the formation of the Earth.[35] According to the computer studies, this same process may also occur around other stars that acquire planets.[35]

Observations suggest that the majority of organic compounds introduced on Earth by interstellar dust particles are considered principal agents in the formation of complex molecules, thanks to their peculiar surface-catalytic activities.[217][218] Studies reported in 2008, based on $^{12}C/^{13}C$ isotopic ratios of organic compounds found in the Murchison meteorite, suggested that the RNA component uracil and related molecules, including xanthine, were formed extraterrestrially.[219][220] On 8 August 2011, a report based on NASA studies of meteorites found on Earth was published suggesting DNA components (adenine, guanine and related organic molecules) were made in outer space.[217][221][222] Scientists also found that the cosmic dust permeating the Universe contains complex organics ("amorphous organic solids with a mixed aromatic–aliphatic structure") that could be created naturally, and rapidly, by stars.[223][224][225] Sun Kwok of The University of Hong Kong suggested that these compounds may have been related to the development of life on Earth said that "If this is the case, life on Earth may have had an easier time getting started as these organics can serve as basic ingredients for life."[223]

Glycolaldehyde, the first example of an interstellar sugar molecule, was detected in the star-forming region near the center of our galaxy. It was discovered in 2000 by Jes Jørgensen and Jan M. Hollis.[226] In 2012, Jørgensen's team reported the detection of glycolaldehyde in a distant star system. The molecule was found around the protostellar binary IRAS 16293-2422 400 light years from Earth.[227][228][229] Glycolaldehyde is needed to form RNA, which is similar in function to DNA. These findings suggest that complex organic molecules may form in stellar systems prior to the formation of planets, eventually arriving on young planets early in their formation.[230] Because sugars are associated with both metabolism and the genetic code, two of the most basic aspects of life, it is thought the discovery of extraterrestrial sugar increases the likelihood that life may exist elsewhere in our galaxy.[226]

NASA announced in 2009 that scientists had identified another fundamental chemical building block of life in a comet for the first time, glycine, an amino acid, which was detected in material ejected from comet Wild 2 in 2004 and grabbed by NASA's *Stardust* probe. Glycine has been detected in meteorites before. Carl Pilcher, who leads the NASA Astrobiology Institute commented that "The discovery of glycine in a comet supports the idea that the fundamental building blocks of life are prevalent in space, and strengthens the argument that life in the Universe may be common rather than rare."[231] Comets are encrusted with outer layers of dark material, thought to be a tar-like substance composed of complex organic material formed from simple carbon compounds after reactions initiated mostly by ionizing radiation. It is possible that a rain of material from comets could have brought significant quantities of such complex organic molecules to Earth.[232][233][234] Amino acids which were formed extraterrestrially may also have arrived on Earth via comets.[39] It is estimated that during the Late Heavy Bombardment, meteorites may have delivered up to five million tons of organic prebiotic elements to Earth per year.[39]

Polycyclic aromatic hydrocarbons (PAH) are the most common and abundant of the known polyatomic molecules in the observable universe, and are considered a likely constituent of the primordial sea.[235][236][237] In 2010, PAHs, along with fullerenes (or "buckyballs"), have been detected in nebulae.[238][239]

In March 2015, NASA scientists reported that, for the first time, complex DNA and RNA organic compounds of life, including uracil, cytosine and thymine, have been formed in the laboratory under outer space conditions, using starting chemicals, such as pyrimidine, found in meteorites. Pyrimidine, like PAHs, the most carbon-rich chemical found in the Universe, may have been formed in red giant stars or in interstellar dust and gas clouds.[240]

9.7.5 Lipid world

Main article: Gard model

The lipid world theory postulates that the first self-replicating object was lipid-like.[241][242] It is known that phospholipids form lipid bilayers in water while under agitation—the same structure as in cell membranes. These molecules were not present on early Earth, but other amphiphilic long-chain molecules also form membranes. Furthermore, these bodies may expand (by insertion of additional lipids), and under excessive expansion may undergo spontaneous splitting which preserves the same size and composition of lipids in the two progenies. The main idea in this theory is that the molecular composition of the lipid bodies is the preliminary way for information storage, and evolution led to the appearance of polymer entities such as RNA or DNA that may store information favorably. Studies on vesicles from potentially prebiotic amphiphiles have so far been limited to systems containing one or two types of amphiphiles. This in contrast to the output of simulated prebiotic chemical reactions, which typically produce very heterogeneous mixtures of compounds.[133] Within the hypothesis of a lipid bilayer membrane composed of a mixture of various distinct amphiphilic compounds there is the opportunity of a huge number of theoretically possible combinations in the arrangements of these amphiphiles in the membrane. Among all these potential combinations, a specific local arrangement of the membrane would have favored the constitution of an hypercycle,[243][244] actually a positive feedback composed of two mutual catalysts represented by a membrane site and a specific compound trapped in the vesicle. Such site/compound pairs are transmissible to the daughter vesicles leading to the emergence of distinct lineages of vesicles which would have allowed Darwinian natural selection.[245]

9.7.6 Polyphosphates

A problem in most scenarios of abiogenesis is that the thermodynamic equilibrium of amino acid versus peptides is in the direction of separate amino acids. What has been missing is some force that drives polymerization. The resolution of this problem may well be in the properties of polyphosphates.[246][247] Polyphosphates are formed by polymerization of ordinary monophosphate ions PO_4^{-3}. Several mechanisms for such polymerization have been suggested. Polyphosphates cause polymerization of amino acids into peptides. They are also logical precursors in the synthesis of such key biochemical compounds as adenosine triphosphate (ATP). A key issue seems to be that calcium reacts with soluble phosphate to form insoluble calcium phosphate (apatite), so some plausible mechanism must be found to keep calcium ions from causing precipitation of phosphate. There has been much work on this topic over the years, but an interesting new idea is that meteorites may have introduced reactive phosphorus species on the early Earth.[248]

9.7.7 PAH world hypothesis

Main article: PAH world hypothesis

Polycyclic aromatic hydrocarbons (PAH) are known to be abundant in the Universe,[235][236][237] including in the interstellar medium, in comets, and in meteorites, and are some of the most complex molecules so far found in space.[213]

Other sources of complex molecules have been postulated, including extraterrestrial stellar or interstellar origin. For example, from spectral analyses, organic molecules are known to be present in comets and meteorites. In 2004, a team detected traces of PAHs in a nebula.[249] In 2010, another team also detected PAHs, along with fullerenes, in nebulae.[238] The use of PAHs has also been proposed as a precursor to the RNA world in the PAH world hypothesis. The Spitzer Space Telescope has detected a star, HH 46-IR, which is forming by a process similar to that by which the Sun formed. In the disk of material surrounding the star, there is a very large range of molecules, including cyanide compounds, hydrocarbons, and carbon monoxide. In September 2012, NASA scientists reported that PAHs, subjected to interstellar medium conditions, are transformed, through hydrogenation, oxygenation and hydroxylation, to more complex organics—"a step along the path toward amino acids and nucleotides, the raw materials of proteins and DNA, respectively."[250][251] Further, as a result of these transformations, the PAHs lose their spectroscopic signature which could be one of the reasons "for the lack of PAH detection in interstellar ice grains, particularly the outer regions of cold, dense clouds or the upper molecular layers of protoplanetary disks."[250][251]

NASA maintains a database for tracking PAHs in the Universe.[213][252] More than 20% of the carbon in the Universe may be associated with PAHs,[213][213] possible starting materials for the formation of life. PAHs seem to have been formed shortly after the Big Bang, are widespread throughout the Universe,[235][236][237] and are associated with new stars and exoplanets.[213]

9.7.8 Radioactive beach hypothesis

Zachary Adam claims that tidal processes that occurred during a time when the Moon was much closer may have concentrated grains of uranium and other radioactive elements at the high-water mark on primordial beaches, where they may have been responsible for generating life's building blocks.[253] According to computer models reported in *Astrobiology*,[254] a deposit of such radioactive materials could show the same self-sustaining nuclear reaction as that found in the Oklo uranium ore seam in Gabon. Such radioactive beach sand might have provided sufficient energy to generate organic molecules, such as amino acids and sugars from acetonitrile in water. Radioactive monazite material also has released soluble phosphate into the regions between sand-grains, making it biologically "accessible." Thus amino acids, sugars, and soluble phosphates might have been produced simultaneously, according to Adam. Radioactive actinides, left behind in some concentration by the reaction, might have formed part of organometallic complexes. These complexes could have been important early catalysts to living processes.

John Parnell has suggested that such a process could provide part of the "crucible of life" in the early stages of any early wet rocky planet, so long as the planet is large enough to have generated a system of plate tectonics which brings radioactive minerals to the surface. As the early Earth is thought to have had many smaller plates, it might have provided a suitable environment for such processes.[255]

9.7.9 Thermodynamic dissipation

Karo Michaelian from the National Autonomous University of Mexico (UNAM) points out that any model for the origin of life must take into account the fact that life is an irreversible thermodynamic process and, like all irreversible processes, its origin and persistence as a "self-organized" system is due to its dissipation an imposed generalized chemical potential, i.e., the production of entropy. That is, entropy production is not incidental to the process of life, but rather the fundamental reason for its existence. Present day life augments the entropy production of Earth in its solar environment by dissipating ultraviolet and visible photons into heat through organic pigments in water. This heat then catalyzes a host of secondary dissipative processes such as the water cycle, ocean and wind currents, hurricanes, etc.[256][257] Michaelian argues that if the thermodynamic function of life today is to produce entropy through photon dissipation, then this probably was its function at its very beginnings.[258] It turns out that both RNA and DNA when in water solution are very strong absorbers and extremely rapid dissipaters of UV light within the 230–290 nm wavelength region, which is a part of the Sun's spectrum that could have penetrated the prebiotic atmosphere.[259] The amount of ultraviolet (UV-C) light reaching the Earth's surface within this spectral range in the Archean could have been on the order of 4 W/m^2,[260] or some 31 orders of magnitude greater than it is today at 260 nm where RNA and DNA absorb most strongly.[259] In fact, not only RNA and DNA, but many fundamental molecules of life (those common to all three domains of life, archea, bacteria, and eucaryote) are also pigments that absorb in the UV-C, and many of these also have a chemical affinity to RNA and DNA.[261] Nucleic acids may thus have acted as acceptor molecules to the UV-C photon excited antenna pigment donor molecules by providing an ultrafast channel for dissipation. Michaelian has shown that there would have existed a non-linear, non-equilibrium thermodynamic imperative to the abiogenic UV-C photochemical synthesis [171] and proliferation of these pigments over the entire Earth surface if they augmented the solar photon dissipation rate.[262]

A simple mechanism to explain enzyme-less replication of RNA and DNA can be given within the same dissipative thermodynamic framework by assuming that life arose when the temperature of the primitive seas had cooled to somewhat below the denaturing temperature of RNA or DNA. The ratio of $^{18}O/^{16}O$ found in cherts of the Barberton greenstone belt of South Africa indicates that the Earth's surface temperature was around 80 °C at 3.8 Ga,[263][264] falling to 70±15 °C about 3.5 to 3.2 Ga,[265] suggestively close to RNA or DNA denaturing (uncoiling and separation) temperatures. During the night, the surface water temperature would drop below the denaturing temperature and single strand RNA/DNA could act as extension template for the formation of double strand RNA/DNA. During the daylight hours, RNA and DNA would absorb UV-C light and convert this directly into heat at the ocean surface, thereby raising the local temperature enough to allow for denaturing of RNA and DNA. Direct experimental evidence for the denaturing of DNA through UV-C light dissipation has now been obtained.[266]

The copying process would have been repeated with each diurnal cycle.[258] Such an ultraviolet and temperature assisted RNA/DNA reproduction (UVTAR) bears similarity to polymerase chain reaction (PCR), a routine laboratory procedure employed to multiply DNA segments. Since denaturation would be most probable in the late afternoon when the Archean sea surface temperature would be highest, and since late afternoon submarine sunlight is somewhat circularly polarized, the homochirality of the organic molecules of life can also be explained within the proposed thermodynamic framework.[258][267]

The fact that the aromatic amino acids have been shown to have chemical affinity to their codons, or anti-codons, and that they also absorb strongly in the UV-C, suggests that they might have originally acted as antenna pigments to increase dissipation and to provide more local heat for UVTAR replication of RNA and DNA as the sea surface temperature cooled. The accumulation of information, e.g., coding for the aromatic amino acids, in RNA or DNA would thus be related to reproductive success under this mechanism. Michaelian suggests that the traditional origin of life research, that expects to describe the emergence of life without overwhelming reference to entropy production through dissipation, is erroneous and that imposed environmental potentials, such as the solar photon flux, and the dissipation of this flux, must be considered to understand the emergence, proliferation, and evolution of life.

9.7.10 Multiple genesis

Different forms of life with variable origin processes may have appeared quasi-simultaneously in the early history of Earth.[268] The other forms may be extinct (having left distinctive fossils through their different biochemistry—(e.g., hypothetical types of biochemistry). It has been proposed that:

> The first organisms were self-replicating iron-rich clays which fixed carbon dioxide into oxalic and other dicarboxylic acids. This system of replicating clays and their metabolic phenotype then evolved into the sulfide rich region of the hotspring acquiring the ability to fix nitrogen. Finally phosphate was incorporated into the evolving system which allowed the synthesis of nucleotides and phospholipids. If biosynthesis recapitulates biopoiesis, then the synthesis of amino acids preceded the synthesis of the purine and pyrimidine bases. Furthermore the polymerization of the amino acid thioesters into polypeptides preceded the directed polymerization of amino acid esters by polynucleotides.[269]

9.7.11 Fluctuating hydrothermal pools on volcanic islands

Bruce Damer and David Deamer have come to the conclusion that cell membranes can not be formed in salty seawater, and must therefore have originated in freshwater. Before the continents formed, the only dry land on earth would be volcanic islands, where rainwater would form ponds where lipids could form the first stages towards cell membranes. These predecessors of true cells are assumed to have behaved more like a superorganism rather than individuals structures, where the porous membranes would house molecules which would leak out and enter other protocells. Only when true cells had evolved would they gradually adapt to saltier environments and enter the ocean.[270]

9.8 See also

- Anthropic principle
- Artificial cell
- Astrochemistry
- Biological immortality
- Common descent
- Emergence
- Entropy and life
- GADV protein world
- Mediocrity principle
- Mycoplasma laboratorium
- Nexus for Exoplanet System Science
- Planetary habitability
- Rare Earth hypothesis
- Shadow biosphere
- Stromatolite

9.9 Notes

[1] The reactions are:

$$FeS + H_2S \rightarrow FeS_2 + 2H^+ + 2e^-$$
$$FeS + H_2S \rightarrow FeS_2 + HCOOH$$

[2] The reactions are:
Reaction 1: Fayalite + water → magnetite + aqueous silica + hydrogen

$$3Fe_2SiO_4 + 2H_2O \rightarrow 2Fe_3O_4 + 3SiO_2 + 2H_2$$

Reaction 2: *Forsterite + aqueous silica → serpentine*

$$3Mg_2SiO_4 + SiO_2 + 4H_2O \rightarrow 2Mg_3Si_2O_5(OH)_4$$

Reaction 3: *Forsterite + water → serpentine + brucite*

$2Mg_2SiO_4 + 3H_2O \rightarrow Mg_3Si_2O_5(OH)_4 + Mg(OH)_2$

Reaction 3 describes the hydration of olivine with water only to yield serpentine and $Mg(OH)_2$ (brucite). Serpentine is stable at high pH in the presence of brucite like calcium silicate hydrate, (C-S-H) phases formed along with portlandite ($Ca(OH)_2$) in hardened Portland cement paste after the hydration of belite (Ca_2SiO_4), the artificial calcium equivalent of forsterite. Analogy of reaction 3 with belite hydration in ordinary Portland cement: *Belite + water → C-S-H phase + portlandite*

$2\ Ca_2SiO_4 + 4\ H_2O \rightarrow 3\ CaO \cdot 2\ SiO_2 \cdot 3\ H_2O + Ca(OH)_2$

9.10 References

[1] Pronunciation: "/ˌeɪbʌɪə(ʊ)'dʒɛnɪsɪs/". Pearsall, Judy; Hanks, Patrick, eds. (1998). "abiogenesis". *The New Oxford Dictionary of English* (1st ed.). Oxford, UK: Oxford University Press. p. 3. ISBN 0-19-861263-X.

[2] OED On-line (2003)

[3] Bernal 1960, p. 30

[4] Oparin 1953, p. vi

[5] Warmflash, David; Warmflash, Benjamin (November 2005). "Did Life Come from Another World?". *Scientific American* (Stuttgart: Georg von Holtzbrinck Publishing Group) **293** (5): 64–71. doi:10.1038/scientificamerican1105-64. ISSN 0036-8733.

[6] Yarus 2010, p. 47

[7] Peretó, Juli (2005). "Controversies on the origin of life" (PDF). *International Microbiology* (Barcelona: Spanish Society for Microbiology) **8** (1): 23–31. ISSN 1139-6709. PMID 15906258. Retrieved 2015-06-01.

[8] Voet & Voet 2004, p. 29

[9] Dyson 1999

[10]
- Copley, Shelley D.; Smith, Eric; Morowitz, Harold J. (December 2007). "The origin of the RNA world: Co-evolution of genes and metabolism" (PDF). *Bioorganic Chemistry* (Amsterdam, the Netherlands: Elsevier) **35** (6): 430–443. doi:10.1016/j.bioorg.2007.08.001. ISSN 0045-2068. PMID 17897696. Retrieved 2015-06-08. The proposal that life on Earth arose from an RNA world is widely accepted.
- Orgel, Leslie E. (April 2003). "Some consequences of the RNA world hypothesis". *Origins of Life and Evolution of the Biosphere* (Kluwer Academic Publishers) **33** (2): 211–218. doi:10.1023/A:1024616317965. ISSN 0169-6149. PMID 12967268. It now seems very likely that our familiar DNA/RNA/protein world was preceded by an RNA world...
- Robertson & Joyce 2012: "There is now strong evidence indicating that an RNA World did indeed exist before DNA- and protein-based life."
- Neveu, Kim & Benner 2013: "[The RNA world's existence] has broad support within the community today."

[11] Robertson, Michael P.; Joyce, Gerald F. (May 2012). "The origins of the RNA world". *Cold Spring Harbor Perspectives in Biology* (Cold Spring Harbor, NY: Cold Spring Harbor Laboratory Press) **4** (5): a003608. doi:10.1101/cshperspect.a003608. ISSN 1943-0264. PMC 3331698. PMID 20739415.

[12] Cech, Thomas R. (July 2012). "The RNA Worlds in Context". *Cold Spring Harbor Perspectives in Biology* (Cold Spring Harbor, NY: Cold Spring Harbor Laboratory Press) **4** (7): a006742. doi:10.1101/cshperspect.a006742. ISSN 1943-0264. PMC 3385955. PMID 21441585.

[13] Keller, Markus A.; Turchyn, Alexandra V.; Ralser, Markus (25 March 2014). "Non-enzymatic glycolysis and pentose phosphate pathway-like reactions in a plausible Archean ocean". *Molecular Systems Biology* (Heidelberg, Germany: EMBO Press on behalf of the European Molecular Biology Organization) **10** (725). doi:10.1002/msb.20145228. ISSN 1744-4292. PMC 4023395. PMID 24771084.

[14] Ehrenfreund, Pascale; Cami, Jan (December 2010). "Cosmic carbon chemistry: from the interstellar medium to the early Earth.". *Cold Spring Harbor Perspectives in Biology* (Cold Spring Harbor, NY: Cold Spring Harbor Laboratory Press) **2** (12): a002097. doi:10.1101/cshperspect.a002097. ISSN 1943-0264. PMC 2982172. PMID 20554702.

[15] Perkins, Sid (8 April 2015). "Organic molecules found circling nearby star". *Science* (News) (Washington, D.C.: American Association for the Advancement of Science). ISSN 1095-9203. Retrieved 2015-06-02.

[16] King, Anthony (14 April 2015). "Chemicals formed on meteorites may have started life on Earth". *Chemistry World* (News) (London: Royal Society of Chemistry). ISSN 1473-7604. Retrieved 2015-04-17.

[17] Saladino, Raffaele; Carota, Eleonora; Botta, Giorgia; et al. (13 April 2015). "Meteorite-catalyzed syntheses of nucleosides and of other prebiotic compounds from formamide under proton irradiation". *Proc. Natl. Acad. Sci. U.S.A.* (Washington, D.C.: National Academy of Sciences) **112** (21): E2746–E2755. doi:10.1073/pnas.1422225112. ISSN 1091-6490. PMID 25870268.

[18] Rampelotto, Pabulo Henrique (26 April 2010). *Panspermia: A Promising Field Of Research* (PDF). Astrobiology Science Conference 2010. Houston, TX: Lunar and Planetary Institute. p. 5224. Bibcode:2010LPICo1538.5224R. Retrieved 2014-12-03. Conference held at League City, TX

[19] Loeb, Abraham (October 2014). "The habitable epoch of the early Universe". *International Journal of Astrobiology* (Cambridge, UK: Cambridge University Press) **13** (4): 337–339. arXiv:1312.0613. Bibcode:2014IJAsB..13..337L. doi:10.1017/S1473550414000196.ISSN 1473-5504.

- Loeb, Abraham (3 June 2014). "The Habitable Epoch of the Early Universe". arXiv:1312.0613v3 [astro-ph.CO].

[20] Dreifus, Claudia (2 December 2014). "Much-Discussed Views That Go Way Back". *The New York Times* (New York: The New York Times Company). p. D2. ISSN 0362-4331. Retrieved 2014-12-03.

[21] Graham, Robert W. (February 1990). "Extraterrestrial Life in the Universe" (PDF) (NASA Technical Memorandum 102363). Lewis Research Center, Cleveland, Ohio: NASA. Retrieved 2015-06-02.

[22] Altermann 2009, p. xvii

[23] "Age of the Earth". United States Geological Survey. 9 July 2007. Retrieved 2006-01-10.

[24] Dalrymple 2001, pp. 205–221

[25] Manhesa, Gérard; Allègre, Claude J.; Dupréa, Bernard; Hamelin, Bruno (May 1980). "Lead isotope study of basic-ultrabasic layered complexes: Speculations about the age of the earth and primitive mantle characteristics". *Earth and Planetary Science Letters* (Amsterdam, the Netherlands: Elsevier) **47** (3): 370–382. Bibcode:1980E&PSL..47..370M. doi:10.1016/0012-821X(80)90024-2. ISSN 0012-821X.

[26] Schopf, J. William; Kudryavtsev, Anatoliy B.; Czaja, Andrew D.; Tripathi, Abhishek B. (5 October 2007). "Evidence of Archean life: Stromatolites and microfossils". *Precambrian Research* (Amsterdam, the Netherlands: Elsevier) **158** (3–4): 141–155. doi:10.1016/j.precamres.2007.04.009. ISSN 0301-9268.

[27] Schopf, J. William (29 June 2006). "Fossil evidence of Archaean life". *Philosophical Transactions of the Royal Society B* (London: Royal Society) **361** (1470): 869–885. doi:10.1098/rstb.2006.1834. ISSN 0962-8436. PMC 1578735. PMID 16754604.

[28] Raven & Johnson 2002, p. 68

[29] Borenstein, Seth (13 November 2013). "Oldest fossil found: Meet your microbial mom". *Excite* (Yonkers, NY: Mindspark Interactive Network). Associated Press. Retrieved 2015-06-02.

[30] Pearlman, Jonathan (13 November 2013). "'Oldest signs of life on Earth found'". *The Daily Telegraph* (London: Telegraph Media Group). Retrieved 2014-12-15.

[31] Noffke, Nora; Christian, Daniel; Wacey, David; Hazen, Robert M. (16 November 2013). "Microbially Induced Sedimentary Structures Recording an Ancient Ecosystem in the *ca.* 3.48 Billion-Year-Old Dresser Formation, Pilbara, Western Australia". *Astrobiology* (New Rochelle, NY: Mary Ann Liebert, Inc.) **13** (12): 1103–1124. Bibcode:2013AsBio..13.1103N. doi:10.1089/ast.2013.1030. ISSN 1531-1074. PMC 3870916. PMID 24205812.

[32] Ohtomo, Yoko; Kakegawa, Takeshi; Ishida, Akizumi; et al. (January 2014). "Evidence for biogenic graphite in early Archaean Isua metasedimentary rocks". *Nature Geoscience* (London: Nature Publishing Group) **7** (1): 25–28. Bibcode:2014NatGe...7...25O.doi:10.1038/ngeo2025. ISSN 1752-0894.

[33] Borenstein, Seth (19 October 2015). "Hints of life on what was thought to be desolate early Earth". *Excite* (Yonkers, NY: Mindspark Interactive Network). Associated Press. Retrieved 2015-10-20.

[34] Bell, Elizabeth A.; Boehnike, Patrick; Harrison, T. Mark; et al. (19 October 2015). "Potentially biogenic carbon preserved in a 4.1 billion-year-old zircon" (PDF). *Proc. Natl. Acad. Sci. U.S.A.* (Washington, D.C.: National Academy of Sciences). doi:10.1073/pnas.1517557112. ISSN 1091-6490. Retrieved 2015-10-20. Early edition, published online before print.

[35] Moskowitz, Clara (29 March 2012). "Life's Building Blocks May Have Formed in Dust Around Young Sun". *Space.com* (Salt Lake City, UT: Purch). Retrieved 2012-03-30.

[36] Fesenkov 1959, p. 9

[37] Kasting, James F. (12 February 1993). "Earth's Early Atmosphere" (PDF). *Science* (Washington, D.C.: American Association for the Advancement of Science) **259** (5097): 922. doi:10.1126/science.11536547. ISSN 0036-8075. PMID 11536547. Retrieved 2015-07-28.

[38] Russell 2010

[39] Follmann, Hartmut; Brownson, Carol (November 2009). "Darwin's warm little pond revisited: from molecules to the origin of life". *Naturwissenschaften* (Berlin: Springer-Verlag) **96** (11): 1265–1292. Bibcode:doi=10.1007/s00114-009-0602-1 2009NW.....96.1265F doi=10.1007/s00114-009-0602-1. ISSN 0028-1042. PMID 19760276.

[40] Morse, John W.; MacKenzie, Fred T. (1998). "Hadean Ocean Carbonate Geochemistry". *Aquatic Geochemistry* (Kluwer Academic Publishers) **4** (3–4): 301–319. doi:10.1023/A:1009632230875. ISSN 1380-6165.

[41] Wilde, Simon A.; Valley, John W.; Peck, William H.; Graham, Colin M. (11 January 2001). "Evidence from detrital zircons for the existence of continental crust and oceans on the Earth 4.4 Gyr ago" (PDF). *Nature* (London: Nature Publishing Group) **409** (6817): 175–178. doi:10.1038/35051550. ISSN 0028-0836. PMID 11196637. Retrieved 2015-06-03.

[42] Rosing, Minik T.; Bird, Dennis K.; Sleep, Norman H.; et al. (22 March 2006). "The rise of continents—An essay on the geologic consequences of photosynthesis" (PDF). *Palaeogeography, Palaeoclimatology, Palaeoecology* (Amsterdam, the Netherlands: Elsevier) **232** (2–4): 99–113. doi:10.1016/j.palaeo.2006.01.007. ISSN 0031-0182. Retrieved 2015-06-08.

[43] Sleep, Norman H.; Zahnle, Kevin J.; Kasting, James F.; et al. (9 November 1989). "Annihilation of ecosystems by large asteroid impacts on early Earth". *Nature* (London: Nature Publishing Group) **342** (6246): 139–142. Bibcode:1989Natur.342..139S. doi:10.1038/342139a0. ISSN 0028-0836. PMID 11536616.

[44] Davies 1999

[45] O'Donoghue, James (21 August 2011). "Oldest reliable fossils show early life was a beach". *New Scientist* (London: Reed Business Information). ISSN 0262-4079. Retrieved 2014-10-13.

- Wacey, David; Kilburn, Matt R.; Saunders, Martin; et al. (October 2011). "Microfossils of sulphur-metabolizing cells in 3.4-billion-year-old rocks of Western Australia". *Nature Geoscience* (London: Nature Publishing Group) **4** (10): 698–702. Bibcode:2011NatGe...4..698W. doi:10.1038/ngeo1238. ISSN 1752-0894.

[46] Wolpert, Stuart (19 October 2015). "Life on Earth likely started at least 4.1 billion years ago — much earlier than scientists had thought". ULCA. Retrieved 20 October 2015.

[47] Gomes, Rodney; Levison, Hal F.; Tsiganis, Kleomenis; Morbidelli, Alessandro (26 May 2005). "Origin of the cataclysmic Late Heavy Bombardment period of the terrestrial planets". *Nature* (London: Nature Publishing Group) **435** (7041): 466–469. Bibcode:2005Natur.435..466G. doi:10.1038/nature03676. ISSN 0028-0836. PMID 15917802.

[48] Davies 2007, pp. 61–73

[49] Maher, Kevin A.; Stevenson, David J. (18 February 1988). "Impact frustration of the origin of life". *Nature* (London: Nature Publishing Group) **331** (6157): 612–614. Bibcode:1988Natur.331..612M. doi:10.1038/331612a0. ISSN 0028-0836. PMID 11536595.

[50] Sheldon 2005

[51] Vartanian 1973, pp. 307–312

[52] Lennox 2001, pp. 229–258

[53] Balme, D. M. (1962). "Development of Biology in Aristotle and Theophrastus: Theory of Spontaneous Generation". *Phronesis* (Leiden, the Netherlands: Brill Publishers) **7** (1–2): 91–104. doi:10.1163/156852862X00052. ISSN 0031-8868.

[54] Ross 1652

[55] Dobell 1960

[56] Bondeson 1999

[57] Bernal 1967

[58] "Biogenesis". *Hmolpedia*. Ancaster, Ontario, Canada: WikiFoundry, Inc. Retrieved 2014-05-19.

[59] Huxley 1968

[60] Bastian 1871

[61] Bastian 1871, p. xi–xii

[62] Oparin 1953, p. 196

[63] Tyndall 1905, IV, XII (1876), XIII (1878)

[64] Bernal 1967, p. 139

[65] Priscu, John C. "Origin and Evolution of Life on a Frozen Earth". Arlington County, VA: National Science Foundation. Retrieved 2014-03-01.

[66] Darwin 1887, p. 18: "It is often said that all the conditions for the first production of a living organism are now present, which could ever have been present. But if (and oh! what a big if!) we could conceive in some warm little pond, with all sorts of ammonia and phosphoric salts, light, heat, electricity, &c., present, that a proteine compound was chemically formed ready to undergo still more complex changes, at the present day such matter would be instantly devoured or absorbed, which would not have been the case before living creatures were formed." — Charles Darwin, 1 February 1871

[67] Bernal 1967, *The Origin of Life* (A. I. Oparin, 1924), pp. 199–234

[68] Oparin 1953

[69] Shapiro 1987, p. 110

[70] Bryson 2004, pp. 300–302

[71] Miller, Stanley L. (15 May 1953). "A Production of Amino Acids Under Possible Primitive Earth Conditions". *Science* (Washington, D.C.: American Association for the Advancement of Science) **117** (3046): 528–529. Bibcode:1953Sci...117..528M. doi:10.1126/science.117.3046.528. ISSN 0036-8075. PMID 13056598.

[72] Parker, Eric T.; Cleaves, Henderson J.; Dworkin, Jason P.; et al. (5 April 2011). "Primordial synthesis of amines and amino acids in a 1958 Miller H_2S-rich spark discharge experiment" (PDF). *Proc. Natl. Acad. Sci. U.S.A.* (Washington, D.C.: National Academy of Sciences) **108** (14): 5526–5531. Bibcode:2011PNAS..108.5526P. doi:10.1073/pnas.1019191108. ISSN 0027-8424. PMC 3078417. PMID 21422282. Retrieved 2015-06-08.

[73] Bernal 1967, p. 143

[74] Walsh, J. Bruce (1995). "Part 4: Experimental studies of the origins of life". *Origins of life* (Lecture notes). Tucson, AZ: University Of Arizona. Archived from the original on 2008-01-13. Retrieved 2015-06-08.

[75] Woodward 1969, p. 287

[76] Bahadur, Krishna (1973). "Photochemical Formation of Self–sustaining Coacervates" (PDF). *Proceedings of the Indian National Science Academy* (New Delhi: Indian National Science Academy) **39B** (4): 455–467. ISSN 0370-0046.

- Bahadur, Krishna (1975). "Photochemical Formation of Self-Sustaining Coacervates". *Zentralblatt für Bakteriologie, Parasitenkunde, Infektionskrankheiten und Hygiene* (Jena, Germany: Gustav Fischer Verlag) **130** (3): 211–218. doi:10.1016/S0044-4057(75)80076-1. OCLC 641018092. PMID 1242552.

[77] Kasting 1993, p. 922

[78] Kasting 1993, p. 920

[79] Bernal 1951

[98] Oró, Joan (16 September 1961). "Mechanism of Synthesis of Adenine from Hydrogen Cyanide under Possible Primitive Earth Conditions". *Nature* (London: Nature Publishing Group) **191** (4794): 1193–1194. Bibcode:1961Natur.191.1193O. doi:10.1038/1911193a0. ISSN 0028-0836. PMID 13731264.

[99] Basile, Brenda; Lazcano, Antonio; Oró, Joan (1984). "Prebiotic syntheses of purines and pyrimidines". *Advances in Space Research* (Amsterdam, the Netherlands: Elsevier) **4** (12): 125–131. Bibcode:1984AdSpR...4..125B. doi:10.1016/0273-1177(84)90554-4. ISSN 0273-1177. PMID 11537766.

[100] Orgel, Leslie E. (August 2004). "Prebiotic Adenine Revisited: Eutectics and Photochemistry". *Origins of Life and Evolution of Biospheres* (Kluwer Academic Publishers) **34** (4): 361–369. Bibcode:2004OLEB...34..361O. doi:10.1023/B:ORIG.0000029882.52156.c2.ISSN 0169-6149. PMID15279171.

[101] Robertson, Michael P.; Miller, Stanley L. (29 June 1995). "An efficient prebiotic synthesis of cytosine and uracil". *Nature* (London: Nature Publishing Group) **375** (6534): 772–774. Bibcode:1995Natur.375..772R. doi:10.1038/375772a0. ISSN 0028-0836. PMID 7596408.

[102] Fox, Douglas (February 2008). "Did Life Evolve in Ice?". *Discover* (Waukesha, WI: Kalmbach Publishing). ISSN 0274-7529. Retrieved 2008-07-03.

[103] Levy, Matthew; Miller, Stanley L.; Brinton, Karen; Bada, Jeffrey L. (June 2000). "Prebiotic Synthesis of Adenine and Amino Acids Under Europa-like Conditions". *Icarus* (Amsterdam, the Netherlands: Elsevier) **145** (2): 609–613. Bibcode:2000Icar..145..609L.doi:10.1006/icar.2000.6365. ISSN 0019-1035. PMID 11543508.

[104] Menor-Salván, César; Ruiz-Bermejo, Marta; Guzmán, Marcelo I.; Osuna-Esteban, Susana; Veintemillas-Verdaguer, Sabino (20 April 2009). "Synthesis of Pyrimidines and Triazines in Ice: Implications for the Prebiotic Chemistry of Nucleobases". *Chemistry: A European Journal* (Weinheim, Germany: Wiley-VCH on behalf of ChemPubSoc Europe) **15** (17): 4411–4418. doi:10.1002/chem.200802656. ISSN 0947-6539. PMID 19288488.

[105] Roy, Debjani; Najafian, Katayoun; von Ragué Schleyer, Paul (30 October 2007). "Chemical evolution: The mechanism of the formation of adenine under prebiotic conditions". *Proc. Natl. Acad. Sci. U.S.A.* (Washington, D.C.: National Academy of Sciences) **104** (44): 17272–17277. Bibcode:2007PNAS..10417272R. doi:10.1073/pnas.0708434104. ISSN 0027-8424. PMC 2077245. PMID 17951429.

[106] Cleaves, H. James; Chalmers, John H.; Lazcano, Antonio; et al. (April 2008). "A Reassessment of Prebiotic Organic Synthesis in Neutral Planetary Atmospheres". *Origins of Life and Evolution of Biospheres* (Dordrecht, the Netherlands: Springer) **38** (2): 105–115. Bibcode:2008OLEB...38..105C. doi:10.1007/s11084-007-9120-3. ISSN 0169-6149. PMID 18204914.

[107] Chyba, Christopher F. (13 May 2005). "Rethinking Earth's Early Atmosphere". *Science* (Washington, D.C.: American Association for the Advancement of Science) **308** (5724): 962–963. doi:10.1126/science.1113157. ISSN 0036-8075. PMID 15890865.

[108] Barton et al. 2007, pp. 93–95

[109] Bada & Lazcano 2009, pp. 56–57

[110] Bada, Jeffrey L.; Lazcano, Antonio (2 May 2003). "Prebiotic Soup--Revisiting the Miller Experiment" (PDF). *Science* (Washington, D.C.: American Association for the Advancement of Science) **300** (5620): 745–746. doi:10.1126/science.1085145. ISSN 0036-8075. PMID 12730584. Retrieved 2015-06-13.

[111] Oró, Joan; Kimball, Aubrey P. (February 1962). "Synthesis of purines under possible primitive earth conditions: II. Purine intermediates from hydrogen cyanide". *Archives of Biochemistry and Biophysics* (Amsterdam, the Netherlands: Elsevier) **96** (2): 293–313. doi:10.1016/0003-9861(62)90412-5. ISSN 0003-9861. PMID 14482339.

[112] Ahuja, Mukesh, ed. (2006). "Origin of Life". *Life Science* **1**. Delhi: Isha Books. p. 11. ISBN 81-8205-386-2. OCLC 297208106.

[113] Service, Robert F. (16 March 2015). "Researchers may have solved origin-of-life conundrum". *Science* (News) (Washington, D.C.: American Association for the Advancement of Science). ISSN 1095-9203. Retrieved 2015-07-26.

[114] Patel, Bhavesh H.; Percivalle, Claudia; Ritson, Dougal J.; Duffy, Colm D.; Sutherland, John D. (April 2015). "Common origins of RNA, protein and lipid precursors in a cyanosulfidic protometabolism". *Nature Chemistry* (London: Nature Publishing Group) **7** (4): 301–307. Bibcode:2015NatCh...7..301P. doi:10.1038/nchem.2202. ISSN 1755-4330. PMID 25803468. Retrieved 2015-07-22.

[115] Patel et al. 2015, p. 302

[116] Paul, Natasha; Joyce, Gerald F. (December 2004). "Minimal self-replicating systems". *Current Opinion in Chemical Biology* (Amsterdam, the Netherlands: Elsevier) **8** (6): 634–639. doi:10.1016/j.cbpa.2004.09.005. ISSN 1367-5931. PMID 15556408.

[117] Bissette, Andrew J.; Fletcher, Stephen P. (2 December 2013). "Mechanisms of Autocatalysis". *Angewandte Chemie International Edition* (Weinheim, Germany: Wiley-VCH on behalf of the German Chemical Society) **52** (49): 12800–12826. doi:10.1002/anie.201303822. ISSN 1433-7851. PMID 24127341.

[118] Kauffman 1993, chpt. 7

[119] Dawkins 2004

[120] Tjivikua, T.; Ballester, Pablo; Rebek, Julius, Jr. (January 1990). "Self-replicating system". *Journal of the American Chemical Society* (Washington, D.C.: American Chemical Society) **112** (3): 1249–1250. doi:10.1021/ja00159a057. ISSN 0002-7863.

[121] Browne, Malcolm W. (30 October 1990). "Chemists Make Molecule With Hint of Life". *The New York Times* (New York: The New York Times Company). ISSN 0362-4331. Retrieved 2015-07-14.

[122] Eigen & Schuster 1979

[123] Hoffmann, Geoffrey W. (25 June 1974). "On the origin of the genetic code and the stability of the translation apparatus". *Journal of Molecular Biology* (Amsterdam, the Netherlands: Elsevier) **86** (2): 349–362. doi:10.1016/0022-2836(74)90024-2. ISSN 0022-2836. PMID 4414916.

[124] Orgel, Leslie E. (April 1963). "The Maintenance of the Accuracy of Protein Synthesis and its Relevance to Ageing". *Proc. Natl. Acad. Sci. U.S.A.* (Washington, D.C.: National Academy of Sciences) **49** (4): 517–521. Bibcode:1963PNAS...49..517O. doi:10.1073/pnas.49.4.517. ISSN 0027-8424. PMC 299893. PMID 13940312.

[125] Hoffmann, Geoffrey W. (October 1975). "The Stochastic Theory of the Origin of the Genetic Code". *Annual Review of Physical Chemistry* (Palo Alto, CA: Annal Reviews) **26**: 123–144. Bibcode:1975ARPC...26..123H. doi:10.1146/annurev.pc.26.100175 .001011.ISSN 0066-426X.

[126] Chaichian, Rojas & Tureanu 2014, pp. 353–364

[127] Plasson, Raphaël; Kondepudi, Dilip K.; Bersini, Hugues; et al. (August 2007). "Emergence of homochirality in far-from-equilibrium systems: Mechanisms and role in prebiotic chemistry". *Chirality* (Hoboken, NJ: John Wiley & Sons) **19** (8): 589–600. doi:10.1002/chir.20440. ISSN 0899-0042. PMID 17559107. "Special Issue: Proceedings from the Eighteenth International Symposium on Chirality (ISCD-18), Busan, Korea, 2006"

[128] Clark, Stuart (July–August 1999). "Polarized Starlight and the Handedness of Life". *American Scientist* (Research Triangle Park, NC: Sigma Xi) **87** (4): 336. Bibcode:1999AmSci..87..336C. doi:10.1511/1999.4.336. ISSN 0003-0996.

[129] Shibata, Takanori; Morioka, Hiroshi; Hayase, Tadakatsu; et al. (17 January 1996). "Highly Enantioselective Catalytic Asymmetric Automultiplication of Chiral Pyrimidyl Alcohol". *Journal of the American Chemical Society* (Washington, D.C.: American Chemical Society) **118** (2): 471–472. doi:10.1021/ja953066g. ISSN 0002-7863.

[130] Soai, Kenso; Sato, Itaru; Shibata, Takanori (2001). "Asymmetric autocatalysis and the origin of chiral homogeneity in organic compounds". *The Chemical Record* (Hoboken, NJ: John Wiley & Sons on behalf of The Japan Chemical Journal Forum) **1** (4): 321–332. doi:10.1002/tcr.1017. ISSN 1528-0691. PMID 11893072.

[131] Hazen 2005

[132] Mullen, Leslie (5 September 2005). "Building Life from Star-Stuff". *Astrobiology Magazine* (New York: NASA). Retrieved 2015-06-15.

[133] Chen, Irene A.; Walde, Peter (July 2010). "From Self-Assembled Vesicles to Protocells" (PDF). *Cold Spring Harbor Perspectives in Biology* (Cold Spring Harbor, NY: Cold Spring Harbor Laboratory Press) **2** (7): a002170. doi:10.1101/cshperspect.a002170. ISSN 1943-0264. PMC 2890201. PMID 20519344. Retrieved 2015-06-15.

[134] "Exploring Life's Origins: Protocells". *Exploring Life's Origins: A Virtual Exhibit.* Arlington County, VA: National Science Foundation. Retrieved 2014-03-18.

[135] Chen, Irene A. (8 December 2006). "The Emergence of Cells During the Origin of Life". *Science* (Washington, D.C.: American Association for the Advancement of Science) **314** (5805): 1558–1559. doi:10.1126/science.1137541. ISSN 0036-8075. PMID 17158315. Retrieved 2015-06-15.

[136] Zimmer, Carl (26 June 2004). "What Came Before DNA?". *Discover* (Waukesha, WI: Kalmbach Publishing). ISSN 0274-7529.

[137] Shapiro, Robert (June 2007). "A Simpler Origin for Life". *Scientific American* (Stuttgart: Georg von Holtzbrinck Publishing Group) **296** (6): 46–53. doi:10.1038/scientificamerican0607-46. ISSN 0036-8733. PMID 17663224. Retrieved 2015-06-15.

[138] Chang 2007

[139] Switek, Brian (13 February 2012). "Debate bubbles over the origin of life". *Nature* (London: Nature Publishing Group). doi:10.1038/nature.2012.10024. ISSN 0028-0836.

[140] Grote, Mathias (September 2011). "*Jeewanu*, or the 'particles of life'" (PDF). *Journal of Biosciences* (Bangalore, India: Indian Academy of Sciences; Springer) **36** (4): 563–570. doi:10.1007/s12038-011-9087-0. ISSN 0250-5991. PMID 21857103. Retrieved 2015-06-15.

[141] Gupta, V. K.; Rai, R. K. (August 2013). "Histochemical localisation of RNA-like material in photochemically formed self-sustaining, abiogenic supramolecular assemblies 'Jeewanu'". *International Research Journal of Science & Engineering* (Amravati, India) **1** (1): 1–4. ISSN 2322-0015. Retrieved 2015-06-15.

[142] Welter, Kira (10 August 2015). "Peptide glue may have held first protocell components together". *Chemistry World* (News) (London: Royal Society of Chemistry). ISSN 1473-7604. Retrieved 2015-08-29.

- Kamat, Neha P.; Tobé, Sylvia; Hill, Ian T.; Szostak, Jack W. (29 July 2015). "Electrostatic Localization of RNA to Protocell Membranes by Cationic Hydrophobic Peptides". *Angewandte Chemie International Edition* (Weinheim, Germany: Wiley-VCH on behalf of the German Chemical Society). doi:10.1002/anie.201505742. ISSN 1433-7851. "Early View (Online Version of Record published before inclusion in an issue)"

[143] Wimberly, Brian T.; Brodersen, Ditlev E.; Clemons, William M., Jr.; et al. (21 September 2000). "Structure of the 30S ribosomal subunit". *Nature* (London: Nature Publishing Group) **407** (6802): 327–339. doi:10.1038/35030006. ISSN 0028-0836. PMID 11014182.

[144] Zimmer, Carl (25 September 2014). "A Tiny Emissary From the Ancient Past". *The New York Times* (New York: The New York Times Company). ISSN 0362-4331. Retrieved 2014-09-26.

[145] Wade, Nicholas (4 May 2015). "Making Sense of the Chemistry That Led to Life on Earth". *The New York Times* (New York: The New York Times Company). ISSN 0362-4331. Retrieved 2015-05-10.

[146] Yarus, Michael (April 2011). "Getting Past the RNA World: The Initial Darwinian Ancestor". *Cold Spring Harbor Perspectives in Biology* (Cold Spring Harbor, NY: Cold Spring Harbor Laboratory Press) **3** (4): a003590. doi:10.1101/cshperspect.a003590. ISSN 1943-0264. PMC 3062219. PMID 20719875.

[147] Neveu, Marc; Kim, Hyo-Joong; Benner, Steven A. (22 April 2013). "The 'Strong' RNA World Hypothesis: Fifty Years Old". *Astrobiology* (New Rochelle, NY: Mary Ann Liebert, Inc.) **13** (4): 391–403. Bibcode:2013AsBio..13..391N. doi:10.1089/ast.2012.0868.ISSN 1531-1074. PMID23551238.

[148] Gilbert, Walter (20 February 1986). "Origin of life: The RNA world". *Nature* (London: Nature Publishing Group) **319** (6055): 618. Bibcode:1986Natur.319..618G. doi:10.1038/319618a0. ISSN 0028-0836.

[149] Noller, Harry F. (April 2012). "Evolution of protein synthesis from an RNA world.". *Cold Spring Harbor Perspectives in Biology* (Cold Spring Harbor, NY: Cold Spring Harbor Laboratory Press) **4** (4): a003681. doi:10.1101/cshperspect.a003681. ISSN 1943-0264. PMC 3312679. PMID 20610545.

[150] Koonin, Eugene V. (31 May 2007). "The cosmological model of eternal inflation and the transition from chance to biological evolution in the history of life". *Biology Direct* (London: BioMed Central) **2**: 15. doi:10.1186/1745-6150-2-15. ISSN 1745-6150. PMC 1892545. PMID 17540027.

[151] Yates, Diana (25 September 2015). "Study adds to evidence that viruses are alive" (Press release). Champaign, IL: University of Illinois at Urbana–Champaign. Retrieved 2015-10-20.

[152] Caetano-Anollés, Gustavo; Nasir, Arshan (6 September 2012). "Benefits of using molecular structure and abundance in phylogenomic analysis". *Frontiers in Genetics* (Lausanne, Switzerland: Frontiers Media) **3** (172). doi:10.3389/fgene.2012.00172. ISSN 1664-8021. PMC 3434437.

[153] Arshan, Nasir; Caetano-Anollés, Gustavo (25 September 2015). "A phylogenomic data-driven exploration of viral origins and evolution". *Science Advances* (Washington, D.C.: American Association for the Advancement of Science) **1** (8): e1500527. doi:10.1126/sciadv.1500527. ISSN 2375-2548.

[154] Nasir, Arshan; Naeem, Aisha; Jawad Khan, Muhammad; et al. (December 2011). "Annotation of Protein Domains Reveals Remarkable Conservation in the Functional Make up of Proteomes Across Superkingdoms". *Genes* (Basel, Switzerland: MDPI) **2** (4): 869–911. doi:10.3390/genes2040869. ISSN 2073-4425. PMC 3927607. PMID 24710297.

[155] Zimmer, Carl (12 September 2013). "A Far-Flung Possibility for the Origin of Life". *The New York Times* (New York: The New York Times Company). ISSN 0362-4331. Retrieved 2015-06-15.

[156] Webb, Richard (29 August 2013). "Primordial broth of life was a dry Martian cup-a-soup". *New Scientist* (London: Reed Business Information). ISSN 0262-4079. Retrieved 2015-06-16.

[157] Wentao Ma; Chunwu Yu; Wentao Zhang; et al. (November 2007). "Nucleotide synthetase ribozymes may have emerged first in the RNA world". *RNA* (Cold Spring Harbor, NY: Cold Spring Harbor Laboratory Press on behalf of the RNA Society) **13** (11): 2012–2019. doi:10.1261/rna.658507. ISSN 1355-8382. PMC 2040096. PMID 17878321.

[158] Orgel, Leslie E. (October 1994). "The origin of life on Earth". *Scientific American* (Stuttgart: Georg von Holtzbrinck Publishing Group) **271** (4): 76–83. doi:10.1038/scientificamerican1094-76. ISSN 0036-8733. PMID 7524147.

[159] Johnston, Wendy K.; Unrau, Peter J.; Lawrence, Michael S.; et al. (18 May 2001). "RNA-Catalyzed RNA Polymerization: Accurate and General RNA-Templated Primer Extension". *Science* (Washington, D.C.: American Association for the Advancement of Science) **292** (5520): 1319–1325. Bibcode:2001Sci...292.1319J. doi:10.1126/science.1060786. ISSN 0036-8075. PMID 11358999.

[160] Szostak, Jack W. (5 February 2015). "The Origins of Function in Biological Nucleic Acids, Proteins, and Membranes". Chevy Chase (CDP), MD: Howard Hughes Medical Institute. Retrieved 2015-06-16.

[161] Lincoln, Tracey A.; Joyce, Gerald F. (27 February 2009). "Self-Sustained Replication of an RNA Enzyme". *Science* (Washington, D.C.: American Association for the Advancement of Science) **323** (5918): 1229–1232. Bibcode:2009Sci...323.1229L. doi:10.1126/science.1167856. ISSN 0036-8075. PMC 2652413. PMID 19131595.

[162] Joyce, Gerald F. (2009). "Evolution in an RNA world" (PDF). *Cold Spring Harbor Perspectives in Biology* (Cold Spring Harbor, NY: Cold Spring Harbor Laboratory Press) **74** (Evolution: The Molecular Landscape): 17–23. doi:10.1101/sqb.2009.74.004. ISSN 1943-0264. PMC 2891321. PMID 19667013. Retrieved 2015-06-16.

[163] Bernstein, Harris; Byerly, Henry C.; Hopf, Frederick A.; et al. (June 1983). "The Darwinian Dynamic". *The Quarterly Review of Biology* (Chicago, IL: University of Chicago Press) **58** (2): 185–207. doi:10.1086/413216. ISSN 0033-5770. JSTOR 2828805.

[164] Michod 1999

[165] Orgel, Leslie E. (17 November 2000). "A Simpler Nucleic Acid". *Science* (Washington, D.C.: American Association for the Advancement of Science) **290** (5495): 1306–1307. doi:10.1126/science.290.5495.1306. ISSN 0036-8075. PMID 11185405.

[166] Nelson, Kevin E.; Levy, Matthew; Miller, Stanley L. (11 April 2000). "Peptide nucleic acids rather than RNA may have been the first genetic molecule". *Proc. Natl. Acad. Sci. U.S.A.* (Washington, D.C.: National Academy of Sciences) **97** (8): 3868–3871. Bibcode:2000PNAS...97.3868N. doi:10.1073/pnas.97.8.3868. ISSN 0027-8424. PMC 18108. PMID 10760258.

[167] Larralde, Rosa; Robertson, Michael P.; Miller, Stanley L. (29 August 1995). "Rates of Decomposition of Ribose and Other Sugars: Implications for Chemical Evolution" (PDF). *Proc. Natl. Acad. Sci. U.S.A.* (Washington, D.C.: National Academy of Sciences) **92** (18): 8158–8160. Bibcode:1995PNAS...92.8158L. doi:10.1073/pnas.92.18.8158. ISSN 0027-8424. PMC 41115. PMID 7667262.

[168] Lindahl, Tomas (22 April 1993). "Instability and decay of the primary structure of DNA". *Nature* (London: Nature Publishing Group) **362** (6422): 709–715. Bibcode:1993Natur.362..709L. doi:10.1038/362709a0. ISSN 0028-0836. PMID 8469282.

[169] Anastasi, Carole; Crowe, Michael A.; Powner, Matthew W.; Sutherland, John D. (18 September 2006). "Direct Assembly of Nucleoside Precursors from Two- and Three-Carbon Units". *Angewandte Chemie International Edition* (Weinheim, Germany: Wiley-VCH on behalf of the German Chemical Society) **45** (37): 6176–6179. doi:10.1002/anie.200601267. ISSN 1433-7851. PMID 16917794.

[170] Powner, Matthew W.; Sutherland, John D. (13 October 2008). "Potentially Prebiotic Synthesis of Pyrimidine β-D-Ribonucleotides by Photoanomerization/Hydrolysis of α-D-Cytidine-2′-Phosphate". *ChemBioChem* (Weinheim, Germany: Wiley-VCH) **9** (15): 2386–2387. doi:10.1002/cbic.200800391. ISSN 1439-4227. PMID 18798212.

[171] Powner, Matthew W.; Gerland, Béatrice; Sutherland, John D. (14 May 2009). "Synthesis of activated pyrimidine ribonucleotides in prebiotically plausible conditions". *Nature* (London: Nature Publishing Group) **459** (7244): 239–242. Bibcode:2009 Natur.459..239P.doi:10.1038/nature08013. ISSN 0028-0836. PMID19444213.

[172] Senthilingam, Meera (25 April 2014). "Metabolism May Have Started in Early Oceans Before the Origin of Life" (Press release). Wellcome Trust. EurekAlert!. Retrieved 2015-06-16.

[173] Yue-Ching Ho, Eugene (July–September 1990). "Evolutionary Epistemology and Sir Karl Popper's Latest Intellectual Interest: A First-Hand Report". *Intellectus* (Hong Kong: Hong Kong Institute of Economic Science) **15**: 1–3. OCLC 26878740. Retrieved 2012-08-13.

[174] Wade, Nicholas (22 April 1997). "Amateur Shakes Up Ideas on Recipe for Life". *The New York Times* (New York: The New York Times Company). ISSN 0362-4331. Retrieved 2015-06-16.

[175] Popper, Karl R. (29 March 1990). "Pyrite and the origin of life". *Nature* (London: Nature Publishing Group) **344** (6265): 387. Bibcode:1990Natur.344..387P. doi:10.1038/344387a0. ISSN 0028-0836.

[176] Huber, Claudia; Wächtershäuser, Günter (31 July 1998). "Peptides by Activation of Amino Acids with CO on (Ni,Fe)S Surfaces: Implications for the Origin of Life". *Science* (Washington, D.C.: American Association for the Advancement of Science) **281** (5377): 670–672. Bibcode:1998Sci...281..670H. doi:10.1126/science.281.5377.670. ISSN 0036-8075. PMID 9685253.

[177] Lane 2009

[178] Musser, George (23 September 2011). "How Life Arose on Earth, and How a Singularity Might Bring It Down". *Observations* (Blog). *Scientific American*. ISSN 0036-8733. Retrieved 2015-06-17.

[179] Carroll, Sean (10 March 2010). "Free Energy and the Meaning of Life". *Cosmic Variance* (Blog). *Discover*. ISSN 0274-7529. Retrieved 2015-06-17.

[180] Wolchover, Natalie (22 January 2014). "A New Physics Theory of Life". *Quanta Magazine* (New York: Simons Foundation). Retrieved 2015-06-17.

[181] England, Jeremy L. (28 September 2013). "Statistical physics of self-replication" (PDF). *Journal of Chemical Physics* (College Park, MD: American Institute of Physics) **139**: 121923. arXiv:1209.1179. Bibcode:2013JChPh.13911923E. doi:10.1063/1.48 18538.ISSN 0021-9606. Retrieved2015-06-18.

[182] Orgel, Leslie E. (7 November 2000). "Self-organizing biochemical cycles". *Proc. Natl. Acad. Sci. U.S.A.* (Washington, D.C.: National Academy of Sciences) **97** (23): 12503–12507. Bibcode:2000PNAS...9712503O. doi:10.1073/pnas.220406697. ISSN 0027-8424. PMC 18793. PMID 11058157.

[183] Mulkidjanian, Armen Y. (24 August 2009). "On the origin of life in the zinc world: 1. Photosynthesizing, porous edifices built of hydrothermally precipitated zinc sulfide as cradles of life on Earth". *Biology Direct* (London: BioMed Central) **4**: 26. doi:10.1186/1745-6150-4-26. ISSN 1745-6150.

[184] Wächtershäuser, Günter (December 1988). "Before Enzymes and Templates: Theory of Surface Metabolism" (PDF). *Microbio logicalReviews* (Washington, D.C.: American Society for Microbiology)**52** (4): 452–484. ISSN 0146-0749. PMC 373159. PMID307 0320.

[185] Mulkidjanian, Armen Y.; Galperin, Michael Y. (24 August 2009). "On the origin of life in the zinc world. 2. Validation of the hypothesis on the photosynthesizing zinc sulfide edifices as cradles of life on Earth". *Biology Direct* (London: BioMed Central) **4**: 27. doi:10.1186/1745-6150-4-27. ISSN 1745-6150.

[186] Macallum, A. B. (1 April 1926). "The Paleochemistry of the body fluids and tissues". *Physiological Reviews* (Bethesda, MD: American Physiological Society) **6** (2): 316–357. ISSN 0031-9333. Retrieved 2015-06-18.

[187] Mulkidjanian, Armen Y.; Bychkov, Andrew Yu.; Dibrova, Daria V.; et al. (3 April 2012). "Origin of first cells at terrestrial, anoxic geothermal fields". *Proc. Natl. Acad. Sci. U.S.A.* (Washington, D.C.: National Academy of Sciences) **109** (14): E821–E830. Bibcode:2012PNAS..109E.821M. doi:10.1073/pnas.1117774109. ISSN 1091-6490. PMC 3325685. PMID 22331915.

[188] For a deeper integrative version of this hypothesis, see in particular Lankenau 2011, pp. 225–286, interconnecting the "Two RNA worlds" concept and other detailed aspects; and Davidovich, Chen; Belousoff, Matthew; Bashan, Anat; Yonath, Ada (September 2009). "The evolving ribosome: from non-coded peptide bond formation to sophisticated translation machinery". *Research in Microbiology* (Amsterdam, the Netherlands: Elsevier) **160** (7): 487–492. doi:10.1016/j.resmic.2009.07.004. ISSN 1769-7123. PMID 19619641.

[189] Schirber, Michael (24 June 2014). "Hydrothermal Vents Could Explain Chemical Precursors to Life". *NASA Astrobiology: Life in the Universe*. NASA. Retrieved 2015-06-19.

[190] Martin, William; Russell, Michael J. (29 January 2003). "On the origins of cells: a hypothesis for the evolutionary transitions from abiotic geochemistry to chemoautotrophic prokaryotes, and from prokaryotes to nucleated cells". *Philosophical Transactions of the Royal Society B* (London: Royal Society) **358** (1429): 59–83; discussion 83–85. doi:10.1098/rstb.2002.1183. ISSN 0962-8436. PMC 1693102. PMID 12594918.

[191] Ignatov, Ignat; Mosin, Oleg V. (2013). "Possible Processes for Origin of Life and Living Matter with modeling of Physiological Processes of Bacterium *Bacillus Subtilis* in Heavy Water as Model System". *Journal of Natural Sciences Research* (New York: International Institute for Science, Technology and Education) **3** (9): 65–76. ISSN 2225-0921.

[192] Calvin 1969

[193] Schirber, Michael (1 March 2010). "First Fossil-Makers in Hot Water". *Astrobiology Magazine* (New York: NASA). Retrieved 2015-06-19.

[194] Kurihara, Kensuke; Tamura, Mieko; Shohda, Koh-ichiroh; et al. (October 2011). "Self-Reproduction of supramolecular giant vesicles combined with the amplification of encapsulated DNA". *Nature Chemistry* (London: Nature Publishing Group) **3** (10): 775–781. Bibcode:2011NatCh...3..775K. doi:10.1038/nchem.1127. ISSN 1755-4330. PMID 21941249.

[195] Usher, Oli (27 April 2015). "Chemistry of seabed's hot vents could explain emergence of life" (Press release). University College London. Retrieved 2015-06-19.

[196] Roldan, Alberto; Hollingsworth, Nathan; Roffey, Anna; Islam, Husn-Ubayda; et al. (May 2015). "Bio-inspired CO2 conversion by iron sulfide catalysts under sustainable conditions" (PDF). *Chemical Communications* (London: Royal Society of Chemistry) **51** (35): 7501–7504. doi:10.1039/C5CC02078F. ISSN 1359-7345. PMID 25835242. Retrieved 2015-06-19.

[197] Muller, Anthonie W. J. (7 August 1985). "Thermosynthesis by biomembranes: Energy gain from cyclic temperature changes". *Journal of Theoretical Biology* (Amsterdam, the Netherlands: Elsevier) **115** (3): 429–453. doi:10.1016/S0022-5193(85)80202-2. ISSN 0022-5193. PMID 3162066.

[198] Muller, Anthonie W. J. (1995). "Were the first organisms heat engines? A new model for biogenesis and the early evolution of biological energy conversion". *Progress in Biophysics and Molecular Biology* (Oxford, UK; New York: Pergamon Press) **63** (2): 193–231. doi:10.1016/0079-6107(95)00004-7. ISSN 0079-6107. PMID 7542789.

[199] Muller, Anthonie W. J.; Schulze-Makuch, Dirk (1 April 2006). "Sorption heat engines: Simple inanimate negative entropy generators". *Physica A: Statistical Mechanics and its Applications* (Utrecht, the Netherlands: Elsevier) **362** (2): 369–381. arXiv:physics/0507173. Bibcode:2006PhyA..362..369M. doi:10.1016/j.physa.2005.12.003. ISSN 0378-4371.

[200] Orgel 1987, pp. 9–16

[201] Perry, Caroline (7 February 2011). "Clay-armored bubbles may have formed first protocells" (Press release). Cambridge, MA: Harvard University. EurekAlert!. Retrieved 2015-06-20.

[202] Dawkins 1996, pp. 148–161

[203] Wenhua Huang; Ferris, James P. (12 July 2006). "One-Step, Regioselective Synthesis of up to 50-mers of RNA Oligomers by Montmorillonite Catalysis". *Journal of the American Chemical Society* (Washington, D.C.: American Chemical Society) **128** (27): 8914–8919. doi:10.1021/ja061782k. ISSN 0002-7863. PMID 16819887.

[204] Moore, Caroline (16 July 2007). "Crystals as genes?". *Highlights in Chemical Science* (London: Royal Society of Chemistry). ISSN 2041-5818. Retrieved 2015-06-21.

- Bullard, Theresa; Freudenthal, John; Avagyan, Serine; et al. (2007). "Test of Cairns-Smith's 'crystals-as-genes' hypothesis". *Faraday Discussions* **136**: 231–245. Bibcode:2007FaDi..136..231B. doi:10.1039/b616612c. ISSN 1359-6640.

[205] Clark, Stuart (25 September 2002). "Tough Earth bug may be from Mars". *New Scientist* (London: Reed Business Information). ISSN 0262-4079. Retrieved 2015-06-21.

[206] "Exobiology and Radiation Assembly (ERA)". *NSSDC Master Catalog*. NASA. NSSDC ID: 1992-049B-03. Retrieved 2015-06-21. Experiment carried on board the European Retrievable Carrier (EURECA).

[207] Horneck, Gerda; Klaus, David M.; Mancinelli, Rocco L. (March 2010). "Space Microbiology". *Microbiology and Molecular Biology Reviews* (Washington, D.C.: American Society for Microbiology) **74** (1): 121–156. doi:10.1128/MMBR.00016-09. ISSN 1092-2172. PMC 2832349. PMID 20197502.

[208] Clancy, Brack & Horneck 2005

[209] Rabbow, Elke; Horneck, Gerda; Rettberg, Petra; et al. (December 2009). "EXPOSE, an Astrobiological Exposure Facility on the International Space Station - from Proposal to Flight". *Origins of Life and Evolution of Biospheres* (Dordrecht, the Netherlands: Springer) **39** (6): 581–598. Bibcode:2009OLEB...39..581R. doi:10.1007/s11084-009-9173-6. ISSN 0169-6149. PMID 19629743.

[210] Onofri, Silvano; de la Torre, Rosa; de Vera, Jean-Pierre; et al. (May 2012). "Survival of Rock-Colonizing Organisms After 1.5 Years in Outer Space". *Astrobiology* (New Rochelle, NY: Mary Ann Liebert, Inc.) **12** (5): 508–516. Bibcode:2012AsBio..12..5 08O.doi:10.1089/ast.2011.0736. ISSN 1531-1074. PMID 22680696.

[211] Amos, Jonathan (23 August 2010). "Beer microbes live 553 days outside ISS". *BBC News* (London: BBC). Retrieved 2015-06-22.

[212] "biological abundance of elements". *Encyclopedia of Science*. Dundee, Scotland: David Darling Enterprises. Retrieved 2008-10-09.

[213] Hoover, Rachel (21 February 2014). "Need to Track Organic Nano-Particles Across the Universe? NASA's Got an App for That". *Ames Research Center*. Mountain View, CA: NASA. Retrieved 2015-06-22.

[214] Chang, Kenneth (18 August 2009). "From a Distant Comet, a Clue to Life". *The New York Times* (New York: The New York Times Company). p. A18. ISSN 0362-4331. Retrieved 2015-06-22.

[215] Goncharuk, Vladislav V.; Zui, O. V. (February 2015). "Water and carbon dioxide as the main precursors of organic matter on Earth and in space". *Journal of Water Chemistry and Technology* (Dordrecht, the Netherlands: Springer on behalf of Allerton Press) **37** (1): 2–3. doi:10.3103/S1063455X15010026. ISSN 1063-455X.

[216] Abou Mrad, Ninette; Vinogradoff, Vassilissa; Duvernay, Fabrice; et al. (2015). "Laboratory experimental simulations: Chemical evolution of the organic matter from interstellar and cometary ice analogs" (PDF). *Bulletin de la Société Royale des Sciences de Liège* (Liège, Belgium: Société royale des sciences de Liège) **84**: 21–32. Bibcode:2015BSRSL..84...21A. ISSN 0037-9565. Retrieved 2015-04-06.

[217] Gallori, Enzo (June 2011). "Astrochemistry and the origin of genetic material". *Rendiconti Lincei* (Milan, Italy: Springer) **22** (2): 113–118. doi:10.1007/s12210-011-0118-4. ISSN 2037-4631. "Paper presented at the Symposium 'Astrochemistry: molecules in space and time' (Rome, 4–5 November 2010), sponsored by Fondazione 'Guido Donegani', Accademia Nazionale dei Lincei."

[218] Martins, Zita (February 2011). "Organic Chemistry of Carbonaceous Meteorites". *Elements* (Chantilly, VA: Mineralogical Society of America et al.) **7** (1): 35–40. doi:10.2113/gselements.7.1.35. ISSN 1811-5209.

[219] Martins, Zita; Botta, Oliver; Fogel, Marilyn L.; et al. (15 June 2008). "Extraterrestrial nucleobases in the Murchison meteorite". *Earth and Planetary Science Letters* (Amsterdam, the Netherlands: Elsevier) **270** (1–2): 130–136. arXiv:0806.2286. Bibcode:2008E&PSL.270..130M. doi:10.1016/j.epsl.2008.03.026. ISSN 0012-821X.

[220] "We may all be space aliens: study". *ABC News* (Sydney: Australian Broadcasting Corporation). AFP. 14 June 2008. Retrieved 2015-06-22.

[221] Callahan, Michael P.; Smith, Karen E.; Cleaves, H. James, II; et al. (23 August 2011). "Carbonaceous meteorites contain a wide range of extraterrestrial nucleobases". *Proc. Natl. Acad. Sci. U.S.A.* (Washington, D.C.: National Academy of Sciences) **108** (34): 13995–13998. Bibcode:2011PNAS..10813995C. doi:10.1073/pnas.1106493108. ISSN 0027-8424. PMC 3161613. PMID 21836052.

[222] Steigerwald, John (8 August 2011). "NASA Researchers: DNA Building Blocks Can Be Made in Space". *Goddard Space Flight Center*. Greenbelt, MD: NASA. Retrieved 2015-06-23.

[223] Chow, Denise (26 October 2011). "Discovery: Cosmic Dust Contains Organic Matter from Stars". *Space.com* (Ogden, UT: Purch). Retrieved 2015-06-23.

[224] "Astronomers Discover Complex Organic Matter Exists Throughout the Universe". Rockville, MD: ScienceDaily, LLC. 26 October 2011. Retrieved 2015-06-23. Post is reprinted from materials provided by The University of Hong Kong.

[225] Sun Kwok; Yong Zhang (3 November 2011). "Mixed aromatic–aliphatic organic nanoparticles as carriers of unidentified infrared emission features". *Nature* (London: Nature Publishing Group) **479** (7371): 80–83. Bibcode:2011Natur.479...80K. doi:10.1038/nature10542. ISSN 0028-0836. PMID 22031328.

[226] Clemence, Lara; Cohen, Jarrett (7 February 2005). "Space Sugar's a Sweet Find". *Goddard Space Flight Center*. Greenbelt, MD: NASA. Retrieved 2015-06-23.

[227] Than, Ker (30 August 2012). "Sugar Found In Space: A Sign of Life?". *National Geographic News* (Washington, D.C.: National Geographic Society). Retrieved 2015-06-23.

[228] "Sweet! Astronomers spot sugar molecule near star". *Excite* (Yonkers, NY: Mindspark Interactive Network). Associated Press. 29 August 2012. Retrieved 2015-06-23.

[229] "Building blocks of life found around young star". *News & Events*. Leiden, the Netherlands: Leiden University. 30 September 2012. Retrieved 2013-12-11.

[230] Jørgensen, Jes K.; Favre, Cécile; Bisschop, Suzanne E.; et al. (20 September 2012). "Detection of the simplest sugar, glycolaldehyde, in a solar-type protostar with ALMA" (PDF). *The Astrophysical Journal Letters* (Bristol, England: IOP Publishing for the American Astronomical Society) **757** (1). arXiv:1208.5498. Bibcode:2012ApJ...757L...4J. doi:10.1088/2041-8205/757/1/L4. ISSN 2041-8213. L4. Retrieved 2015-06-23.

[231] "'Life chemical' detected in comet". *BBC News* (London: BBC). 18 August 2009. Retrieved 2015-06-23.

[232] Thompson, William Reid; Murray, B. G.; Khare, Bishun Narain; Sagan, Carl (30 December 1987). "Coloration and darkening of methane clathrate and other ices by charged particle irradiation: Applications to the outer solar system". *Journal of Geophysical Research* (Washington, D.C.: American Geophysical Union) **92** (A13): 14933–14947. Bibcode:1987JGR....9214933T. doi:10.1029/JA092iA13p14933. ISSN 0148-0227. PMID 11542127.

[233] Stark, Anne M. (5 June 2013). "Life on Earth shockingly comes from out of this world". Livermore, CA: Lawrence Livermore National Laboratory. Retrieved 2015-06-23.

[234] Goldman, Nir; Tamblyn, Isaac (20 June 2013). "Prebiotic Chemistry within a Simple Impacting Icy Mixture". *Journal of Physical Chemistry A* (Washington, D.C.: American Chemical Society) **117** (24): 5124–5131. doi:10.1021/jp402976n. ISSN 1089-5639. PMID 23639050.

[235] Carey, Bjorn (18 October 2005). "Life's Building Blocks 'Abundant in Space'". *Space.com* (Watsonville, CA: Imaginova). Retrieved 2015-06-23.

[236] Hudgins, Douglas M.; Bauschlicher, Charles W., Jr.; Allamandola, Louis J. (10 October 2005). "Variations in the Peak Position of the 6.2 μm Interstellar Emission Feature: A Tracer of N in the Interstellar Polycyclic Aromatic Hydrocarbon Population" (PDF). *The Astrophysical Journal* (Bristol, England: IOP Publishing for the American Astronomical Society) **632** (1): 316–332. Bibcode:2005ApJ...632..316H. doi:10.1086/432495. ISSN 0004-637X.

[237] Des Marais, David J.; Allamandola, Louis J.; Sandford, Scott; et al. (2009). "Cosmic Distribution of Chemical Complexity". *Ames Research Center*. Mountain View, CA: NASA. Retrieved 2015-06-24. See the Ames Research Center 2009 annual team report to the NASA Astrobiology Institute here.

[238] García-Hernández, Domingo. A.; Manchado, Arturo; García-Lario, Pedro; et al. (20 November 2010). "Formation of Fullerenes in H-Containing Planetary Nebulae". *The Astrophysical Journal Letters* (Bristol, England: IOP Publishing for the American Astronomical Society) **724** (1): L39–L43. arXiv:1009.4357. Bibcode:2010ApJ...724L..39G. doi:10.1088/2041-8205/724/1/L39. ISSN 2041-8213.

[239] Atkinson, Nancy (27 October 2010). "Buckyballs Could Be Plentiful in the Universe". *Universe Today* (Courtenay, British Columbia: Fraser Cain). Retrieved 2015-06-24.

[240] Marlaire, Ruth, ed. (3 March 2015). "NASA Ames Reproduces the Building Blocks of Life in Laboratory". *Ames Research Center*. Moffett Field, CA: NASA. Retrieved 2015-03-05.

[241] Lancet, Doron (30 December 2014). "Systems Prebiology-Studies of the origin of Life". *The Lancet Lab*. Rehovot, Israel: Department of Molecular Genetics; Weizmann Institute of Science. Retrieved 2015-06-26.

[242] Segré, Daniel; Ben-Eli, Dafna; Deamer, David W.; Lancet, Doron (February 2001). "The Lipid World" (PDF). *Origins of Life and Evolution of the Biosphere* (Kluwer Academic Publishers) **31** (1–2): 119–145. doi:10.1023/A:1006746807104. ISSN 0169-6149. PMID 11296516. Retrieved 2008-09-11.

[243] Eigen, Manfred; Schuster, Peter (November 1977). "The Hypercycle. A Principle of Natural Self-Organization. Part A: Emergence of the Hypercycle" (PDF). *Naturwissenschaften* (Berlin: Springer-Verlag) **64** (11): 541–565. Bibcode:1977NW.....64..541E.doi:10.1007/bf00450633. ISSN 0028-1042. PMID593400. Retrieved2015-06-13.

- Eigen, Manfred; Schuster, Peter (1978). "The Hypercycle. A Principle of Natural Self-Organization. Part B: The Abstract Hypercycle" (PDF). *Naturwissenschaften* (Berlin: Springer-Verlag) **65**: 7–41. Bibcode:1978NW.....65....7E. doi:10.1007/bf00420631. ISSN 0028-1042. Retrieved 2015-06-13.
- Eigen, Manfred; Schuster, Peter (July 1978). "The Hypercycle. A Principle of Natural Self-Organization. Part C: The Realistic Hypercycle" (PDF). *Naturwissenschaften* (Berlin: Springer-Verlag) **65** (7): 341–369. Bibcode:1978NW.....65..341E.doi:10.1007/bf00439699. ISSN 0028-1042. Retrieved2015-06-13.

[244] Markovitch, Omer; Lancet, Doron (Summer 2012). "Excess Mutual Catalysis Is Required for Effective Evolvability" (PDF). *Artificial Life* (Cambridge, MA: MIT Press) **18** (3): 243–266. doi:10.1162/artl_a_00064. ISSN 1064-5462. PMID 22662913. Retrieved 2015-06-26.

[245] Tessera, Marc (2011). "Origin of Evolution *versus* Origin of Life: A Shift of Paradigm". *International Journal of Molecular Sciences* (Basel, Switzerland: MDPI) **12** (6): 3445–3458. doi:10.3390/ijms12063445. ISSN 1422-0067. PMC 3131571. PMID 21747687. Special Issue: "Origin of Life 2011"

[246] Brown, Michael R. W.; Kornberg, Arthur (16 November 2004). "Inorganic polyphosphate in the origin and survival of species". *Proc. Natl. Acad. Sci. U.S.A.* (Washington, D.C.: National Academy of Sciences) **101** (46): 16085–16087. Bibcode:2004PNAS..10116085B. doi:10.1073/pnas.0406909101. ISSN 0027-8424. PMC 528972. PMID 15520374.

[247] Clark, David P. (3 August 1999). "The Origin of Life". *Microbiology 425: Biochemistry and Physiology of Microorganism* (Lecture). Carbondale, IL: College of Science; Southern Illinois University Carbondale. Archived from the original on 2000-10-02. Retrieved 2015-06-26.

[248] Pasek, Matthew A. (22 January 2008). "Rethinking early Earth phosphorus geochemistry". *Proc. Natl. Acad. Sci. U.S.A.* (Washington, D.C.: National Academy of Sciences) **105** (3): 853–858. Bibcode:2008PNAS..105..853P. doi:10.1073/pnas.0708205105.ISSN 0027-8424. PMC2242691. PMID 18195373.

[249] Witt, Adolf N.; Vijh, Uma P.; Gordon, Karl D. (2003). "Discovery of Blue Fluorescence by Polycyclic Aromatic Hydrocarbon Molecules in the Red Rectangle". *Bulletin of the American Astronomical Society* (Washington, D.C.: American Astronomical Society) **35**: 1381. Bibcode:2003AAS...20311017W. Archived from the original on 2003-12-19. Retrieved 2015-06-26. American Astronomical Society Meeting 203, #110.17, January 2004.

[250] "NASA Cooks Up Icy Organics to Mimic Life's Origins". *Space.com*. Ogden, UT: Purch. 20 September 2012. Retrieved 2015-06-26.

[251] Gudipati, Murthy S.; Rui Yang (1 September 2012). "In-situ Probing of Radiation-induced Processing of Organics in Astrophysical Ice Analogs—Novel Laser Desorption Laser Ionization Time-of-flight Mass Spectroscopic Studies". *The Astrophysical Journal Letters* (Bristol, England: IOP Publishing for the American Astronomical Society) **756** (1). Bibcode:2012ApJ...756L..24G.doi:10.1088/2041-8205/756/1/L24. ISSN 2041-8213. L24.

[252] "NASA Ames PAH IR Spectroscopic Database". NASA. Retrieved 2015-06-17.

[253] Dartnell, Lewis (12 January 2008). "Did life begin on a radioactive beach?". *New Scientist* (London: Reed Business Information) (2638): 8. ISSN 0262-4079. Retrieved 2015-06-26.

[254] Adam, Zachary (2007). "Actinides and Life's Origins". *Astrobiology* (New Rochelle, NY: Mary Ann Liebert, Inc.) **7** (6): 852–872. Bibcode:2007AsBio...7..852A. doi:10.1089/ast.2006.0066. ISSN 1531-1074. PMID 18163867.

[255] Parnell, John (December 2004). "Mineral Radioactivity in Sands as a Mechanism for Fixation of Organic Carbon on the Early Earth". *Origins of Life and Evolution of Biospheres* (Kluwer Academic Publishers) **34** (6): 533–547. Bibcode:2004OLEB...34..533P.doi:10.1023/B:ORIG.0000043132.23966.a1. ISSN 0169-6149. PMID15570707.

[256] Michaelian, Karo (30 June 2009). "Thermodynamic Function of Life". arXiv:0907.0040 [physics.gen-ph].

[257] Michaelian, Karo (25 January 2011). "Biological catalysis of the hydrological cycle: life's thermodynamic function". *Hydrology and Earth System Sciences Discussions* (Göttingen, Germany: Copernicus Publications on behalf of the European Geosciences Union) **8**: 1093–1123. Bibcode:2011HESSD...8.1093M. doi:10.5194/hessd-8-1093-2011. ISSN 1812-2116.

[258] Michaelian, Karo (11 March 2011). "Thermodynamic Dissipation Theory for the Origin of Life" (PDF). *Earth System Dynamics* (Göttingen, Germany: Copernicus Publications on behalf of the European Geosciences Union) **2**: 37–51. arXiv:0907.0042. Bibcode:2011ESD.....2...37M. doi:10.5194/esd-2-37-2011. ISSN 2190-4987. Retrieved 2015-06-28.

[259] Cnossen, Ingrid; Sanz-Forcada, Jorge; Favata, Fabio; et al. (February 2007). "Habitat of early life: Solar X-ray and UV radiation at Earth's surface 4–3.5 billion years ago". *Journal of Geophysical Research* (Washington, D.C.: American Geophysical Union) **112** (E2): E02008. arXiv:astro-ph/0702529. Bibcode:2007JGRE..112.2008C. doi:10.1029/2006JE002784. ISSN 0148-0227.

[260] Sagan, Carl (April 1973). "Ultraviolet Selection Pressure on the Earliest Organisms". *Journal of Theoretical Biology* (Amsterdam, the Netherlands: Elsevier) **39** (1): 195–200. doi:10.1016/0022-5193(73)90216-6. ISSN 0022-5193. PMID 4741712.

[261] Michaelian, Karo; Simeonov, Aleksandar (16 May 2014). "Fundamental Molecules of Life are Pigments which Arose and Evolved to Dissipate the Solar Spectrum". arXiv:1405.4059 [physics.bio-ph].

[262] Michaelian, Karo (2013). "A non-linear irreversible thermodynamic perspective on organic pigment proliferation and biological evolution" (PDF). *Journal of Physics: Conference Series* (Bristol, England: IOP Publishing) **475** (conference 1): 012010. arXiv:1307.5924. Bibcode:2013JPhCS.475a2010M. doi:10.1088/1742-6596/475/1/012010. ISSN 1742-6596. "4th National Meeting in Chaos, Complex System and Time Series 29 November to 2 December 2011, Xalapa, Veracruz, Mexico"

[263] Knauth 1992, pp. 123–152

[264] Knauth, L. Paul; Lowe, Donald R. (May 2003). "High Archean climatic temperature inferred from oxygen isotope geochemistry of cherts in the 3.5 Ga Swaziland group, South Africa". *Geological Society of America Bulletin* (Boulder, CO: Geological Society of America) **115**: 566–580. Bibcode:2003GSAB..115..566K. doi:10.1130/0016-7606(2003)115<0566:hactif>2.0.co;2. ISSN 0016-7606.

[265] Lowe, Donald R.; Tice, Michael M. (June 2004). "Geologic evidence for Archean atmospheric and climatic evolution: Fluctuating levels of CO_2, CH_4, and O_2 with an overriding tectonic control". *Geology* (Boulder, CO: Geological Society of America) **32** (6): 493–496. Bibcode:2004Geo....32..493L. doi:10.1130/G20342.1. ISSN 0091-7613.

[266] Michaelian, Karo; Santillán Padilla, Norberto (24 November 2014). "DNA Denaturing through UV-C Photon Dissipation: A Possible Route to Archean Non-enzymatic Replication" (PDF). *bioRxiv* (Cold Spring Harbor, NY: Cold Spring Harbor Laboratory). doi:10.1101/009126. Retrieved 2015-06-29.

[267] Michaelian, Karo (2010). "Homochirality Through Photon-Induced Melting of RNA/DNA: Thermodynamic Dissipation Theory Of The Origin Of Life". *WebmedCentral* (Durham, UK: Webmed Limited, UK) **1** (10): WMC00924. doi:10.9754/journal.wmc.2010.00924.

[268] Davies, Paul (December 2007). "Are Aliens Among Us?" (PDF). *Scientific American* (Stuttgart: Georg von Holtzbrinck Publishing Group) **297** (6): 62–69. doi:10.1038/scientificamerican1207-62. ISSN 0036-8733. Retrieved 2015-07-16. ...if life does emerge readily under terrestrial conditions, then perhaps it formed many times on our home planet. To pursue this possibility, deserts, lakes and other extreme or isolated environments have been searched for evidence of "alien" life-forms—organisms that would differ fundamentally from known organisms because they arose independently.

[269] Hartman, Hyman (October 1998). "Photosynthesis and the Origin of Life". *Origins of Life and Evolution of Biospheres* (Kluwer Academic Publishers) **28** (4–6): 515–521. Bibcode:1998OLEB...28..515H. doi:10.1023/A:1006548904157. ISSN 0169-6149. PMID 11536891.

[270] Damer, Bruce; Deamer, David (13 March 2015). "Coupled Phases and Combinatorial Selection in Fluctuating Hydrothermal Pools: A Scenario to Guide Experimental Approaches to the Origin of Cellular Life". *Life* (Basel, Switzerland: MDPI) **5** (1): 872–887. doi:10.3390/life5010872. ISSN 2075-1729. PMC 4390883. PMID 25780958.

9.11 Bibliography

- Altermann, Wladyslaw (2009). "From Fossils to Astrobiology – A Roadmap to Fata Morgana?" (PDF). In Seckbach, Joseph; Walsh, Maud. *From Fossils to Astrobiology: Records of Life on Earth and the Search for Extraterrestrial Biosignatures*. Cellular Origin, Life in Extreme Habitats and Astrobiology **12**. Dordrecht, the Netherlands; London: Springer Science+Business Media. ISBN 978-1-4020-8836-0. LCCN 2008933212. Retrieved 2015-06-05.

- Bada, Jeffrey L.; Lazcano, Antonio (2009). "The Origin of Life". In Ruse, Michael; Travis, Joseph. *Evolution: The First Four Billion Years*. Foreword by Edward O. Wilson. Cambridge, MA: Belknap Press of Harvard University Press. ISBN 978-0-674-03175-3. LCCN 2008030270. OCLC 225874308.

- Barton, Nicholas H.; Briggs, Derek E. G.; Eisen, Jonathan A.; et al. (2007). *Evolution*. Cold Spring Harbor, NY: Cold Spring Harbor Laboratory Press. ISBN 978-0-87969-684-9. LCCN 2007010767. OCLC 86090399.

- Bastian, H. Charlton (1871). *The Modes of Origin of Lowest Organisms*. London; New York: Macmillan and Company. LCCN 11004276. OCLC 42959303. Retrieved 2015-06-06.

- Bernal, J. D. (1951). *The Physical Basis of Life*. London: Routledge & Kegan Paul. LCCN 51005794.

- Bernal, J. D. (1960). "The Problem of Stages in Biopoesis". In Florkin, M. *Aspects of the Origin of Life*. International Series of Monographs on Pure and Applied Biology. Oxford, UK; New York: Pergamon Press. ISBN 978-1-4831-3587-8. LCCN 60013823.

- Bernal, J. D. (1967) [Reprinted work by A. I. Oparin originally published 1924; Moscow: The Moscow Worker]. *The Origin of Life*. The Weidenfeld and Nicolson Natural History. Translation of Oparin by Ann Synge. London: Weidenfeld & Nicolson. LCCN 67098482.

- Bock, Gregory R.; Goode, Jamie A., eds. (1996). *Evolution of Hydrothermal Ecosystems on Earth (and Mars?)*. Ciba Foundation Symposium **202**. Chichester, UK; New York: John Wiley & Sons. ISBN 0-471-96509-X. LCCN 96031351.

- Bondeson, Jan (1999). *The Feejee Mermaid and Other Essays in Natural and Unnatural History*. Ithaca, NY: Cornell University Press. ISBN 0-8014-3609-5. LCCN 98038295.

- Bryson, Bill (2004). *A Short History of Nearly Everything*. London: Black Swan. ISBN 978-0-552-99704-1. OCLC 55589795.

- Calvin, Melvin (1969). *Chemical Evolution: Molecular Evolution Towards the Origin of Living Systems on the Earth and Elsewhere*. Oxford, UK: Clarendon Press. ISBN 0-19-855342-0. LCCN 70415289. OCLC 25220.

- Chaichian, Masud; Rojas, Hugo Perez; Tureanu, Anca (2014). "Physics and Life". *Basic Concepts in Physics: From the Cosmos to Quarks*. Undergraduate Lecture Notes in Physics. Berlin; Heidelberg: Springer Berlin Heidelberg. doi:10.1007/978-3-642-19598-3_12. ISBN 978-3-642-19597-6. ISSN 2192-4791. LCCN 2013950482. OCLC 900189038.

- Chang, Thomas Ming Swi (2007). *Artificial Cells: Biotechnology, Nanomedicine, Regenerative Medicine, Blood Substitutes, Bioencapsulation, and Cell/Stem Cell Therapy*. Regenerative Medicine, Artificial Cells and Nanomedicine **1**. Hackensack, NJ: World Scientific. ISBN 978-981-270-576-1. LCCN 2007013738. OCLC 173522612.

- Clancy, Paul; Brack, André; Horneck, Gerda (2005). *Looking for Life, Searching the Solar System*. Cambridge, UK: Cambridge University Press. ISBN 978-0-521-82450-7. LCCN 2006271630. OCLC 57574490.

- Dalrymple, G. Brent (2001). "The age of the Earth in the twentieth century: a problem (mostly) solved". In Lewis, C. L. E.; Knell, S. J. *The Age of the Earth: from 4004 BC to AD 2002*. Geological Society Special Publication **190**. London: Geological Society of London. Bibcode:2001GSLSP.190..205D. doi:10.1144/gsl.sp.2001.190.01.14. ISBN 1-86239-093-2. ISSN 0305-8719. LCCN 2003464816. OCLC 48570033.

- Darwin, Charles (1887). Darwin, Francis, ed. *The Life and Letters of Charles Darwin, Including an Autobiographical Chapter* **3** (3rd ed.). London: John Murray. OCLC 834491774.

- Davies, Geoffrey F. (2007). "Chapter 2.3 Dynamics of the Hadean and Archaean Mantle". In van Kranendonk, Martin J.; Smithies, R. Hugh; Bennett, Vickie C. *Earth's Oldest Rocks*. Developments in Precambrian Geology **15**. Amsterdam, the Netherlands; Boston: Elsevier. doi:10.1016/S0166-2635(07)15023-4. ISBN 978-0-444-52810-0. LCCN 2009525003.

- Davies, Paul (1999). *The Fifth Miracle: The Search for the Origin of Life*. London: Penguin Books. ISBN 0-14-028226-2.

- Dawkins, Richard (1996). *The Blind Watchmaker* (Reissue with a new introduction ed.). New York: W. W. Norton & Company. ISBN 0-393-31570-3. LCCN 96229669. OCLC 35648431.

- Dawkins, Richard (2004). *The Ancestor's Tale: A Pilgrimage to the Dawn of Evolution*. Boston, MA: Houghton Mifflin. ISBN 0-618-00583-8. LCCN 2004059864. OCLC 56617123.

- Dobell, Clifford (1960) [Originally published 1932; New York: Harcourt, Brace & Company]. *Antony van Leeuwenhoek and His 'Little Animals'*. New York: Dover Publications. LCCN 60002548.

- Dyson, Freeman (1999). *Origins of Life* (Revised ed.). Cambridge, UK; New York: Cambridge University Press. ISBN 0-521-62668-4. LCCN 99021079.

- Eigen, M.; Schuster, P. (1979). *The Hypercycle: A Principle of Natural Self-Organization*. Berlin; New York: Springer-Verlag. ISBN 0-387-09293-5. LCCN 79001315. OCLC 4665354.

- Fesenkov, V. G. (1959). "Some Considerations about the Primaeval State of the Earth". In Oparin, A. I.; et al. *The Origin of Life on the Earth*. I.U.B. Symposium Series **1**. Edited for the International Union of Biochemistry by Frank Clark and R. L. M. Synge (English-French-German ed.). London; New York: Pergamon Press. ISBN 978-1-4832-2240-0. LCCN 59012060. Retrieved 2015-06-03. International Symposium on the Origin of Life on the Earth (held at Moscow, 19–24 August 1957)

- Hazen, Robert M. (2005). *Genesis: The Scientific Quest for Life's Origin*. Washington, D.C.: Joseph Henry Press. ISBN 0-309-09432-1. LCCN 2005012839. OCLC 60321860.

- Huxley, Thomas Henry (1968) [Originally published 1897]. "VIII Biogenesis and Abiogenesis [1870]". *Discourses, Biological and Geological*. Collected Essays **VIII** (Reprint ed.). New York: Greenwood Press. LCCN 70029958. Retrieved 2014-05-19.

- Kauffman, Stuart (1993). *The Origins of Order: Self-Organization and Selection in Evolution*. New York: Oxford University Press. ISBN 978-0-19-507951-7. LCCN 91011148. OCLC 23253930.

- Kauffman, Stuart (1995). *At Home in the Universe: The Search for Laws of Self-Organization and Complexity*. New York: Oxford University Press. ISBN 0-19-509599-5. LCCN 94025268.

- Klyce, Brig (22 January 2001). Kingsley, Stuart A.; Bhathal, Ragbir, eds. *Panspermia Asks New Questions*. The Search for Extraterrestrial Intelligence (SETI) in the Optical Spectrum III **4273**. Bellingham, WA: SPIE. doi:10.1117/12.435366. ISBN 0-8194-3951-7. LCCN 2001279159. Retrieved 2015-06-09. Proceedings of the SPIE held at San Jose, CA, 22–24 January 2001

- Knauth, L. Paul (1992). "Origin and diagenesis of cherts: An isotopic perspective". In Clauer, Norbert; Chaudhuri, Sambhu. *Isotopic Signatures and Sedimentary Records*. Lecture Notes in Earth Sciences **43**. Berlin; New York: Springer-Verlag. doi:10.1007/BFb0009863. ISBN 3-540-55828-4. ISSN 0930-0317. LCCN 92025372. OCLC 26262469.

- Lane, Nick (2009). *Life Ascending: The 10 Great Inventions of Evolution* (1st American ed.). New York: W. W. Norton & Company. ISBN 978-0-393-06596-1. LCCN 2009005046. OCLC 286488326.

- Lankenau, Dirk-Henner (2011). "Two RNA Worlds: Toward the Origin of Replication, Genes, Recombination and Repair". In Egel, Richard; Lankenau, Dirk-Henner; Mulkidjanian,, Armen Y. *Origins of Life: The Primal Self-Organization*. Heidelberg: Springer. doi:10.1007/978-3-642-21625-1. ISBN 978-3-642-21624-4. LCCN 2011935879. OCLC 733245537.

- Lennox, James G. (2001). *Aristotle's Philosophy of Biology: Studies in the Origins of Life Science*. Cambridge Studies in Philosophy and Biology. Cambridge, UK; New York: Cambridge University Press. ISBN 0-521-65976-0. LCCN 00026070.

- McKinney, Michael L. (1997). "How do rare species avoid extinction? A paleontological view". In Kunin, William E.; Gaston, Kevin J. *The Biology of Rarity: Causes and consequences of rare—common differences* (1st ed.). London; New York: Chapman & Hall. ISBN 0-412-63380-9. LCCN 96071014. OCLC 36442106.

- Michod, Richard E. (1999). "Darwinian Dynamics: Evolutionary Transitions in Fitness and Individuality". Princeton, NJ: Princeton University Press. ISBN 0-691-02699-8. LCCN 98004166. OCLC 38948118.

- Miller, G. Tyler; Spoolman, Scott E. (2012). *Environmental Science* (14th ed.). Belmont, CA: Brooks/Cole. ISBN 978-1-111-98893-7. LCCN 2011934330. OCLC 741539226.

- Oparin, A. I. (1953) [Originally published 1938; New York: The Macmillan Company]. *The Origin of Life*. Translation and new introduction by Sergius Morgulis (2nd ed.). Mineola, NY: Dover Publications. ISBN 0-486-49522-1. LCCN 53010161.

- Orgel, Leslie E. (1987). "Evolution of the Genetic Apparatus: A Review". *Evolution of Catalytic Function*. Cold Spring Harbor Symposia on Quantitative Biology **52**. Cold Spring Harbor, NY: Cold Spring Harbor Laboratory Press. doi:10.1101/SQB.1987.052.01.004. ISBN 0-87969-054-2. OCLC 19850881. "Proceedings of a symposium held at Cold Spring Harbor Laboratory in 1987"

- Raven, Peter H.; Johnson, George B. (2002). *Biology* (6th ed.). Boston, MA: McGraw-Hill. ISBN 0-07-112261-3. LCCN 2001030052. OCLC 45806501.

- Ross, Alexander (1652). *Arcana Microcosmi*. Book II. London. Retrieved 2015-07-07.

- Russell, Michael, ed. (2010). *Origins, Abiogenesis and the Search for Life*. Cambridge, MA: Cosmology Science Publishers. ISBN 978-0-9829552-1-5.

- Shapiro, Robert (1987). *Origins: A Skeptic's Guide to the Creation of Life on Earth*. Toronto; New York: Bantam Books. ISBN 0-553-34355-6.

- Sheldon, Robert B. (22 September 2005). Hoover, Richard B.; Levin, Gilbert V.; Rozanov, Alexei Y.; Gladstone, G. Randall, eds. *Historical Development of the Distinction between Bio- and Abiogenesis* (PDF). Astrobiology and Planetary Missions **5906**. Bellingham, WA: SPIE. doi:10.1117/12.663480. ISBN 978-0-8194-5911-4. LCCN 2005284378. Retrieved 2015-04-13. Proceedings of the SPIE held at San Diego, CA, 31 July–2 August 2005

- Stearns, Beverly Peterson; Stearns, Stephen C. (1999). *Watching, from the Edge of Extinction*. New Haven, CT: Yale University Press. ISBN 0-300-07606-1. LCCN 98034087. OCLC 47011675.

- Tyndall, John (1905) [Originally published 1871; London; New York: Longmans, Green & Co.; D. Appleton and Company]. *Fragments of Science* **2** (6th ed.). New York: P.F. Collier & Sons. OCLC 726998155. Retrieved 2015-06-06.

- Vartanian, Aram (1973). "Spontaneous Generation". In Wiener, Philip P. *Dictionary of the History of Ideas* **IV**. New York: Charles Scribner's Sons. ISBN 0-684-13293-1. LCCN 72007943. Retrieved 2015-06-05.

- Voet, Donald; Voet, Judith G. (2004). *Biochemistry* **1** (3rd ed.). New York: John Wiley & Sons. ISBN 0-471-19350-X. LCCN 2003269978.

- Woodward, Robert J., ed. (1969). *Our Amazing World of Nature: Its Marvels & Mysteries*. Pleasantville, NY: Reader's Digest Association. ISBN 0-340-13000-8. LCCN 69010418.

- Yarus, Michael (2010). *Life from an RNA World: The Ancestor Within*. Cambridge, MA: Harvard University Press. ISBN 978-0-674-05075-4. LCCN 2009044011.

9.12 Further reading

- Arrhenius, Gustaf O.; Sales, Brian C.; Mojzsis, Stephen J.; et al. (21 August 1997). "Entropy and Charge in Molecular Evolution—the Case of Phosphate" (PDF). *Journal of Theoretical Biology* (Amsterdam, the Netherlands: Elsevier) **187** (4): 503–522. doi:10.1006/jtbi.1996.0385. ISSN 0022-5193. PMID 9299295.
- Cavalier-Smith, Thomas (June 2006). "Cell evolution and Earth history: stasis and revolution". *Philosophical Transactions of the Royal Society B* (London: Royal Society) **361** (1470): 969–1006. doi:10.1098/rstb.2006.1842. ISSN 0962-8436. PMC 1578732. PMID 16754610.
- de Duve, Christian (1995). *Vital Dust: Life As A Cosmic Imperative* (1st ed.). New York: Basic Books. ISBN 0-465-09044-3. LCCN 94012964. OCLC 30624716.
- Fernando, Chrisantha T.; Rowe, Jonathan (7 July 2007). "Natural selection in chemical evolution". *Journal of Theoretical Biology* (Amsterdam, the Netherlands) **247** (1): 152–167. doi:10.1016/j.jtbi.2007.01.028. ISSN 0022-5193. PMID 17399743.
- Gribbin, John (1998). *The Case of the Missing Neutrinos: And other Curious Phenomena of the Universe* (1st Fromm International ed.). New York: Fromm International. ISBN 0-88064-199-1. LCCN 98027948. OCLC 39368356.
- Harris, Henry (2002). *Things Come to Life: Spontaneous Generation Revisited.* Oxford, UK; New York: Oxford University Press. ISBN 0-19-851538-3. LCCN 2001054856. OCLC 48100507.
- Horgan, John (February 1991). "In the Beginning...". *Scientific American* (Stuttgart: Georg von Holtzbrinck Publishing Group) **264** (2): 116–125. doi:10.1038/scientificamerican0291-116. ISSN 0036-8733.
- Ignatov, Ignat; Mosin, Oleg V. (2013). "Modeling of Possible Processes for Origin of Life and Living Matter in Hot Mineral and Seawater with Deuterium". *Journal of Environment and Earth Science* (New York: International Institute for Science, Technology and Education) **3** (14): 103–118. ISSN 2224-3216. Retrieved 2015-06-29.
- Jortner, Joshua (October 2006). "Conditions for the emergence of life on the early Earth: summary and reflections". *Philosophical Transactions of the Royal Society B* (London: Royal Society) **361** (1474): 1877–1891. doi:10.1098/rstb.2006.1909. ISSN 0962-8436. PMC 1664691. PMID 17008225.
- Klotz, Irene (24 February 2012). "Did Life Start in a Pond, Not Oceans?". *Discovery News* (Silver Spring, MD: Discovery Communications). Retrieved 2015-06-29.
- Knoll, Andrew H. (2003). *Life on a Young Planet: The First Three Billion Years of Evolution on Earth.* Princeton, NJ: Princeton University Press. ISBN 0-691-00978-3. LCCN 2002035484. OCLC 50604948.
- Luisi, Pier Luigi (2006). *The Emergence of Life: From Chemical Origins to Synthetic Biology.* Cambridge, UK: Cambridge University Press. ISBN 978-0-521-82117-9. LCCN 2006285720. OCLC 173609999.
- Maynard Smith, John; Szathmáry, Eörs (1999). *The Origins of Life: From the Birth of Life to the Origin of Language.* Oxford, UK; New York: Oxford University Press. ISBN 0-19-850493-4. LCCN 99230990. OCLC 40980149.
- Morowitz, Harold J. (1992). *Beginnings of Cellular Life: Metabolism Recapitulates Biogenesis.* New Haven, CT: Yale University Press. ISBN 0-300-05483-1. LCCN 92006849. OCLC 25316379.
- NASA Astrobiology Institute: Harrison, T. Mark; McKeegan, Kevin D.; Mojzsis, Stephen J. "Earth's Early Environment and Life: When did Earth become suitable for habitation?". Archived from the original on 2012-02-17. Retrieved 2015-06-30.
- NASA Specialized Center of Research and Training in Exobiology: Arrhenius, Gustaf O. (11 September 2002). "Arrhenius". Archived from the original on 2007-12-21. Retrieved 2015-06-30.
- "The physico-chemical basis of life". *What is Life.* Spring Valley, CA: Lukas K. Buehler. Retrieved 27 October 2005.

- Pitsch, Stefan; Krishnamurthy, Ramanarayanan; Arrhenius, Gustaf O. (6 September 2000). "Concentration of Simple Aldehydes by Sulfite-Containing Double-Layer Hydroxide Minerals: Implications for Biopoesis". *Helvetica Chimica Acta* (Hoboken, NJ: John Wiley & Sons) **83** (9): 2398–2411. doi:10.1002/1522-2675(20000906)83:9<2398::AID-HLCA2398>3.0.CO;2-5. ISSN 0018-019X. PMID 11543578.

- Pons, Marie-Laure; Quitté, Ghylaine; Fujii, Toshiyuki; et al. (25 October 2011). "Early Archean Serpentine Mud Volcanoes at Isua, Greenland, as a Niche for Early Life". *Proc. Natl. Acad. Sci. U.S.A.* (Washington, D.C.: National Academy of Sciences) **108** (43): 17639–17643. Bibcode:2011PNAS..10817639P. doi:10.1073/pnas.1108061108. ISSN 0027-8424. PMC 3203773. PMID 22006301.

- Pross, Addy (2012). *What is Life?: How Chemistry Becomes Biology* (1st ed.). Oxford, UK: Oxford University Press. ISBN 978-0-19-964101-7. LCCN 2012538842. OCLC 812020290.

- Roy, Debjani; Schleyer, Paul von Ragué (2010). "Chemical Origin of Life: How do Five HCN Molecules Combine to form Adenine under Prebiotic and Interstellar Conditions". In Matta, Chérif F. *Quantum Biochemistry*. Weinheim, Germany: Wiley-VCH. doi:10.1002/9783527629213.ch6. ISBN 978-3-527-62921-3. LCCN 2011499476. OCLC 905973537.

- Russell, Michael J.; Hall, A. J.; Cairns-Smith, Alexander Graham; et al. (10 November 1988). "Submarine hot springs and the origin of life". *Nature* (London: Nature Publishing Group) **336** (6195): 117. Bibcode:1988Natur.336..117R. doi:10.1038/336117a0. ISSN 0028-0836. PMID 11536607.

- Shock, Everett L. (25 October 1997). "High-temperature life without photosynthesis as a model for Mars" (PDF). *Journal of Geophysical Research* (Washington, D.C.: American Geophysical Union) **102** (E10): 23687–23694. Bibcode:1997JGR...10223687S. doi:10.1029/97je01087. ISSN 0148-0227.

9.13 External links

- "Exploring Life's Origins: A Virtual Exhibit". *Exploring Life's Origins: A Virtual Exhibit.* Arlington County, VA: National Science Foundation. Retrieved 2015-07-02.

- Fields, Helen (October 2010). "The Origins of Life". *Smithsonian* (Washington, D.C.: Smithsonian Institution). ISSN 0037-7333. Retrieved 2015-07-02.

- Fox, Douglas (28 March 2007). "Primordial Soup's On: Scientists Repeat Evolution's Most Famous Experiment". *Scientific American* (Stuttgart: Georg von Holtzbrinck Publishing Group). ISSN 0036-8733. Retrieved 2015-07-02.

- "The Geochemical Origins of Life by Michael J. Russell & Allan J. Hall". Glasgow, Scotland: University of Glasgow. 13 December 2008. Retrieved 2015-07-02.

- Kauffman, Stuart (8 August 1996). "Even peptides do it". *Nature* (London: Nature Publishing Group) **382** (6591): 496–497. Bibcode:1996Natur.382..496K. doi:10.1038/382496a0. ISSN 0028-0836. PMID 8700218. Archived from the original on 2006-10-15. Retrieved 2015-07-02.

- Malory, Marcia. "How life began on Earth". *Earth Facts.* Retrieved 2015-07-02.

- Nowak, Martin A.; Ohtsuki, Hisashi (30 September 2008). "Prevolutionary dynamics and the origin of evolution" (PDF). *Proc. Natl. Acad. Sci. U.S.A.* (Washington, D.C.: National Academy of Sciences) **105** (39): 14924–14927. Bibcode:2008PNAS..10514924N. doi:10.1073/pnas.0806714105. ISSN 0027-8424. PMC 2567469. PMID 18791073.

- "Possible Connections Between Interstellar Chemistry and the Origin of Life on the Earth". *Space Science and Astrobiology at Ames.* NASA. Archived from the original on 2009-07-31. Retrieved 2015-07-02.

- "Research Spotlight: Jack Szostak: Making Life from Scratch". *Origins of Life Initiative.* Cambridge, MA: Harvard University. Retrieved 2015-07-02.

- Schirber, Michael (9 June 2006). "How Life Began: New Research Suggests Simple Approach". *LiveScience* (Ogden, UT: Purch). Retrieved 2015-07-02.
- "Scientists Find Clues That Life Began in Deep Space". *NASA Astrobiology Institute*. Mountain View, CA: NASA. 30 January 2001. Archived from the original on 2013-04-29. Retrieved 2015-07-02.
- "Simple Artificial Cell Created From Scratch To Study Cell Complexity". *Science Daily* (Rockville, MD: ScienceDaily, LLC). 16 May 2008. Retrieved 2015-07-02. Post is reprinted from materials provided by Pennsylvania State University.
- Singer, Emily (19 July 2015). "Chemists Invent New Letters for Nature's Genetic Alphabet". *Wired*. New York: Condé Nast. Retrieved 2015-07-20.
- Swaminathan, Nikhil (10 June 2008). "Scientists Close to Reconstructing First Living Cell". *Scientific American* (News) (Stuttgart: Georg von Holtzbrinck Publishing Group). ISSN 0036-8733. Retrieved 2015-07-02.
- Vasas, Vera; Fernando, Chrisantha; Santos, Mauro; et al. (5 January 2012). "Evolution before genes" (PDF). *Biology Direct* (London: BioMed Central) **7**: 1. doi:10.1186/1745-6150-7-1. ISSN 1745-6150.
- Zlobin, Andrei E. (2013). "Tunguska similar impacts and origin of life". *Modern Scientific Researches and Innovations* (Moscow: International Centre of Science and Innovations Ltd.) (12). Retrieved 2015-07-02.
- Zlobin, Andrei E. (2014). "Symmetry infringement in mathematical metrics of hydrogen atom as illustration of ideas by V.I.Vernadsky concerning origin of life and biosphere" (PDF). *Acta Naturae* (Moscow: Park Media Ltd.) (Special Issue 1): 48. ISSN 2075-8251. Retrieved 2015-07-02.

9.13.1 Video resources

- Hazen, Robert M. (29 April 2014). *The Origins of Life* (Webcast). Baltimore, MD: Space Telescope Science Institute. Retrieved 2015-07-03. — A 2014 Spring Symposium webcast (video; 38m)
- "The Origin of Life" on YouTube — A Royal Institution Discourse lecture given by John Maynard Smith in 1995 (video; 58m)
- "Space Experts Discuss the Search for Life in the Universe at NASA" on YouTube — Panel discussion at NASA headquarters on 14 July 2014 (video; 87m)

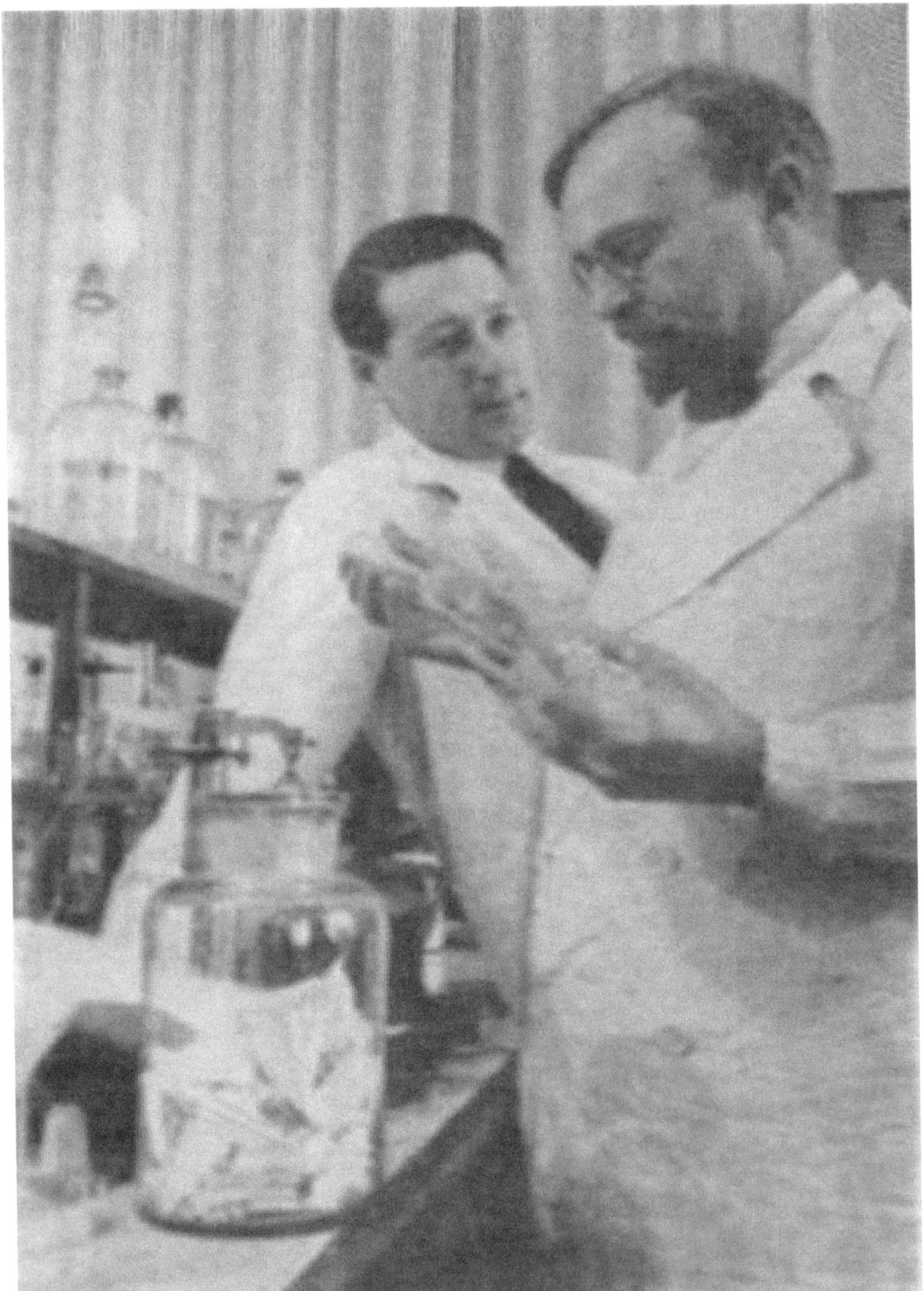

Alexander Oparin (right) at the laboratory

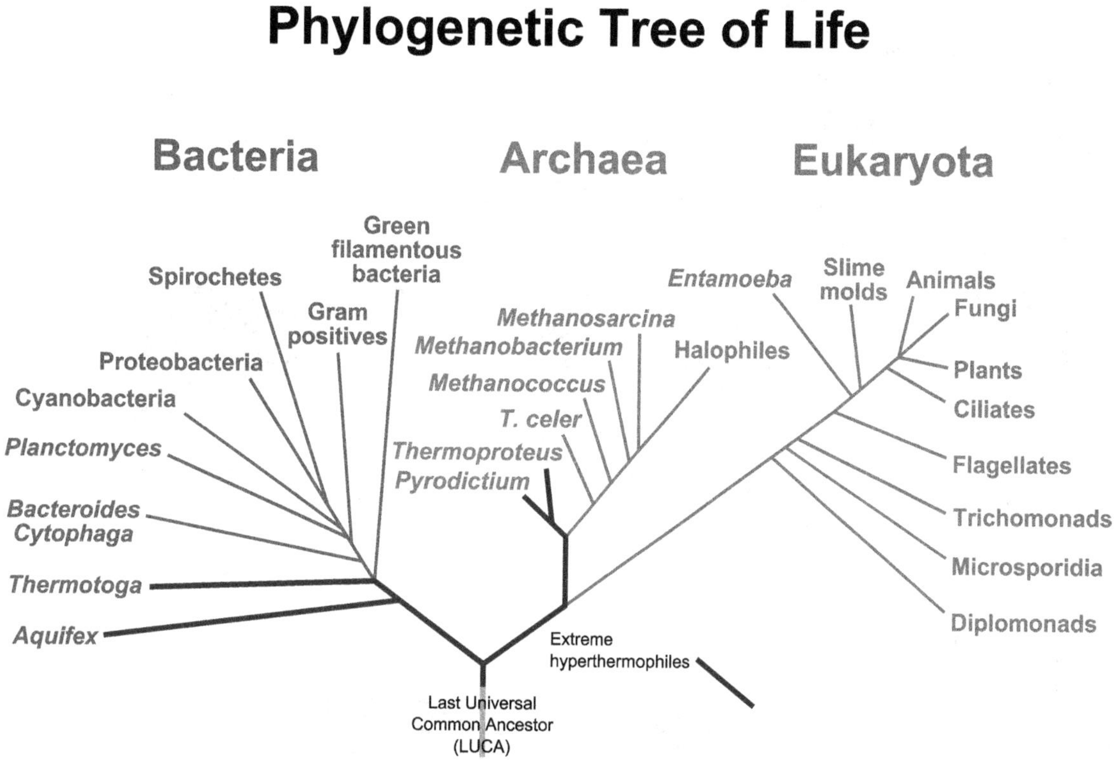

A cladogram demonstrating extreme hyperthermophiles at the base of the phylogenetic tree of life.

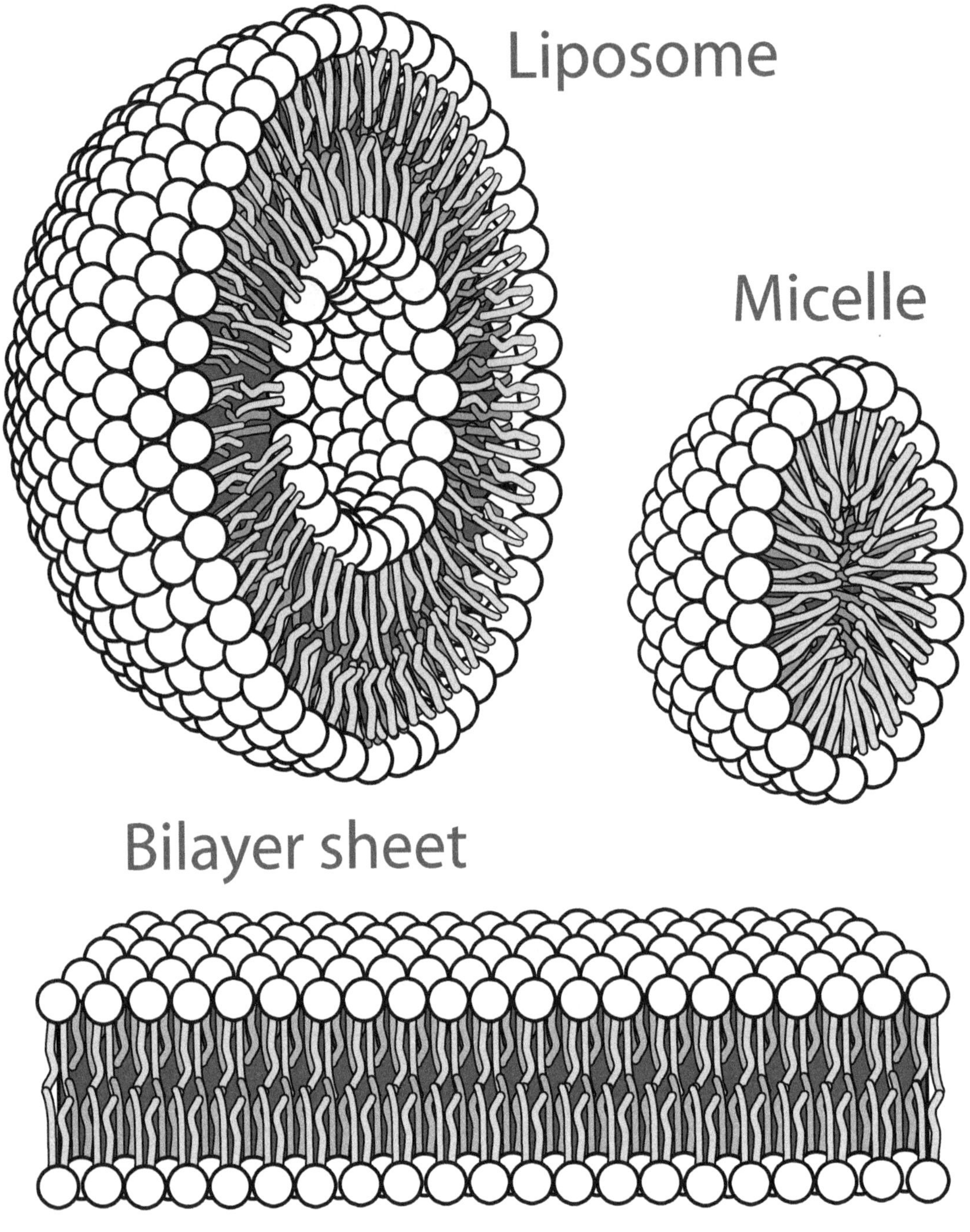

The three main structures phospholipids form spontaneously in solution: the liposome (a closed bilayer), the micelle and the bilayer.

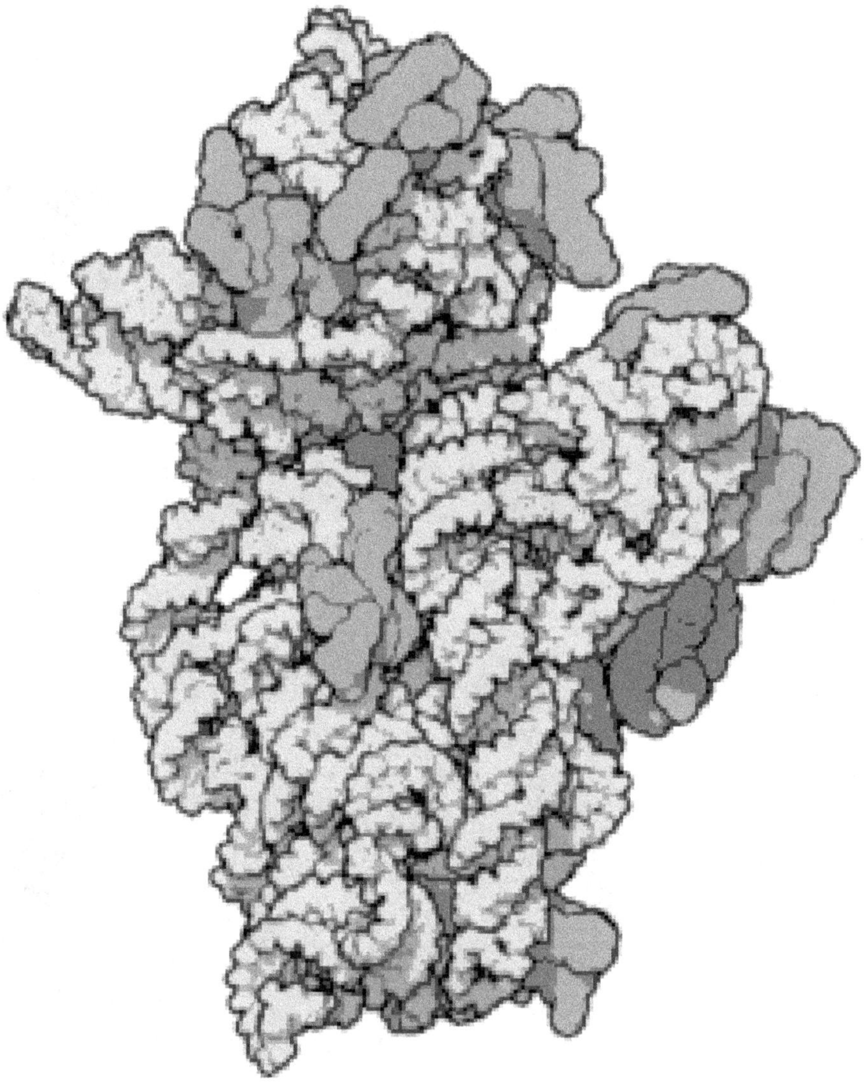

Molecular structure of the ribosome 30S subunit from Thermus thermophilus.[143] *Proteins are shown in blue and the single RNA chain in orange.*

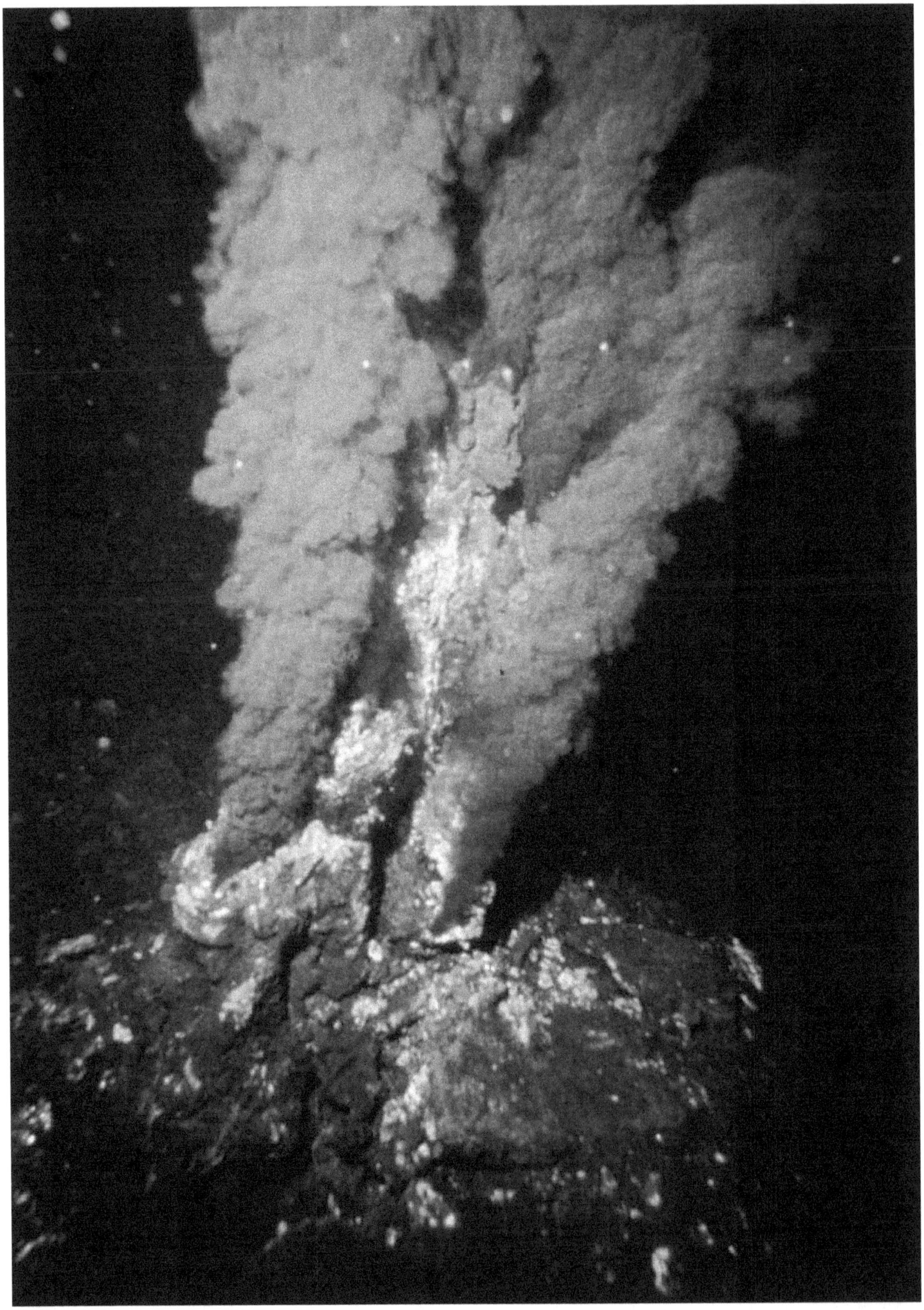

Deep-sea hydrothermal vent or 'black smoker'

White smokers emitting liquid carbon dioxide (CO_2) at the Champagne *vent, Marianas Trench Marine National Monument*

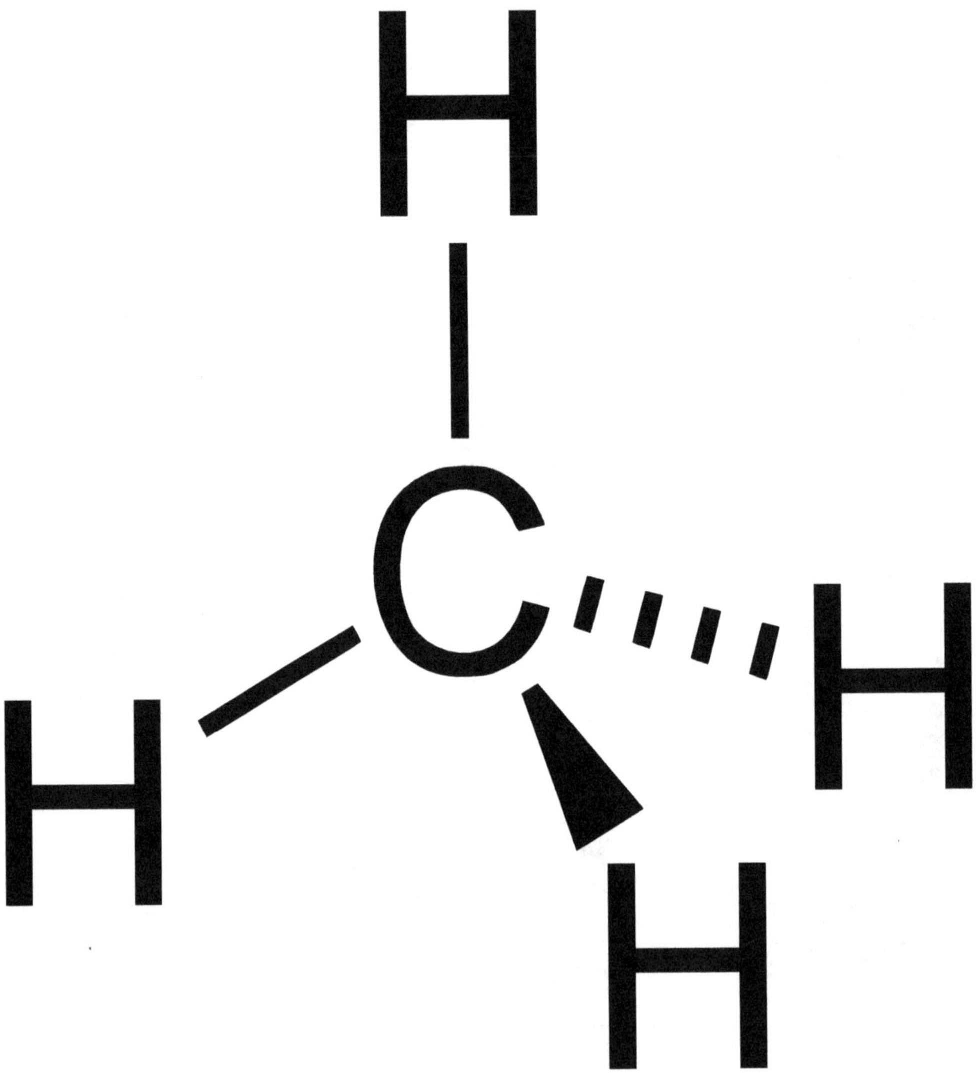

Methane is one of the simplest organic compounds

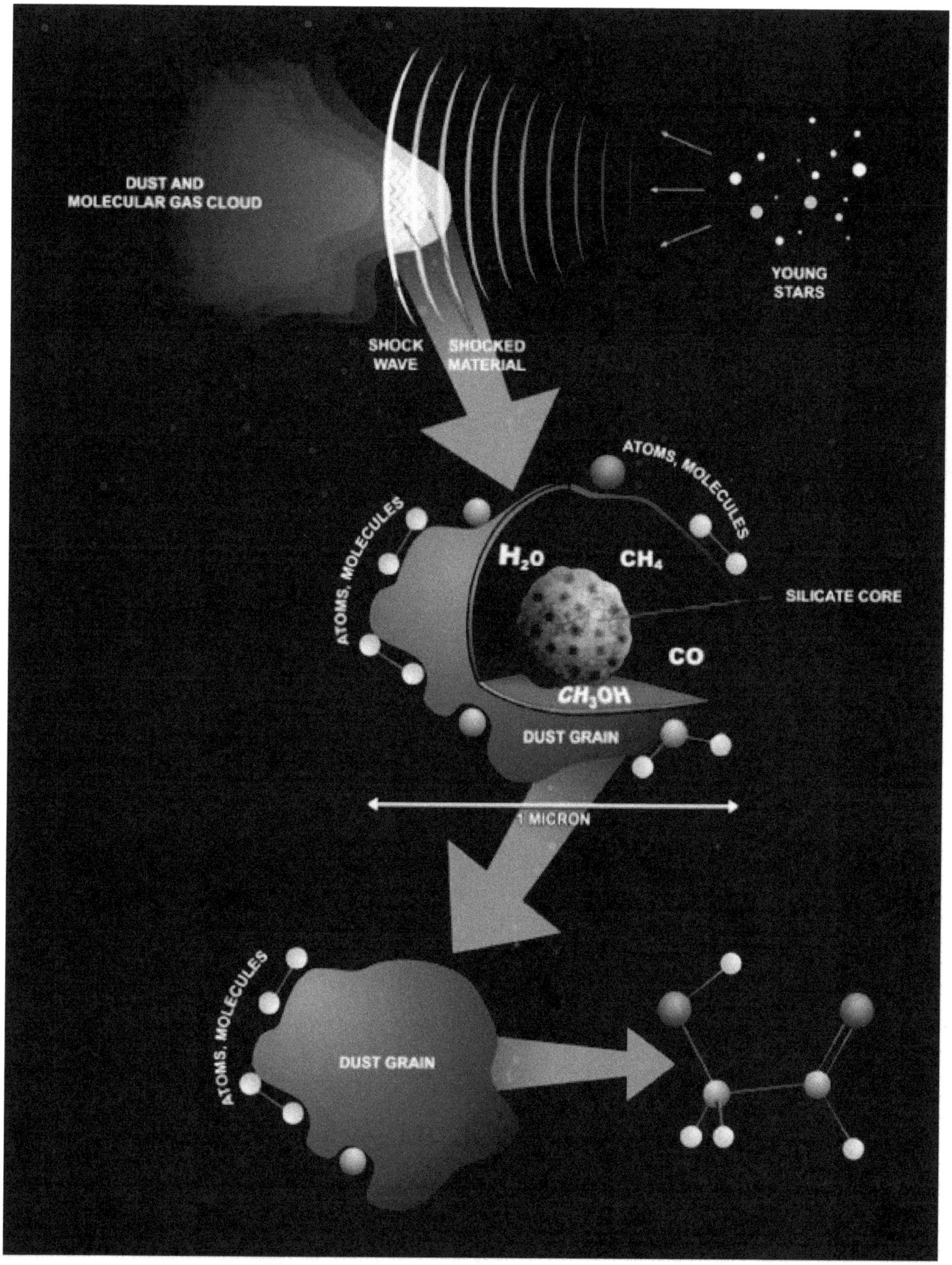

Formation of glycolaldehyde in stardust

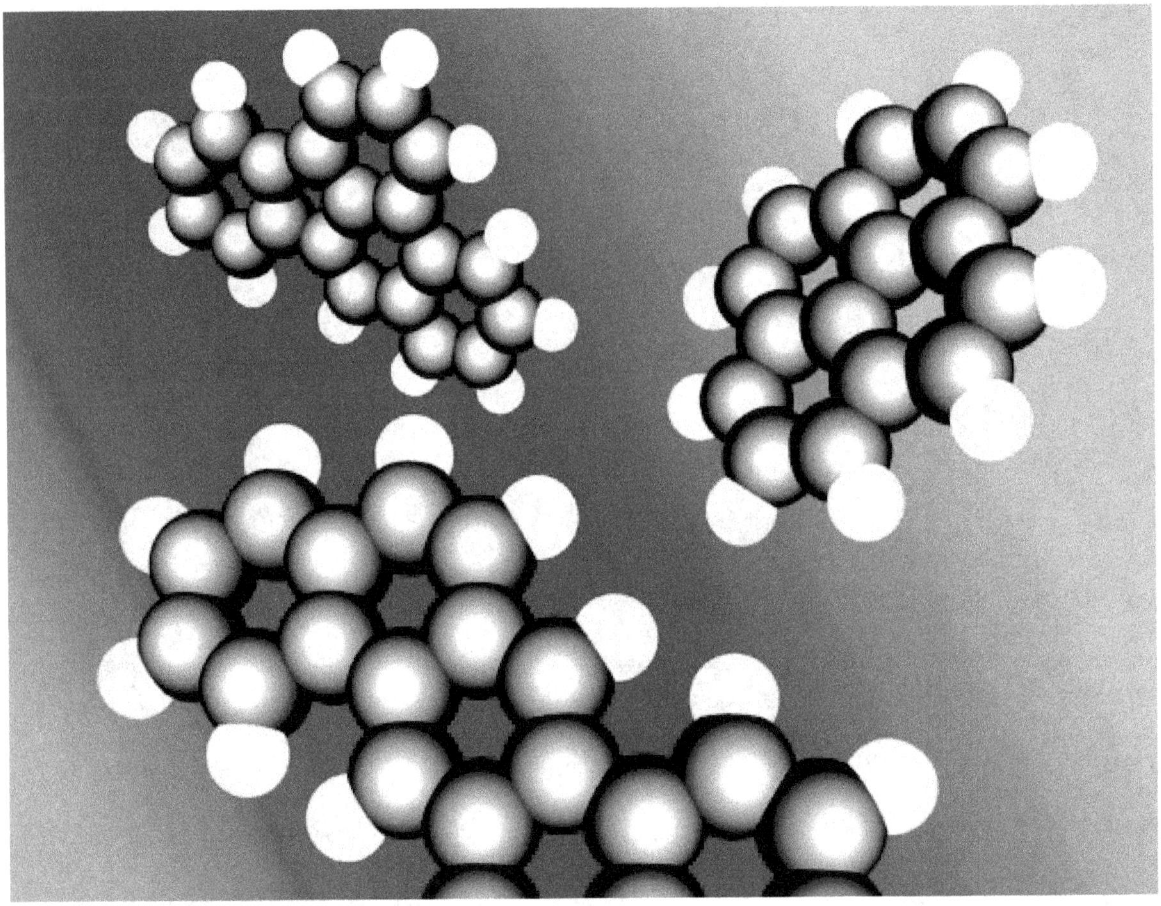

An illustration of typical polycyclic aromatic hydrocarbons. Clockwise from top left: benz(e)acephenanthrylene, pyrene and dibenz(ah)anthracene.

Chapter 10

Biotic material

Biotic material or **biological derived material** is any material that originates from living organisms. Most such materials contain carbon and are capable of decay.

The earliest life on Earth arose at least 3.5 billion years ago.[1][2][3] Earlier physical evidences of life include graphite, a biogenic substance, in 3.7 billion-year-old metasedimentary rocks discovered in southwestern Greenland,[4] as well as, "remains of biotic life" found in 4.1 billion-year-old rocks in Western Australia.[5][6] Earth's biodiversity has expanded continually except when interrupted by mass extinctions.[7] Although scholars estimate that over 99 percent of all species of life (over five billion)[8] that ever lived on Earth are extinct,[9][10] there are still an estimated 10–14 million extant species,[11][12] of which about 1.2 million have been documented and over 86% have not yet been described.[13]

Examples of biotic materials are wood, linoleum, straw, humus, manure, bark, crude oil, cotton, spider silk, chitin, fibrin, and bone.

The use of biotic materials, and processed biotic materials (bio-based material) as alternative natural materials, over synthetics is popular with those who are environmentally conscious because such materials are usually biodegradable, renewable, and the processing is commonly understood and has minimal environmental impact. However, not all biotic materials are used in an environmentally friendly way, such as those that require high levels of processing, are harvested unsustainably, or are used to produce carbon emissions.

When the source of the recently living material has little importance to the product produced, such as in the production of biofuels, biotic material is simply called biomass. Many fuel sources may have biological sources, and may be divided roughly into fossil fuels, and biofuel.

In soil science, biotic material is often referred to as *organic matter*. Biotic materials in soil include glomalin, Dopplerite and humic acid. Some biotic material may not be considered to be organic matter if it is low in organic compounds, such as a clam's shell, which is an essential component of the living organism, but contains little organic carbon.

Examples of the use of biotic materials include:

- Alternative natural materials
- building material, for a stylistic reasons, or to reduce allergic reactions.
- clothing
- energy production
- food
- medicine
- ink
- composting and mulch

10.1 References

[1] Schopf, JW, Kudryavtsev, AB, Czaja, AD, and Tripathi, AB. (2007). *Evidence of Archean life: Stromatolites and microfossils.* Precambrian Research 158:141–155.

[2] Schopf, JW (2006). *Fossil evidence of Archaean life.* Philos Trans R Soc Lond B Biol Sci 29;361(1470) 869-85.

[3] Hamilton Raven, Peter; Brooks Johnson, George (2002). *Biology.* McGraw-Hill Education. p. 68. ISBN 978-0-07-112261-0. Retrieved 7 July 2013.

[4] Ohtomo, Yoko; Kakegawa, Takeshi; Ishida, Akizumi; et al. (January 2014). "Evidence for biogenic graphite in early Archaean Isua metasedimentary rocks". *Nature Geoscience* (London: Nature Publishing Group) **7** (1): 25–28. Bibcode:2014NatGe...7...25O. doi:10.1038/ngeo2025. ISSN 1752-0894.

[5] Borenstein, Seth (19 October 2015). "Hints of life on what was thought to be desolate early Earth". *Excite* (Yonkers, NY: Mindspark Interactive Network). Associated Press. Retrieved 2015-10-20.

[6] Bell, Elizabeth A.; Boehnike, Patrick; Harrison, T. Mark; et al. (19 October 2015). "Potentially biogenic carbon preserved in a 4.1 billion-year-old zircon" (PDF). *Proc. Natl. Acad. Sci. U.S.A.* (Washington, D.C.: National Academy of Sciences). doi:10.1073/pnas.1517557112. ISSN 1091-6490. Retrieved 2015-10-20. Early edition, published online before print.

[7] Sahney, S., Benton, M.J. and Ferry, P.A. (27 January 2010). "Links between global taxonomic diversity, ecological diversity and the expansion of vertebrates on land" (PDF). *Biology Letters* **6** (4): 544–47. doi:10.1098/rsbl.2009.1024. PMC 2936204. PMID 20106856.

[8] Kunin, W.E.; Gaston, Kevin, eds. (31 December 1996). *The Biology of Rarity: Causes and consequences of rare—common differences.* ISBN 978-0412633805. Retrieved 26 May 2015.

[9] Stearns, Beverly Peterson; Stearns, S. C.; Stearns, Stephen C. (1 August 2000). *Watching, from the Edge of Extinction.* Yale University Press. p. 1921. ISBN 978-0-300-08469-6. Retrieved 27 December 2014.

[10] Novacek, Michael J. (8 November 2014). "Prehistory's Brilliant Future". *New York Times.* Retrieved 25 December 2014.

[11] May, Robert M. (1988). "How many species are there on earth?". *Science* **241** (4872): 1441–1449. Bibcode:1988Sci...241.1441M. doi:10.1126/science.241.4872.1441. PMID 17790039.

[12] Miller, G.; Spoolman, Scott (1 January 2012). "Biodiversity and Evolution". *Environmental Science.* Cengage Learning. p. 62. ISBN 1-133-70787-4. Retrieved 27 December 2014.

[13] Mora, C.; Tittensor, D.P.; Adl, S.; Simpson, A.G.; Worm, B. (23 August 2011). "How many species are there on Earth and in the ocean?". *PLOS Biology.* doi:10.1371/journal.pbio.1001127. Retrieved 26 May 2015.

Chapter 11

Directed panspermia

Directed panspermia concerns the deliberate transport of microorganisms in space to be used as introduced species on lifeless planets. Directed panspermia may have been sent to Earth to start life here, or may be sent from Earth to seed exoplanets with life.

Historically, Shklovskii and Sagan (1966) and Crick and Orgel (1973) hypothesized that life on Earth may have been seeded deliberately by other civilizations. Conversely, Mautner and Matloff (1979) and Mautner (1995, 1997) proposed that we ourselves should seed new planetary systems, protoplanetary discs or star-forming clouds with microorganisms, to secure and expand our organic gene/protein life-form. To avoid interference with local life, the targets may be young planetary systems where local life is unlikely. Directed panspermia can be motivated by biotic ethics that value the basic patterns of organic gene/protein life with its unique complexity and unity, and its drive for self-propagation.

Belonging to life then implies panbiotic ethics with a purpose to propagate and expand life in space. Directed panspermia for this purpose is becoming possible due to developments in solar sails, precise astrometry, the discovery of extrasolar planets, extremophiles and microbial genetic engineering. Cosmological projections suggests that life in space can then have an immense future.[1][2]

11.1 History and motivation

An early example of the idea of directed panspermia dates to the early science fiction work *Last and First Men* by Olaf Stapledon, first published in 1930. It details the manner in which the last humans, upon discovering that the Solar System will soon be destroyed, send microscopic "seeds of a new humanity" towards potentially habitable areas of the universe.[3]

In 1966 Shklovskii and Sagan proposed that life on Earth may have been seeded through directed panspermia by other civilisations.[4] and 1973 Crick and Orgel also discussed [5] Conversely, Mautner and Matloff proposed in 1979, and Mautner examined in detail in 1995 and 1997 the technology and motivation to secure and expand our organic gene/protein life-form by directed panspermia missions to new planetary systems, protoplanetary discs and star-forming clouds.[2][6][7][8] Technological aspects include propulsion by solar sails, deceleration by radiation pressure or viscous drag at the target, and capture of the colonizing micro-organisms by planets. A possible objection is potential interference with local life at the targets, but targeting young planetary systems where local life, especially advanced life, could not have started yet, avoids this problem.[8] Directed panspermia may be motivated by the desire to perpetuate the common genetic heritage of all terrestrial life. This motivation was formulated as biotic ethics that value the common gene/protein patterns of our family of organic life,[9] and as panbiotic ethics that aim to secure and expand life in the universe.[7][8]

11.2 Strategies and targets

Directed panspermia may be aimed at nearby young planetary systems such as Alpha PsA (25 ly (light-years) away) and Beta Pictoris (63.4 ly), both of which show accretion discs and signs of comets and planets. More suitable targets may be identified by space telescopes such as the Kepler mission that will identify nearby star systems with habitable planets. Alternatively, directed panspermia may aim at star-forming interstellar clouds such as Rho Ophiuchi cloud complex (427 ly), that contains clusters of new stars too young to originate local life (425 infrared-emitting young stars aged 100,000 to a million years). Such clouds contain zones with various densities (diffuse cloud < dark fragment < dense core < protostellar condensation < accretion disc)[10] that could selectively capture panspermia capsules of various sizes.

Habitable planets or habitable zones about nearby stars may be targeted by large (10 kg) missions where microbial capsules are bundled and shielded. Upon arrival, microbial capsules in the payload may be dispersed in orbit for capture by planets. Alternatively, small microbial capsules may be sent in large swarms to habitable planets, protoplanetary discs, or zones of various density in interstellar clouds. The microbial swarm provides minimal shielding but does not require high precision targeting, especially when aiming at large interstellar clouds.[2]

11.3 Propulsion and launch

Panspermia missions should deliver microorganisms that can grow in the new habitats. They may be sent in 10^{-10} kg, 30 micrometer radius capsules that allow intact atmospheric entry at the target planets, each containing 100,000 diverse microorganisms suited to various environments. Both for bundled large mass missions and microbial capsule swarms, solar sails may provide the most simple propulsion for interstellar transit.[11] Spherical sails will avoid orientation control both at launch and at deceleration at the targets. For bundled shielded missions to nearby star systems, solar sails with thicknesses of 10^{-7} m and areal densities of 0.0001 kg/m^2 seem feasible, and sail/payload mass ratios of 10:1 will allow exit velocities near the maximum possible for such sails. Sails with about 540 m radius and area of 10^6 m^2 can impart 10 kg payloads with interstellar cruise velocities of 0.0005 *c* (1.5×10^5 m/s) when launched from 1 au (astronomical unit). At this speed, voyage to the Alpha PsA star will last 50,000 y, and to the Rho Opiuchus cloud, 824,000 years.

At the targets, the microbial payload would decompose into 10^{11} (100 billion) 30 μm capsules to increase the probability of capture. In the swarm strategy to protoplanetary discs and interstellar clouds, 1 mm radius, 4.2×10^{-6} kg microbial capsules are launched from 1 au using sails of 4.2×10^{-5} kg with radius of 0.37 m and area of 0.42 m^2 to achieve cruising speeds of 0.0005 *c*. At the target, each capsule decomposes into 4,000 delivery microcapsules of 10^{-10} kg and of 30 micrometer radius that allow intact entry to planetary atmospheres.[12] For missions that do not encounter dense gas zones, such as interstellar transit to mature planets or to habitable zones about stars, the microcapsules can be launched directly from 1 au using 10^{-9} kg sails of 1.8 mm radius to achieve velocities of 0.0005 *c* to be decelerated by radiation pressure for capture at the targets. The 1 mm and 30 micrometer radius vehicles and payloads are needed in large numbers for both the bundled and swarm missions. These capsules and the miniature sails for swarm missions can be mass manufactured readily.

11.4 Astrometry and targeting

The panspermia vehicles would be aimed at moving targets whose locations at the time of arrival must be predicted. This can be calculated using their measured proper motions, their distances, and the cruising speeds of the vehicles. The positional uncertainty and size of the target object then allow estimating the probability that the panspermia vehicles will arrive at their targets. The positional uncertainty δy (m) of the target at arrival time is given by equation (1), where $\alpha_{(p)}$ is the resolution of proper motion of the target object (arsec/year), d is the distance (m) and v is the velocity of the vehicle (m/s)

$$\delta y = 1.5 \times 10^{-13}\ \alpha_p d^2 v$$

Given the positional uncertainty, the vehicles may be launched with a scatter in a circle about the predicted position of the target. The probability P_{target} for a capsule to hit the target area with radius r_{target} (m) is the given by the ratio of the targeting scatter and the target area.

$$_{target} = A_{target}/\pi(\delta y)^2 = 4.4 \times 10^{25}\ r_{target}{}^2 v^2/(\alpha_p{}^2 d^4)$$

To apply these equations, the precision of astrometry of star proper motion of 0.00001 parsec/year, and the solar sail vehicle velocity of 0.0005 *c* (1.5 x 10^5 m/s) may be expected within a few decades. For a chosen planetary system, the area A_{target} may be the width of the habitable zone, while for interstellar clouds, it may be the sizes of the various density zones of the cloud.

11.5 Deceleration and capture

Solar sail missions to Sun-like stars can decelerate by radiation pressure in reverse dynamics of the launch. The sails must be properly oriented at arrival, but orientation control may be avoided using spherical sails. The vehicles must approach the target Sun-like stars at radial distances similar to the launch, about 1 au. After the vehicles are captured in orbit, the microbial capsules may be dispersed in a ring orbiting the star, some within the gravitational capture zone of planets. Missions to accretion discs of planets and to star-forming clouds will decelerate by viscous drag at the rate dv/dt as determined by equation (3), where v is the velocity, rc the radius of the spherical capsule, ρc is density of the capsule and ρm is the density of the medium.

$$dv/dt = -(3v^2/2\rho_c)\ \rho_m/r_c$$

A vehicle entering the cloud with a velocity of 0.0005 *c* (1.5×10^5 m/s) will be captured when decelerated to 2,000 m/s, the typical speed of grains in the cloud. The size of the capsules can be designed to stop at zones with various densities in the interstellar cloud. Simulations show that a 35 micron radius capsule will be captured in a dense core, and a 1 mm radius capsule in a protostellar condensation in the cloud. As for approach to accretion discs about stars, a millimetre size capsule entering the 1000 km thick disc face at 0.0005 *c* will be captured at 100 km into the disc. Therefore, 1 mm sized objects may be the best for seeding protoplanetary discs about new stars and protostellar condensations in interstellar clouds.[8]

The captured panspermia capsules will mix with dust. A fraction of the dust and a proportional fraction of the captured capsules will be delivered to planets. Dispersing the payload into delivery microcapsules will increase the chance that some will be delivered to planets. Particles of 0.6 - 60 micron radius can remain cold enough to preserve organic matter during atmospheric entry to planets.[12] Accordingly, each 1 mm, 4.2×10^{-6} kg capsule captured in the viscous medium can be dispersed into 42,000 delivery microcapsules of 30 micron radius, each weighing 10^{-10} kg and containing 100,000 microbes. These objects will not be ejected from the dust cloud by radiation pressure from the star, and will remain mixed with the dust.[13][14] A fraction of the dust, containing the captured microbial capsules, will be captured by planets, or captured in comets and delivered by them later to planets. The probability of capture, $P_{capture}$, can be estimated from similar processes, such as the capture of interplanetary dust particles by planets in our Solar System, where 10^{-5} of the Zodiacal cloud maintained by comet ablation, and also a similar fraction of asteroid fragments, is collected by the Earth.[15][16] The probability of capture of an initially launched capsule by a planet, P_{planet} is given by the equation below, where P_{target} is the probability that the capsule reaches the target accretion disc or cloud zone, and $P_{capture}$ is the probability of capture from this zone by a planet.

$$P_{planet} = P_{target} \times P_{capture}$$

The probability P_{planet} depends on the mixing ratio of the capsules with the dust and on the fraction of the dust delivered to planets. These variables can be estimated for capture in planetary accretion discs or in various zones in the interstellar cloud.

11.6 Biomass requirements

After determining the composition of chosen meteorites, astroecologists performed laboratory experiments that suggest that many colonizing microorganisms and some plants could obtain most of their chemical nutrients from asteroid and cometary materials.[17] However, the scientists noted that phosphate (PO_4) and nitrate (NO_3–N) critically limit nutrition to many terrestrial lifeforms.[17] For successful missions, enough biomass must be launched and captured for a reasonable chance to initiate life at the target planet. An optimistic requirement is the capture by the planet of 100 capsules with 100,000 microorganisms each, for a total of 10 million organisms with a total biomass of 10^{-8} kg.

The required biomass to launch for a successful mission is given by following equation. $m_{biomass}$ (kg) = $10^{-8} / P_{planet}$ Using the above equations for P_{target} with transit velocities of 0.0005 c, the known distances to the targets, and the masses of the dust in the target regions then allows calculating the biomass that needs to be launched for probable success. With these parameters, as little as 1 gram of biomass (10^{12} microorganisms) could seed Alpha PsA and 4.5 gram could seed Beta Pictoris. More biomass needs to be launched to the Rho Ophiuchi cloud complex, mainly because its larger distance. A biomass on the order of 300 tons would need to be launched to seed a protostellar condensation or an accretion disc, but two hundred kilograms would be sufficient to seed a young stellar object in the Rho Ophiuchi cloud complex.

Consequently, as long as the required physical range of tolerance are met (e.g.: growth temperature, cosmic radiation shielding, atmosphere and gravity), lifeforms viable on Earth may be chemically nourished by watery asteroid and planetary materials in this and other planetary systems.[17]

11.7 Biological payload

The seeding organisms need to survive and multiply in the target environments and establish a viable biosphere. Some of the new branches of life may develop intelligent beings who will further expand life in the galaxy. The messenger microorganisms may find diverse environments, requiring extremophile microorganisms with a range of tolerances, including thermophile (high temperature), psychrophile (low temperature), acidophile (high acidity), halophile (high salinity), oligotroph (low nutrient concentration), xerophile (dry environments) and radioresistant (high radiation tolerance) microorganisms. Genetic engineering may produce polyextremophile microorganisms with several tolerances. The target atmospheres will probably lack oxygen, so the colonizers should include anaerobic microorganisms. Colonizing anaerobic cyanobacteria may later establish atmospheric oxygen that is needed for higher evolution, as it happened on Earth. Aerobic organisms in the biological payload may be delivered to the planets later when the conditions are right, by comets that captured and preserved the capsules.

The development of eukaryote microorganisms was a major bottleneck to higher evolution on Earth. Including eukaryote microrganisms in the payload can bypass this barrier. Multicellular organisms are even more desirable, but being much heavier than bacteria, fewer can be sent. Hardy tardigrades (water-bears) may be suitable but they are similar to arthropods and would lead to insects. The body-plan of rotifers could lead to higher animals, if the rotifers can be hardened to survive interstellar transit.

Microorganisms or capsules captured in the accretion disc can be captured along with the dust into asteroids. During aqueous alteration the asteroids contain water, inorganic salts and organics, and astroecology experiments with meteorites showed that algae, bacteria, fungi and plant cultures can grow in the asteroids in these media.[18] Microorganisms can then spread in the accreting solar nebula, and will be delivered to planets in comets and in asteroids. The microorganisms can grow on nutrients in the carrier comets and asteroids in the aqueous planetary environments, until they adapt to the local environments and nutrients on the planets.[17][18][19]

11.7.1 Signal in the genome

A number of publications since 1979 have proposed the idea that directed panspermia could be demonstrated to be the origin of all life on Earth if a distinctive 'signature' message were found, deliberately implanted into either the genome or the genetic code of the first microorganisms by our hypothetical progenitor.[20][21][22][23] In 2013 a team of physicists claimed that they had found mathematical and semiotic patterns in the genetic code which, they believe, is evidence for such a signature.[24][25] Further investigations are needed.

11.8 Advanced missions

Significantly, panspermia missions can be launched by present or near-future technologies. However, more advanced technologies may be also used when these become available. The biological aspects of directed panspermia may be improved by genetic engineering to produce hardy polyextremophile microorganisms and multicellular organisms, suitable to diverse planetary environments. Hardy polyextremophile anaerobic multicellular eukaryots with high radiation resistance, that can form a self-sustaining ecosystem with cyanobacteria, would combine ideally the features needed for survival and higher evolution.

For advanced missions, ion thrusters or solar sails using beam-powered propulsion accelerated by Earth-based lasers can achieve speeds up to 0.01 *c* (3×10^6 m/s). Robots may provide in-course navigation, may control the reviving of the frozen microbes periodically during transit to repair radiation damage, and may also choose suitable targets. These propulsion methods and robotics are under development.

Microbial payloads may be also planted on hyperbolic comets bound for interstellar space. This strategy follows the mechanisms of natural panspermia by comets, as suggested by Hoyle and Wikramasinghe.[26] The microorganisms would be frozen in the comets at interstellar temperatures of a few kelvins and protected from radiation for eons. It is unlikely that an ejected comet will be captured in another planetary system, but the probability can be increased by allowing the microbes to multiply during warm perihelion approach to the Sun, then fragmenting the comet. A 1 km radius comet would yield 4.2×10^{12} one-kg seeded fragments, and rotating the comet would eject these shielded icy objects in random directions into the galaxy. This increases a trilion-fold the probability of capture in another planetary system, compared with transport by a single comet.[2][7][8] Such manipulation of comets is a speculative long-term prospect.

11.9 Motivation and ethics

Directed panspermia aims to secure and expand our family of organic gene/protein life. It may be motivated by the desire to perpetuate the common genetic heritage of all terrestrial life. This motivation was formulated as biotic ethics, that value the common gene/protein patters of organic life,[9] and as panbiotic ethics that aim to secure and expand life in the universe.[7][8]

Molecular biology shows complex patterns common to all cellular life, a common genetic code and a common mechanism to translate it into proteins, which in turn help to reproduce the DNA code. Also, shared are the basic mechanisms of energy use and material transport. These self-propagating patterns and processes are the core of organic gene/protein life. Life is unique because of this complexity, and because of the exact coincidence of the laws of physics that allow life to exist. Also unique to life is the pursuit of self-propagation, which implies a human purpose to secure and expand life. These objectives are best secured in space, suggesting a panbiotic ethics aimed to secure this future.[2][7][8][9]

The longevity of human space-faring technological society is uncertain, and it would be prudent to start a directed panspermia program promptly. This program could secure life and allow it to expand in space and in biodiversity with an immense future for trillions of eons.[1][2]

11.10 Objections and counterarguments

The main objection to directed panspermia is that it may interfere with local life at the targets. The colonizing microorganisms may out-compete local life for resources, or infect and harm local organisms. However, this probability can be minimized by targeting newly forming planetary systems, accretion discs and star-forming clouds, where local life, and especially advanced life, could not have emerged yet. If there is local life that is fundamentally different, the colonizing microorganisms may not harm it. If there is local organic gene/protein life, it may exchange genes with the colonizing microorganisms, increasing galactic biodiversity.

Another objection is that space should be left pristine for scientific studies, a reason for planetary quarantine. However, directed panspermia may reach only a few, at best a few hundred new stars, still leaving a hundred billion pristine for local life and for research. A technical objection is the uncertain survival of the messenger organisms during long interstellar transit. Research by simulations, and the development on hardy colonizers is needed to address this questions.

11.11 See also

- Astrobiology
- List of microorganisms tested in outer space
- Planetary protection

11.12 References

[1] Mautner, Michael N. (2005). "Life in the cosmological future: Resources, biomass and populations" (PDF). *Journal of the British Interplanetary Society* **58**: 167–180. Bibcode:2005JBIS...58..167M.

[2] Mautner, Michael N. (2000). *Seeding the Universe with Life: Securing Our Cosmological Future* (PDF). Washington D. C.: Legacy Books (www.amazon.com). ISBN 047600330X.

[3] Stapledon, Olaf (2008). *Last and first men* (Unabridged republ. ed.). Mineola, N.Y.: Dover Publications. p. 238. ISBN 978-0486466828.

[4] Shklovskii, I. S.; Sagan, C. (1966). *Intelligent life in the universe.* New York: Dell. ISBN 978-1892803023.

[5] Crick, F. H.; Orgel, L. E. (1973). "Directed panspermia". *Icarus* **19**: 341. Bibcode:1973Icar...19..341C. doi:10.1016/0019-1035(73)90110-3.

[6] Mautner, M.; Matloff, G. L. (1979). "A technical and ethical evaluation of seeding nearby solar systems" (PDF). *J. British Interplanetary Soc.* **32**: 419–423.

[7] Mautner, Michael N. (1995). "Directed Panspermia. 2. Technological Advances Toward Seeding Other Solar Systems, and the Foundations of Panbiotic Ethics". *J. British Interplanetary Soc.* **48**: 435–440.

[8] Mautner, Michael N. (1997). "Directed panspermia. 3. Strategies and motivation for seeding star-forming clouds" (PDF). *J. British Interplanetary Soc.* **50**: 93–102.

[9] Mautner, Michael N. (2009). "Life-centered ethics and the human future in space" (PDF). *Bioethics* **23**: 433–440. doi:10.1111/j.1467-8519.2008.00688.x. PMID 19077128.

[10] Mezger, P. G. B. F. Burke, J. H. Rahe and E. E. Roettger, eds. "The search for protostars using millimetre/submillimeter dust emission as a tracer". *Planetary Systems: Formation, Evolution and Detection*: 208–220.

[11] Vulpetti, G.; Johnson, L.; Matloff, G. L. (2008). *Solar Sails : A Novel Approach to Interplanetary Flight.* New York: Springer. ISBN 978-0-387-34404-1.

[12] Anders, E. (1989). "Prebiotic organic matter from comets and asteroids". *Nature* **342** (6247): 255–257. Bibcode:1989Natur.342.. doi:10.1038/342255a0. 255A.

[13] Morrison, D. (1977). "Sizes and albedos of the larger asteroids". *Comets, Asteroids and Meteorites: Interrelations, Evolution and Origins, A. H. Delsemme, ed., U. of Toledo Press*: 177–183.

[14] Sekanina, Z. (1977). "Meteor streams in the making". *Comets, Asteroids and Meteorites: Interrelations, Evolution and Origins, A. H. Delsemme, ed., U. of Toledo Press*: 159–169.

[15] Weatherill, G. W. (1977). "Fragmentation of asteroids and delivery of fragments to Earth". *Comets, Asteroids and Meteorites: Interrelations, Evolution and Origins, A. H. Delsemme, ed., U. of Toledo Press*: 283–291.

[16] Kyte, F. T.; Wasson, J. T. (1989). "Accretion rate of exraterrestrial matter: Iridium deposited 33 to 67 million years ago". *Science* **232**: 1225–1229. Bibcode:1986Sci...232.1225K. doi:10.1126/science.232.4755.1225.

[17] Mautner, Michael N. (2002). "Planetary bioresources and astroecology. 1. Planetary microcosm bioessays of Martian and meteorite materials: soluble electrolytes, nutrients, and algal and plant responses" (PDF). *Icarus* **158**: 72–86. Bibcode:2002Icar..158. ..72M.doi:10.1006/icar.2002.6841. PMID 12449855.

[18] Mautner, Michael N. (2002). "Planetary resources and astroecology. Planetary microcosm models of asteroid and meteorite interiors: electrolyte solutions and microbial growth. Implications for space populations and panspermia" (PDF). *Astrobiology* **2**: 59–76. Bibcode:2002AsBio...2...59M. doi:10.1089/153110702753621349.

[19] Olsson-Francis, Karen; Cockell, Charles S. (2010). "Use of cyanobacteria for in-situ resource use in space applications". *Planetary and Space Science* **58**: 1279–1285. Bibcode:2010P&SS...58.1279O. doi:10.1016/j.pss.2010.05.005.

[20] G. Marx (1979). "Message through time". *Acta Astronautica* **6** (1-2): 221–225. Bibcode:1979AcAau...6..221M. doi:10.1016/0 5765(79)90158-9. 094-

[21] H. Yokoo, T. Oshima (1979). "Is bacteriophage φX174 DNA a message from an extraterrestrial intelligence?". *Icarus* **38** (1): 148–153. Bibcode:1979Icar...38..148Y. doi:10.1016/0019-1035(79)90094-0.

[22] Overbye, Dennis (26 June 2007). "Human DNA, the Ultimate Spot for Secret Messages (Are Some There Now?)". Retrieved 2014-10-09.

[23] Davies, Paul C.W. (2010). *The Eerie Silence: Renewing Our Search for Alien Intelligence.* Boston, Massachusetts: Houghton Mifflin Harcourt. ISBN 978-0-547-13324-9.

[24] V. I. shCherbak, M. A. Makukov (2013). "The "Wow! signal" of the terrestrial genetic code". *Icarus* **224** (1): 228–242. arXiv:1303.6739. Bibcode:2013Icar..224..228S. doi:10.1016/j.icarus.2013.02.017.

[25] M. A. Makukov, V. I. shCherbak (2014). "Space ethics to test directed panspermia". *Life Sciences in Space Research* **3**: 10–17. doi:10.1016/j.lssr.2014.07.003.

[26] Hoyle, F.; Wickramasinghe, C. (1978). *Lifecloud: The Origin of Life in the Universe.* London: J. M. Dent and Sons.

11.13 External links

- Society for Life in Space (SOLIS) Interstellar Panspermia Society, for research and education on directed panspermia. Aims to launch directed panspermia missions in this century.
- Astro-ecology and resources for life in space. Experimental research on nutrients in asteroids and meteorites, and on microorganisms and plants that grow on these in-situ space resources.

Chapter 12

List of interstellar and circumstellar molecules

This is a list of molecules that have been detected in the interstellar medium and circumstellar envelopes, grouped by the number of component atoms. The chemical formula is listed for each detected compound, along with any ionized form that has also been observed.

12.1 Detection

The molecules listed below were detected by spectroscopy. Their spectral features are generated by transitions of component electrons between different energy levels, or by rotational or vibrational spectra. Detection usually occurs in radio, microwave, or infrared portions of the spectrum.[1]

Interstellar molecules are formed by chemical reactions within very sparse interstellar or circumstellar clouds of dust and gas. Usually this occurs when a molecule becomes ionized, often as the result of an interaction with a cosmic ray. This positively charged molecule then draws in a nearby reactant by electrostatic attraction of the neutral molecule's electrons. Molecules can also be generated by reactions between neutral atoms and molecules, although this process is generally slower.[2] The dust plays a critical role of shielding the molecules from the ionizing effect of ultraviolet radiation emitted by stars.[3]

12.1.1 History

The chemistry of life may have begun shortly after the Big Bang, 13.8 billion years ago, during a habitable epoch when the Universe was only 10–17 million years old.[4][5]

The first carbon-containing molecule detected in the interstellar medium was the methylidyne radical (CH) in 1937.[6] From the early 1970s it was becoming evident that interstellar dust consisted of a large component of more complex organic molecules (COMs),[7] probably polymers. Chandra Wickramasinghe proposed the existence of polymeric composition based on the molecule formaldehyde (H_2CO).[8] Fred Hoyle and Chandra Wickramasinghe later proposed the identification of bicyclic aromatic compounds from an analysis of the ultraviolet extinction absorption at 2175A.,[9] thus demonstrating the existence of polycyclic aromatic hydrocarbon molecules in space.

In 2004, scientists reported[10] detecting the spectral signatures of anthracene and pyrene in the ultraviolet light emitted by the Red Rectangle nebula (no other such complex molecules had ever been found before in outer space). This discovery was considered a confirmation of a hypothesis that as nebulae of the same type as the Red Rectangle approach the ends of their lives, convection currents cause carbon and hydrogen in the nebulae's core to get caught in stellar winds, and radiate outward.[11] As they cool, the atoms supposedly bond to each other in various ways and eventually form particles of a million or more atoms. The scientists inferred[10] that since they discovered polycyclic aromatic hydrocarbons (PAHs)

— which may have been vital in the formation of early life on Earth — in a nebula, by necessity they must originate in nebulae.[11]

In 2010, fullerenes (or "buckyballs") were detected in nebulae.[12] Fullerenes have been implicated in the origin of life; according to astronomer Letizia Stanghellini, “It's possible that buckyballs from outer space provided seeds for life on Earth.”[13]

In October 2011, scientists found using spectroscopy that cosmic dust contains complex organic compounds (“amorphous organic solids with a mixed aromatic-aliphatic structure”) that could be created naturally, and rapidly, by stars.[14][15][16] The compounds are so complex that their chemical structures resemble the makeup of coal and petroleum; such chemical complexity was previously thought to arise only from living organisms.[14] These observations suggest that organic compounds introduced on Earth by interstellar dust particles could serve as basic ingredients for life due to their surface-catalytic activities.[17][18] One of the scientists suggested that these compounds may have been related to the development of life on Earth and said that, “If this is the case, life on Earth may have had an easier time getting started as these organics can serve as basic ingredients for life.”[14]

In August 2012, astronomers at Copenhagen University reported the detection of a specific sugar molecule, glycolaldehyde, in a distant star system. The molecule was found around the protostellar binary *IRAS 16293-2422*, which is located 400 light years from Earth.[19][20] Glycolaldehyde is needed to form ribonucleic acid, or RNA, which is similar in function to DNA. This finding suggests that complex organic molecules may form in stellar systems prior to the formation of planets, eventually arriving on young planets early in their formation.[21]

In September 2012, NASA scientists reported that PAHs, subjected to interstellar medium (ISM) conditions, are transformed, through hydrogenation, oxygenation, and hydroxylation, to more complex organics — “a step along the path toward amino acids and nucleotides, the raw materials of proteins and DNA, respectively”.[22][23] Further, as a result of these transformations, the PAHs lose their spectroscopic signature which could be one of the reasons “for the lack of PAH detection in interstellar ice grains, particularly the outer regions of cold, dense clouds or the upper molecular layers of protoplanetary disks.”[22][23]

PAHs are found everywhere in deep space[24] and, in June 2013, PAHs were detected in the upper atmosphere of Titan, the largest moon of the planet Saturn.[25]

In 2013, Dwayne Heard at the University of Leeds suggested[26] that quantum mechanical tunneling could explain a reaction his group observed taking place, at a significantly higher than expected rate, between cold (around 63 Kelvin) hydroxyl and methanol molecules, apparently bypassing intramolecular energy barriers which would have to be overcome by thermal energy or ionization events for the same rate to exist at warmer temperatures. The proposed tunneling mechanism may help explain the common observation of fairly complex molecules (up to tens of atoms) in interstellar space.

A particularly large and rich region for detecting interstellar molecules is Sagittarius B2 (Sgr B2). This giant molecular cloud lies near the center of the Milky Way galaxy and is a frequent target for new searches. About half of the molecules listed below were first found near Sgr B2, and nearly every other molecule has since been detected in this feature.[27] A rich source of investigation for circumstellar molecules is the relatively nearby star CW Leonis (IRC +10216), where about 50 compounds have been identified.[28]

In March 2015, NASA scientists reported that, for the first time, complex DNA and RNA organic compounds of life, including uracil, cytosine and thymine, have been formed in the laboratory under outer space conditions, using starting chemicals, such as pyrimidine, found in meteorites. Pyrimidine, like polycyclic aromatic hydrocarbons (PAHs), the most carbon-rich chemical found in the Universe, may have been formed in red giants or in interstellar dust and gas clouds, according to the scientists.[29]

12.2 Molecules

The following tables list molecules that have been detected in the interstellar medium, grouped by the number of component atoms. If there is no entry in the Molecule column, only the ionized form has been detected. For molecules where no designation was given in the scientific literature, that field is left empty. Mass is given in Atomic mass units. The total number of unique species, including distinct ionization states, is listed in parentheses in each section header.

Most of the molecules detected so far are organic. Only one inorganic species has been observed in molecules which contain at least five atoms, SiH_4.[30] Larger molecules have so far all had at least one carbon atom, with no N-N or O-O bonds.[30]

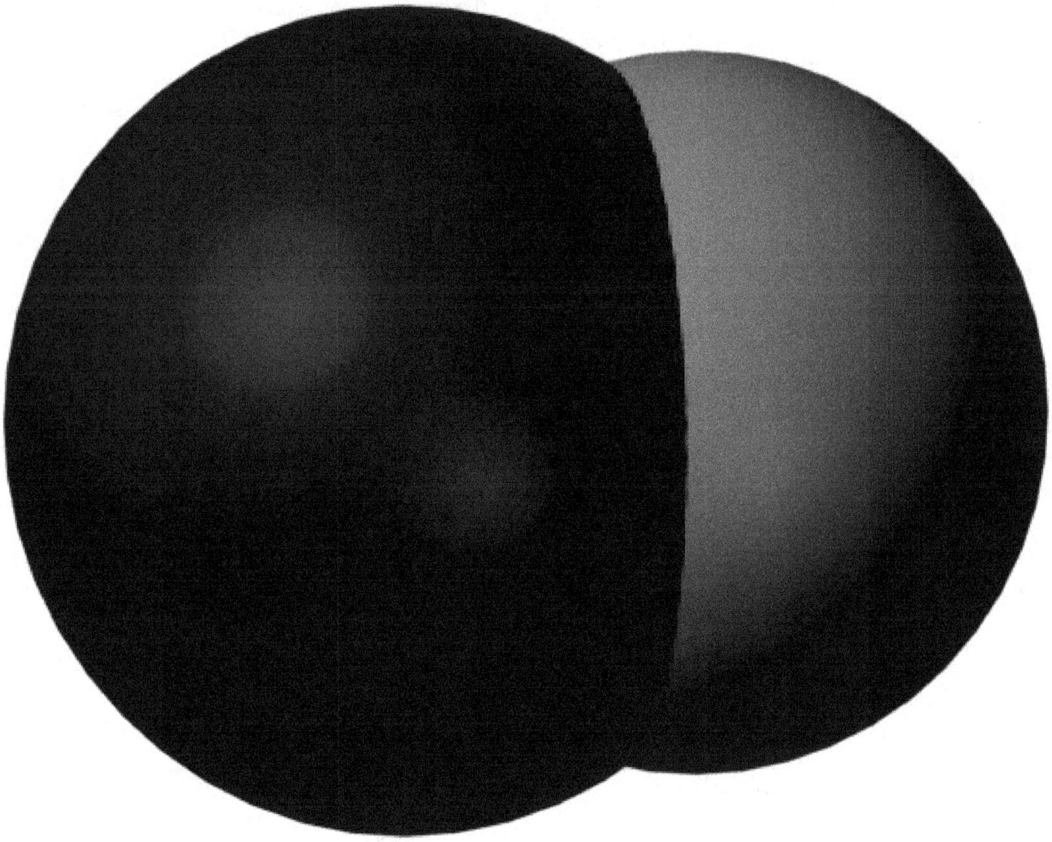

Carbon monoxide is frequently used to trace the distribution of mass in molecular clouds.[31]

The H_3^+ cation is one of the most abundant ions in the universe. It was first detected in 1993.[2][69]

Formaldehyde is an organic molecule that is widely distributed in the interstellar medium.[98]

12.2.1 Diatomic (43)

12.2.2 Triatomic (43)

12.2.3 Four atoms (27)

12.2.4 Five atoms (19)

12.2.5 Six atoms (16)

12.2.6 Seven atoms (10)

12.2.7 Eight atoms (11)

12.2.8 Nine atoms (10)

Methane, the primary component of natural gas, has also been detected on comets and in the atmosphere of several planets in the Solar System.[117]

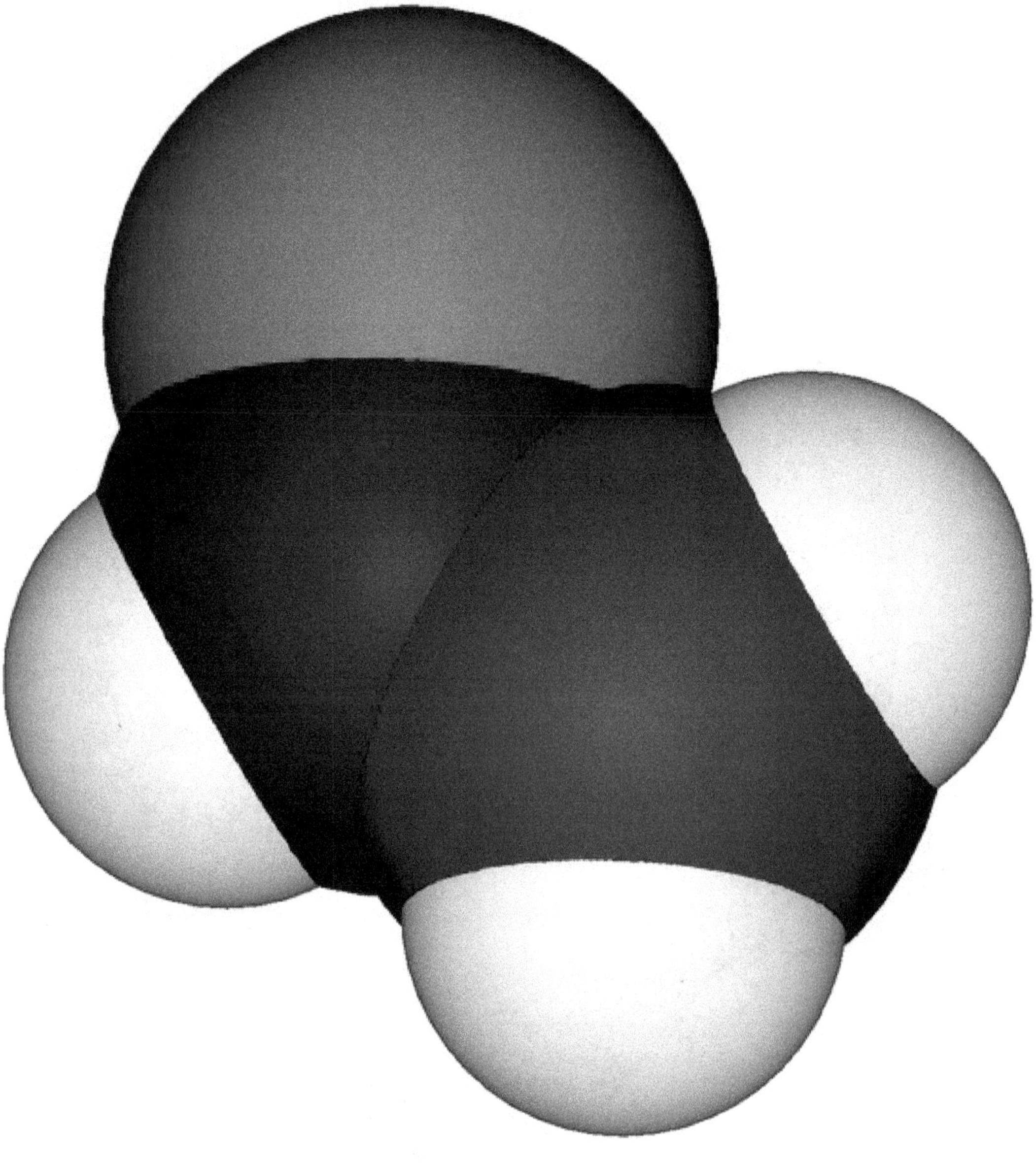

In the ISM, formamide (above) can combine with methylene to form acetamide.[133]

A number of polyyne-derived chemicals are among the heaviest molecules found in the interstellar medium.

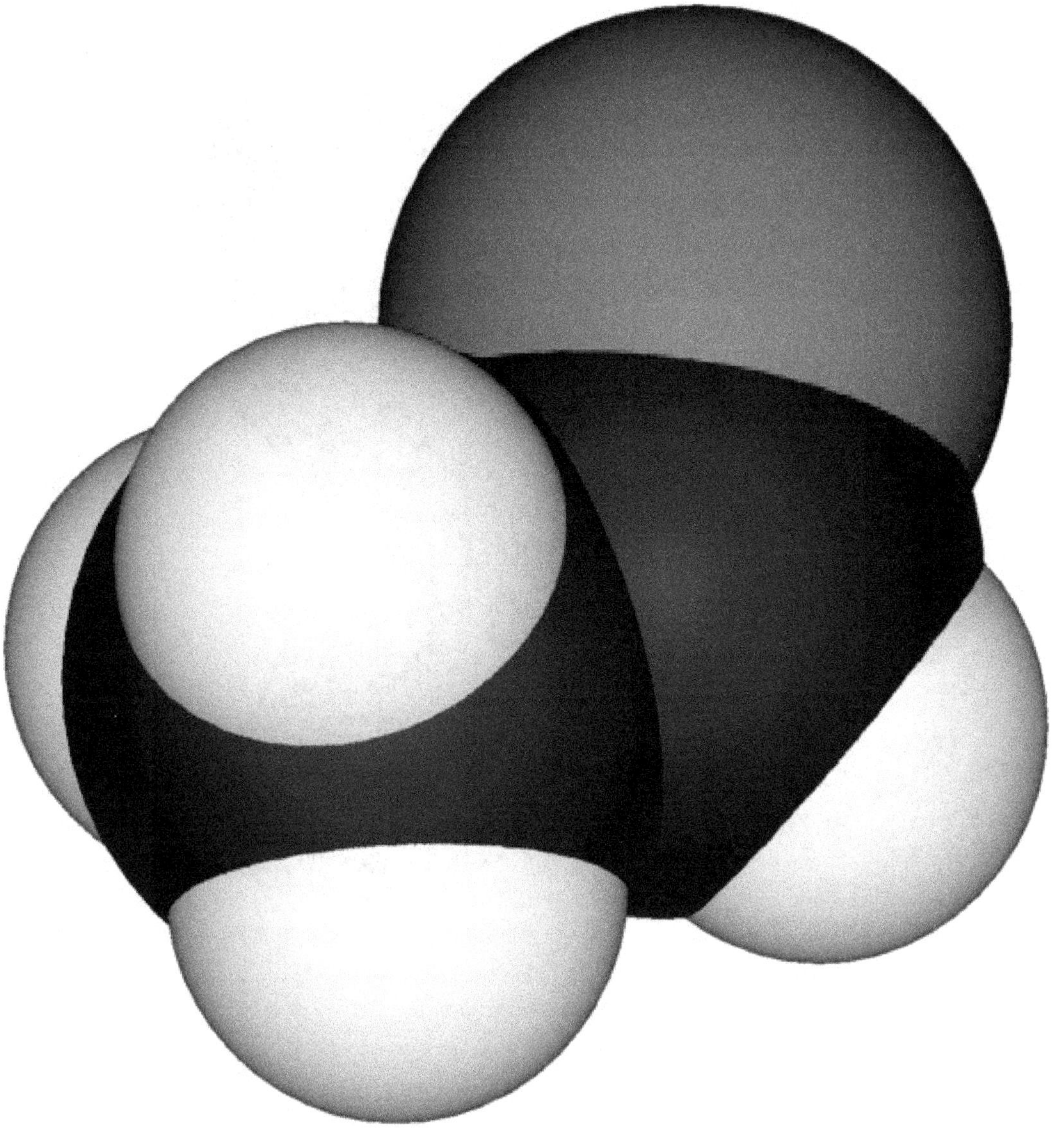

Acetaldehyde (above) and its isomers vinyl alcohol and ethylene oxide have all been detected in interstellar space.[142]

12.2.9 Ten or more atoms (15)

12.3 Deuterated molecules (17)

These molecules all contain one or more deuterium atoms, a heavier isotope of hydrogen.

12.4 Unconfirmed (13)

Evidence for the existence of the following molecules has been reported in scientific literature, but the detections are either described as tentative by the authors, or have been challenged by other researchers. They await independent confirmation.

The radio signature of acetic acid, a compound found in vinegar, was confirmed in 1997.[147]

12.5 See also

- Abiogenesis
- Astrobiology
- Astrochemistry
- Atomic and molecular astrophysics
- Cosmic dust
- Cosmic ray
- Cosmochemistry
- Diffuse interstellar band
- Extraterrestrial liquid water
- Forbidden mechanism

- Intergalactic dust
- Interplanetary medium
- Interstellar medium
- Organic compound
- Outer space
- Panspermia
- Polycyclic aromatic hydrocarbon (PAH)
- Spectroscopy

12.6 References

[1] Shu, Frank H. (1982), *The Physical Universe: An Introduction to Astronomy*, University Science Books, ISBN 0-935702-05-9

[2] Dalgarno, A. (2006), "Interstellar Chemistry Special Feature: The galactic cosmic ray ionization rate", *Proceedings of the National Academy of Sciences* **103** (33): 12269–12273, Bibcode:2006PNAS..10312269D, doi:10.1073/pnas.0602117103, PMC 1567869, PMID 16894166

[3] Brown, Laurie M.; Pais, Abraham; Pippard, A. B. (1995), "The physics of the interstellar medium", *Twentieth Century Physics* (2nd ed.), CRC Press, p. 1765, ISBN 0-7503-0310-7

[4] Loeb, Abraham (October 2014). "The Habitable Epoch of the Early Universe". *International Journal of Astrobiology* **13** (04): 337–339. arXiv:1312.0613. Bibcode:2014IJAsB..13..337L. doi:10.1017/S1473550414000196. Retrieved 15 December 2014.

[5] Dreifus, Claudia (2 December 2014). "Much-Discussed Views That Go Way Back - Avi Loeb Ponders the Early Universe, Nature and Life". *New York Times*. Retrieved 3 December 2014.

[6] Woon, D. E. (May 2005), *Methylidyne radical*, The Astrochemist, retrieved 2007-02-13

[7] Ruaud, M.; Loison, J.C.; Hickson, K.M.; Gratier, P.; Hersant, F.; Wakelam, V. (22 December 2014). "Modeling Complex Organic Molecules in dense regions: Eley-Rideal and complex induced reaction" (PDF). *arXiv*. arXiv:1412.6256v1. Retrieved 26 January 2015.

[8] N.C. Wickramasinghe, Formaldehyde Polymers in Interstellar Space, Nature, 252, 462, 1974

[9] F. Hoyle and N.C. Wickramasinghe, Identification of the lambda 2200A interstellar absorption feature, Nature, 270, 323, 1977

[10] Battersby, S. (2004). "Space molecules point to organic origins". New Scientist. Retrieved 11 December 2009.

[11] Mulas, G.; Malloci, G.; Joblin, C.; Toublanc, D. (2006). "Estimated IR and phosphorescence emission fluxes for specific polycyclic aromatic hydrocarbons in the Red Rectangle". *Astronomy and Astrophysics* **446** (2): 537–549. arXiv:astro-ph/0509586. Bibcode:2006A&A...446..537M. doi:10.1051/0004-6361:20053738.

[12] García-Hernández, D. A.; Manchado, A.; García-Lario, P.; Stanghellini, L.; Villaver, E.; Shaw, R. A.; Szczerba, R.; Perea-Calderón, J. V. (2010-10-28). "Formation Of Fullerenes In H-Containing Planatary Nebulae". *The Astrophysical Journal Letters* **724** (1): L39–L43. arXiv:1009.4357. Bibcode:2010ApJ...724L..39G. doi:10.1088/2041-8205/724/1/L39.

[13] Atkinson, Nancy (2010-10-27). "Buckyballs Could Be Plentiful in the Universe". Universe Today. Retrieved 2010-10-28.

[14] Chow, Denise (26 October 2011). "Discovery: Cosmic Dust Contains Organic Matter from Stars". Space.com. Retrieved 2011-10-26.

[15] ScienceDaily Staff (26 October 2011). "Astronomers Discover Complex Organic Matter Exists Throughout the Universe". ScienceDaily. Retrieved 2011-10-27.

[16] Kwok, Sun; Zhang, Yong (26 October 2011). "Mixed aromatic–aliphatic organic nanoparticles as carriers of unidentified infrared emission features". *Nature* **479** (7371): 80–3. Bibcode:2011Natur.479...80K. doi:10.1038/nature10542. PMID 22031328.

[17] Gallori, Enzo (November 2010). "Astrochemistry and the origin of genetic material". *Rendiconti Lincei* **22** (2): 113–118. doi:10.1007/s12210-011-0118-4. Retrieved 2011-08-11.

[18] Martins, Zita (February 2011). "Organic Chemistry of Carbonaceous Meteorites". *Elements* **7** (1): 35–40. doi:10.2113/gselements.7.1.35. Retrieved 2011-08-11.

[19] Than, Ker (August 29, 2012). "Sugar Found In Space". *National Geographic*. Retrieved August 31, 2012.

[20] Staff (August 29, 2012). "Sweet! Astronomers spot sugar molecule near star". AP News. Retrieved August 31, 2012.

[21] Jørgensen, J. K.; Favre, C.; Bisschop, S.; Bourke, T.; Dishoeck, E.; Schmalzl, M. (2012). "Detection of the simplest sugar, glycolaldehyde, in a solar-type protostar with ALMA" (PDF). *The Astrophysical Journal Letters*. eprint **757**: L4. arXiv:1208.5498. Bibcode:2012ApJ...757L...4J. doi:10.1088/2041-8205/757/1/L4.

[22] Staff (September 20, 2012). "NASA Cooks Up Icy Organics to Mimic Life's Origins". Space.com. Retrieved September 22, 2012.

[23] Gudipati, Murthy S.; Yang, Rui (September 1, 2012). "In-Situ Probing Of Radiation-Induced Processing Of Organics In Astrophysical Ice Analogs—Novel Laser Desorption Laser Ionization Time-Of-Flight Mass Spectroscopic Studies". *The Astrophysical Journal Letters* **756** (1): L24. Bibcode:2012ApJ...756L..24G. doi:10.1088/2041-8205/756/1/L24. Retrieved September 22, 2012.

[24] Clavin, Whitney (10 February 2015). "Why Comets Are Like Deep Fried Ice Cream". *NASA*. Retrieved 10 February 2015.

[25] López-Puertas, Manuel (June 6, 2013). "PAH's in Titan's Upper Atmosphere". *CSIC*. Retrieved June 6, 2013.

[26] http://www.sciencenews.org/view/generic/id/351444/description/Interstellar_chemistry_makes_use_of_quantum_shortcut#351468 comment_

[27] Cummins, S. E.; Linke, R. A.; Thaddeus, P. (1986), "A survey of the millimeter-wave spectrum of Sagittarius B2", *Astrophysical Journal Supplement Series* **60**: 819–878, Bibcode:1986ApJS...60..819C, doi:10.1086/191102

[28] Kaler, James B. (2002), *The hundred greatest stars*, Copernicus Series, Springer, ISBN 0-387-95436-8, retrieved 2011-05-09

[29] Marlaire, Ruth (3 March 2015). "NASA Ames Reproduces the Building Blocks of Life in Laboratory". *NASA*. Retrieved 5 March 2015.

[30] Klemperer, William (2011), "Astronomical Chemistry", *Annual Review of Physical Chemistry* **62**: 173–184, doi:10.1146/annurev-physchem-032210-103332

[31] *The Structure of Molecular Cloud Cores*, Centre for Astrophysics and Planetary Science, University of Kent, retrieved 2007-02-16

[32] Ziurys, Lucy M. (2006), "The chemistry in circumstellar envelopes of evolved stars: Following the origin of the elements to the origin of life", *Proceedings of the National Academy of Sciences* **103** (33): 12274–12279, Bibcode:2006PNAS..10312274Z, doi:10.1073/pnas.0602277103, PMC 1567870, PMID 16894164

[33] Cernicharo, J.; Guelin, M. (1987), "Metals in IRC+10216 - Detection of NaCl, AlCl, and KCl, and tentative detection of AlF", *Astronomy and Astrophysics* **183** (1): L10–L12, Bibcode:1987A&A...183L..10C

[34] Ziurys, L. M.; Apponi, A. J.; Phillips, T. G. (1994), "Exotic fluoride molecules in IRC +10216: Confirmation of AlF and searches for MgF and CaF", *Astrophysical Journal* **433** (2): 729–732, Bibcode:1994ApJ...433..729Z, doi:10.1086/174682

[35] Tenenbaum, E. D.; Ziurys, L. M. (2009), "Millimeter Detection of AlO ($X^2\Sigma^+$): Metal Oxide Chemistry in the Envelope of VY Canis Majoris", *Astrophysical Journal* **694**: L59–L63, Bibcode:2009ApJ...694L..59T, doi:10.1088/0004-637X/694/1/L59

[36] Barlow, M. J.; Swinyard, B. M.; Owen, P. J.; Cernicharo, J.; Gomez, H. L.; Ivison, R. J.; Lim, T. L.; Matsuura, M.; Miller, S.; Olofsson, G.; Polehampton, E. T. (2013), "Detection of a Noble Gas Molecular Ion, ^{36}ArH+, in the Crab Nebula", *Science* **342** (6164): 1343–1345, doi:10.1126/science.124358213

[37] Quenqua, Douglas (13 December 2013). "Noble Molecules Found in Space". *New York Times*. Retrieved 13 December 2013.

[38] Lambert, D. L.; Sheffer, Y.; Federman, S. R. (1995), "Hubble Space Telescope observations of C_2 molecules in diffuse interstellar clouds", *Astrophysical Journal* **438**: 740–749, Bibcode:1995ApJ...438..740L, doi:10.1086/175119

[39] Galazutdinov, G. A.; Musaev, F. A.; Krelowski, J. (2001), "On the detection of the linear C_5 molecule in the interstellar medium", *Monthly Notices of the Royal Astronomical Society* **325** (4): 1332–1334, Bibcode:2001MNRAS.325.1332G, doi:10.1046/j.1365-8711.2001.04388.x

[40] Neufeld, D. A.; et al. (2006), "Discovery of interstellar CF^{+}", *Astronomy and Astrophysics* **454** (2): L37–L40, arXiv:astro-ph/0603201, Bibcode:2006A&A...454L..37N, doi:10.1051/0004-6361:200600015

[41] Adams, Walter S. (1941), "Some Results with the COUDÉ Spectrograph of the Mount Wilson Observatory", *Astrophysical Journal* **93**: 11–23, Bibcode:1941ApJ....93...11A, doi:10.1086/144237

[42] Smith, D. (1988), "Formation and Destruction of Molecular Ions in Interstellar Clouds", *Philosophical Transactions of the Royal Society of London* **324** (1578): 257–273, Bibcode:1988RSPTA.324..257S, doi:10.1098/rsta.1988.0016

[43] Fuente, A.; et al. (2005), "Photon-dominated Chemistry in the Nucleus of M82: Widespread HOC^+ Emission in the Inner 650 Parsec Disk", *Astrophysical Journal* **619** (2): L155–L158, arXiv:astro-ph/0412361, Bibcode:2005ApJ...619L.155F, doi:10.1086/427990

[44] Guelin, M.; Cernicharo, J.; Paubert, G.; Turner, B. E. (1990), "Free CP in IRC + 10216", *Astronomy and Astrophysics* **230**: L9–L11, Bibcode:1990A&A...230L...9G

[45] Dopita, Michael A.; Sutherland, Ralph S. (2003), *Astrophysics of the diffuse universe*, Springer-Verlag, ISBN 3-540-43362-7

[46] Agúndez, M.; et al. (2010-07-30), "Astronomical identification of CN^-, the smallest observed molecular anion", *Astronomy & Astrophysics* **517**: L2, arXiv:1007.0662, Bibcode:2010A&A...517L...2A, doi:10.1051/0004-6361/201015186, retrieved 2010-09-03

[47] Khan, Amina. "Did two planets around nearby star collide? Toxic gas holds hints". *LA Times*. Retrieved March 9, 2014.

[48] Dent, W.R.F.; Wyatt, M.C.;Roberge, A.; Augereau,J.-C.; Casassus, S.;Corder, S.; Greaves, J.S.; de Gregorio-Monsalvo, I; Hales, A.; Jackson, A.P.; Hughes, A. Meredith; Lagrange, A.-M; Matthews, B.; Wilner, D. (March 6, 2014). "Molecular Gas Clumps from the Destruction of Icy Bodies in the β Pictoris Debris Disk". *Science* **343**: 1490–1492. arXiv:1404.1380. Bibcode:2014Sci...343.1490D. doi:10.1126/science.1248726. Retrieved March 9, 2014.

[49] Latter, W. B.; Walker, C. K.; Maloney, P. R. (1993), "Detection of the Carbon Monoxide Ion (CO^+) in the Interstellar Medium and a Planetary Nebula", *Astrophysical Journal Letters* **419**: L97, Bibcode:1993ApJ...419L..97L, doi:10.1086/187146

[50] Furuya, R. S.; et al. (2003), "Interferometric observations of FeO towards Sagittarius B2", *Astronomy and Astrophysics* **409** (2): L21–L24, Bibcode:2003A&A...409L..21F, doi:10.1051/0004-6361:20031304

[51] Adams, Walter S. (1970), "Rocket Observation of Interstellar Molecular Hydrogen", *Astrophysical Journal* **161**: L81–L85, Bibcode:1970ApJ...161L..81C, doi:10.1086/180575

[52] Blake, G. A.; Keene, J.; Phillips, T. G. (1985), "Chlorine in dense interstellar clouds - The abundance of HCl in OMC-1", *Astrophysical Journal, Part 1* **295**: 501–506, Bibcode:1985ApJ...295..501B, doi:10.1086/163394

[53] De Luca, M.; Gupta, H.; Neufeld, D.; Gerin, M.; Teyssier, D.; Drouin, B. J.; Pearson, J. C.; Lis, D. C.; et al. (2012), "Herschel/HIFI Discovery of HCl+ in the Interstellar Medium", *The Astrophysical Journal Letters* **751** (2): L37, Bibcode:2012ApJ...751L..37D,doi:10.1088/2041-8205/751/2/L37

[54] Neufeld, David A.; et al. (1997), "Discovery of Interstellar Hydrogen Fluoride", *Astrophysical Journal Letters* **488** (2): L141–L144, arXiv:astro-ph/9708013, Bibcode:1997ApJ...488L.141N, doi:10.1086/310942

[55] Wyrowski, F.; et al. (2009), "First interstellar detection of OH^{+}", *Astronomy & Astrophysics* **518**: A26, arXiv:1004.2627, Bibcode:2010A&A...518A..26W, doi:10.1051/0004-6361/201014364

[56] Meyer, D. M.; Roth, K. C. (1991), "Discovery of interstellar NH", *Astrophysical Journal, Part 2 - Letters* **376**: L49–L52, Bibcode:1991ApJ...376L..49M, doi:10.1086/186100

[57] Wagenblast, R.; et al. (January 1993), "On the origin of NH in diffuse interstellar clouds", *Monthly Notices of the Royal Astronomical Society* **260** (2): 420–424, Bibcode:1993MNRAS.260..420W, doi:10.1093/mnras/260.2.420

[58] <Please add first missing authors to populate metadata.> (June 9, 2004), *Astronomers Detect Molecular Nitrogen Outside Solar System*, Space Daily, retrieved 2010-06-25

[59] Knauth, D. C; et al. (2004), "The interstellar N_2 abundance towards HD 124314 from far-ultraviolet observations", *Nature* **429** (6992): 636–638, Bibcode:2004Natur.429..636K, doi:10.1038/nature02614, PMID 15190346, retrieved 2010-06-25

[60] McGonagle, D.; et al. (1990), "Detection of nitric oxide in the dark cloud L134N", *Astrophysical Journal, Part 1* **359**: 121–124, Bibcode:1990ApJ...359..121M, doi:10.1086/169040

[61] Whiteoak, J. B.; Gardner, F. F. (1985), "Interstellar NaI absorption towards the stellar association ARA OB1", *Astronomical Society of Australia, Proceedings* (Sydney) **6** (2): 164–171, Bibcode:1985PASAu...6..164W

[62] Staff writers (March 27, 2007), *Elusive oxygen molecule finally discovered in interstellar space*, Physorg.com, retrieved 2007-04-02

[63] Ziurys, L. M. (1987), "Detection of interstellar PN - The first phosphorus-bearing species observed in molecular clouds", *Astrophysical Journal, Part 2 - Letters to the Editor* **321**: L81–L85, Bibcode:1987ApJ...321L..81Z, doi:10.1086/185010

[64] Tenenbaum, E. D.; Woolf, N. J.; Ziurys, L. M. (2007), "Identification of phosphorus monoxide ($X\ ^2 Pi\ _r$) in VY Canis Majoris: Detection of the first PO bond in space", *Astrophysical Journal, Part 2 - Letters to the Editor* **666**: L29–L32, Bibcode:2007ApJ...6 66L..29T,doi:10.1086/521361

[65] Yamamura, S. T.; Kawaguchi, K.; Ridgway, S. T. (2000), "Identification of SH v=1 Ro-vibrational Lines in R Andromedae", *The Astrophysical Journal* **528** (1): L33–L36, arXiv:astro-ph/9911080, Bibcode:2000ApJ...528L..33Y, doi:10.1086/312420, PMID 10587489

[66] Menten, K. M.; et al. (2011), "Submillimeter Absorption from SH^+, a New Widespread Interstellar Radical, $^{13}CH^+$ and HCl", *Astronomy & Astrophysics* **525**: A77, arXiv:1009.2825, Bibcode:2011A&A...525A..77M, doi:10.1051/0004-6361/201014363, retrieved 2010-12-03.

[67] Pascoli, G.; Comeau, M. (1995), "Silicon Carbide in Circumstellar Environment", *Astrophysics and Space Science* **226**: 149–163, Bibcode:1995Ap&SS.226..149P, doi:10.1007/BF00626907

[68] Kamiński, T.; et al. (2013), "Pure rotational spectra of TiO and TiO_2 in VY Canis Majoris", *Astronomy and Astrophysics* **551**: A113, arXiv:1301.4344, Bibcode:2013A&A...551A.113K, doi:10.1051/0004-6361/201220290

[69] Geballe, T. R.; Oka, T. (1996), "Detection of $H_3{}^+$in Interstellar Space",*Nature***384**(6607): 334–335,Bibcode:1996Natur.384.., doi:10.1038/384334a0, PMID 8934516

[70] Tenenbaum, E. D.; Ziurys, L. M. (2010), "Exotic Metal Molecules in Oxygen-rich Envelopes: Detection of AlOH (X1Σ+) in VY Canis Majoris", *Astrophysical Journal* **712**: L93–L97, Bibcode:2010ApJ...712L..93T, doi:10.1088/2041-8205/712/1/L93

[71] Anderson, J. K.; et al. (2014), "Detection of CCN ($X^2\Pi_r$) in IRC+10216: Constraining Carbon-chain Chemistry", *Astrophysical Journal* **795**: L1, Bibcode:2014ApJ...795L...1A, doi:10.1088/2041-8205/795/1/L1

[72] Ohishi, Masatoshi, Masatoshi; et al. (1991), "Detection of a new carbon-chain molecule, CCO", *Astrophysical Journal, Part 2 - Letters* **380**: L39–L42, Bibcode:1991ApJ...380L..39O, doi:10.1086/186168

[73] Irvine, William M.; et al. (1988), "Newly detected molecules in dense interstellar clouds", *Astrophysical Letters and Communications* **26**: 167–180, Bibcode:1988ApL&C..26..167I, PMID 11538461

[74] Halfen, D. T.; Clouthier, D. J.; Ziurys, L. M. (2008), "Detection of the CCP Radical ($X\ ^2\Pi_r$) in IRC +10216: A New Interstellar Phosphorus-containing Species", *Astrophysical Journal* **677** (2): L101–L104, Bibcode:2008ApJ...677L.101H, doi:10.1086/588024

[75] Whittet, D. C. B.; Walker, H. J. (1991), "On the occurrence of carbon dioxide in interstellar grain mantles and ion-molecule chemistry", *Monthly Notices of the Royal Astronomical Society* **252**: 63–67, Bibcode:1991MNRAS.252...63W, doi:10.1093/mnra s/252.1.63

[76] Zack, L. N.; Halfen, D. T.; Ziurys, L. M. (June 2011), "Detection of FeCN ($X\ ^4\Delta_i$) in IRC+10216: A New Interstellar Molecule", *The Astrophysical Journal Letters* **733** (2): L36, Bibcode:2011ApJ...733L..36Z, doi:10.1088/2041-8205/733/2/L36

[77] Lis, D. C.; et al. (2010-10-01), "Herschel/HIFI discovery of interstellar chloronium (H_2Cl^+)", *Astronomy & Astrophysics* **521**: L9, arXiv:1007.1461, Bibcode:2010A&A...521L...9L, doi:10.1051/0004-6361/201014959.

[78] *Europe's space telescope ISO finds water in distant places*, ESO, April 29, 1997, archived from the original on 2006-12-22, retrieved 2007-02-08

[79] Ossenkopf, V.; et al. (2010), "Detection of interstellar oxidaniumyl: Abundant $H2O^+$ towards the star-forming regions DR21, Sgr B2, and NGC6334", *Astronomy & Astrophysics* **518**: L111, arXiv:1005.2521, Bibcode:2010A&A...518L.111O, doi:10.1051/0004-6361/201014577.

[80] Parise, B.; Bergman, P.; Du, F. (2012), "Detection of the hydroperoxyl radical HO_2 toward ρ Ophiuchi A. Additional constraints on the water chemical network", *Astronomy & Astrophysics Letters* **541**: L11–L14, arXiv:1205.0361, Bibcode:2012A&A...541L..11P,doi:10.1051/0004-6361/201219379

[81] Snyder, L. E.; Buhl, D. (1971), "Observations of Radio Emission from Interstellar Hydrogen Cyanide", *Astrophysical Journal* **163**: L47–L52, Bibcode:1971ApJ...163L..47S, doi:10.1086/180664

[82] Schilke, P.; Benford, D. J.; Hunter, T. R.; Lis, D. C., Phillips, T. G.; Phillips, T. G. (2001), "A Line Survey of Orion-KL from 607 to 725 GHz",*Astrophysical Journal Supplement Series***132**(2): 281–364,Bibcode:2001ApJS..132..281S,doi:10.1086/3189

[83] Schenewerk, M. S.; Snyder, L. E.; Hjalmarson, A. (1986), "Interstellar HCO - Detection of the missing 3 millimeter quartet", *Astrophysical Journal, Part 2 - Letters to the Editor* **303**: L71–L74, Bibcode:1986ApJ...303L..71S, doi:10.1086/184655

[84] Kawaguchi, Kentarou; et al. (1994), "Detection of a new molecular ion HC3NH(+) in TMC-1", *Astrophysical Journal* **420**: L95, Bibcode:1994ApJ...420L..95K, doi:10.1086/187171

[85] Agúndez, M.; Cernicharo, J.; Guélin, M. (2007), "Discovery of Phosphaethyne (HCP) in Space: Phosphorus Chemistry in Circumstellar Envelopes", *The Astrophysical Journal* **662** (2): L91, Bibcode:2007ApJ...662L..91A, doi:10.1086/519561, retrieved 2007-06-02

[86] Schilke, P.; Comito, C.; Thorwirth, S. (2003), "First Detection of Vibrationally Excited HNC in Space", *The Astrophysical Journal* **582** (2): L101–L104, Bibcode:2003ApJ...582L.101S, doi:10.1086/367628, retrieved 2008-09-14

[87] Hollis, J. M.; et al. (1991), "Interstellar HNO: Confirming the Identification - Atoms, ions and molecules: New results in spectral line astrophysics", *Atoms* (San Francisco: ASP) **16**: 407–412, Bibcode:1991ASPC...16..407H

[88] van Dishoeck, Ewine F.; et al. (1993), "Detection of the Interstellar NH 2 Radical", *Astrophysical Journal, Part 2 - Letters* **416**: L83–L86, Bibcode:1993ApJ...416L..83V, doi:10.1086/187076

[89] Womack, M.; Ziurys, L. M.; Wyckoff, S. (1992), "A survey of $N_2H(+)$ in dense clouds - Implications for interstellar nitrogen and ion-molecule chemistry", *Astrophysical Journal, Part 1* **387**: 417–429, Bibcode:1992ApJ...387..417W, doi:10.1086/171094

[90] Ziurys, L. M.; et al. (1994), "Detection of interstellar N_2O: A new molecule containing an N-O bond", *Astrophysical Journal, Part 2 - Letters* **436**: L181–L184, Bibcode:1994ApJ...436L.181Z, doi:10.1086/187662

[91] Hollis, J. M.; Rhodes, P. J. (November 1, 1982), "Detection of interstellar sodium hydroxide in self-absorption toward the galactic center", *Astrophysical Journal, Part 2 - Letters to the Editor* **262**: L1–L5, Bibcode:1982ApJ...262L...1H, doi:10.1086/183900

[92] Goldsmith, P. F.; Linke, R. A. (1981), "A study of interstellar carbonyl sulfide", *Astrophysical Journal, Part 1* **245**: 482–494, Bibcode:1981ApJ...245..482G, doi:10.1086/158824

[93] Phillips, T. G.; Knapp, G. R. (1980), "Interstellar Ozone",*American Astronomical Society Bulletin***12**: 440,Bibcode:1980BAAS

[94] Johansson, L. E. B.; et al. (1984), "Spectral scan of Orion A and IRC+10216 from 72 to 91 GHz", *Astronomy and Astrophysics* **130** (2): 227–256, Bibcode:1984A&A...130..227J

[95] Cernicharo, José; et al. (2015), "Discovery of SiCSi in IRC+10216: a Missing Link Between Gas and Dust Carriers OF Si–C Bonds", *Astrophysical Journal Letters* **806**: L3, Bibcode:2015ApJ...806L.3C, doi:10.1088/2041-8025

[96] Guélin, M.; et al. (2004), "Astronomical detection of the free radical SiCN", *Astronomy and Astrophysics* **363**: L9–L12, Bibcode:2000A&A...363L...9G

[97] Guélin, M.; et al. (2004), "Detection of the SiNC radical in IRC+10216", *Astronomy and Astrophysics* **426** (2): L49–L52, Bibcode:2004A&A...426L..49G, doi:10.1051/0004-6361:200400074

[98] Snyder, Lewis E.; et al. (1999), "Microwave Detection of Interstellar Formaldehyde", *Physical Review Letters* **61** (2): 77–115, Bibcode:1969PhRvL..22..679S, doi:10.1103/PhysRevLett.22.679

[99] Feuchtgruber, H.; et al. (June 2000), "Detection of Interstellar CH_3",*The Astrophysical Journal***535**(2): L111–L114,arXiv:ast ph/0005273, Bibcode:2000ApJ...535L.111F, doi:10.1086/312711, PMID 10835311

[100] Irvine, W. M.; et al. (1984), "Confirmation of the Existence of Two New Interstellar Molecules: C_3H and C_3O", *Bulletin of the American Astronomical Society* **16**: 877, Bibcode:1984BAAS...16..877I

[101] Pety, J.; et al. (2012), "The IRAM-30 m line survey of the Horsehead PDR. II. First detection of the l-C_3MH^+ hydrocarbon cation", *Astronomy & Astrophysica* **548**: A68, arXiv:1210.8178, Bibcode:2012A&A...548A..68P, doi:10.1051/0004-6361/201220062

[102] Mangum, J. G.; Wootten, A. (1990), "Observations of the cyclic C_3H radical in the interstellar medium", *Astronomy and Astrophysics* **239**: 319–325, Bibcode:1990A&A...239..319M

[103] Wootten, Alwyn; et al. (1991), "Detection of interstellar H_3O(+) - A confirming line", *Astrophysical Journal, Part 2 - Letters* **380**: L79–L83, Bibcode:1991ApJ...380L..79W, doi:10.1086/186178

[104] Ridgway, S. T.; et al. (1976), "Circumstellar acetylene in the infrared spectrum of IRC+10216", *Nature* **264**: 345, 346, Bibcode:1976Natur.264..345R, doi:10.1038/264345a0

[105] Ohishi, Masatoshi; et al. (1994), "Detection of a new interstellar molecule, H_2CN", *Astrophysical Journal, Part 2 - Letters* **427**: L51–L54, Bibcode:1994ApJ...427L..51O, doi:10.1086/187362

[106] Minh, Y. C.; Irvine, W. M.; Brewer, M. K. (1991), "H2CS abundances and ortho-to-para ratios in interstellar clouds", *Astronomy and Astrophysics* **244**: 181–189, Bibcode:1991A&A...244..181M, PMID 11538284

[107] Guelin, M.; Cernicharo, J. (1991), "Astronomical detection of the HCCN radical - Toward a new family of carbon-chain molecules?", *Astronomy and Astrophysics* **244**: L21–L24, Bibcode:1991A&A...244L..21G

[108] Agúndez, M.; et al. (2015), "Discovery of interstellar ketenyl (HCCO), a surprisingly abundant radical", *Astronomy and Astrophysics* **577**: A5, doi:10.1051/0004-6361

[109] Minh, Y. C.; Irvine, W. M.; Ziurys, L. M. (1988), "Observations of interstellar HOCO(+) - Abundance enhancements toward the Galactic center", *Astrophysical Journal, Part 1* **334**: 175–181, Bibcode:1988ApJ...334..175M, doi:10.1086/166827

[110] Marcelino, Núria; et al. (2009), "Discovery of fulminic acid, HCNO, in dark clouds", *Astrophysical Journal* **690**: L27–L30, arXiv:0811.2679, Bibcode:2009ApJ...690L..27M, doi:10.1088/0004-637X/690/1/L27

[111] Brünken, S.; et al. (2010-07-22), "Interstellar HOCN in the Galactic center region", *Astronomy & Astrophysics* **516**: A109, arXiv:1005.2489, Bibcode:2010A&A...516A.109B, doi:10.1051/0004-6361/200912456

[112] Bergman; Parise; Liseau; Larsson; Olofsson; Menten; Güsten (2011), "Detection of interstellar hydrogen peroxide", *Astronomy & Astrophysics* **531**: L8, arXiv:1105.5799, Bibcode:2011A&A...531L...8B, doi:10.1051/0004-6361/201117170.

[113] Frerking, M. A.; Linke, R. A.; Thaddeus, P. (1979), "Interstellar isothiocyanic acid", *Astrophysical Journal, Part 2 - Letters to the Editor* **234**: L143–L145, Bibcode:1979ApJ...234L.143F, doi:10.1086/183126

[114] Nguyen-Q-Rieu; Graham, D.; Bujarrabal, V. (1984), "Ammonia and cyanotriacetylene in the envelopes of CRL 2688 and IRC + 10216", *Astronomy and Astrophysics* **138** (1): L5–L8, Bibcode:1984A&A...138L...5N

[115] Halfen, D. T.; et al. (September 2009), "Detection of a New Interstellar Molecule: Thiocyanic Acid HSCN", *The Astrophysical Journal Letters* **702** (2): L124–L127, Bibcode:2009ApJ...702L.124H, doi:10.1088/0004-637X/702/2/L124

[116] Cabezas, C.; et al. (2013), "Laboratory and Astronomical Discovery of Hydromagnesium Isocyanide", *Astrophysical Journal* **775**: 133, arXiv:1309.0371, Bibcode:2013ApJ...775..133C, doi:10.1088/0004-637X/775/2/133

[117] Butterworth, Anna L.; et al. (2004), "Combined element (H and C) stable isotope ratios of methane in carbonaceous chondrites", *Monthly Notices of the Royal Astronomical Society* **347** (3): 807–812, Bibcode:2004MNRAS.347..807B, doi:10.1111/j.1365-2966.2004.07251.x

[118] http://www.astro.uni-koeln.de/site/vorhersagen/molecules/ism/Ammonium.html

[119] http://iopscience.iop.org/2041-8205/771/1/L10/

[120] Cernicharo, J.; Marcelino, N.; Roueff, E.; Gerin, M.; Jiménez-Escobar, A.; Muñoz Caro, G. M. (2012), "Discovery of the Methoxy Radical, CH_3O, toward B1: Dust Grain and Gas-phase Chemistry in Cold Dark Clouds", *The Astrophysical Journal Letters* **759** (2): L43–L46, Bibcode:2012ApJ...759L..43C, doi:10.1088/2041-8205/759/2/L43

[121] Finley, Dave (August 7, 2006), *Researchers Use NRAO Telescope to Study Formation Of Chemical Precursors to Life*, National Radio Astronomy Observatory, retrieved 2006-08-10

[122] Fossé, David; et al. (2001), "Molecular Carbon Chains and Rings in TMC-1", *Astrophysical Journal* **552** (1): 168–174, arXiv:astro-ph/0012405, Bibcode:2001ApJ...552..168F, doi:10.1086/320471, retrieved 2008-09-14

[123] Dickens, J. E.; et al. (1997), "Hydrogenation of Interstellar Molecules: A Survey for Methylenimine (CH_2NH)", *Astrophysical Journal* **479** (1 Pt 1): 307–12, Bibcode:1997ApJ...479..307D, doi:10.1086/303884, PMID 11541227

[124] McGuire, B.A.; et al. (2012), "Interstellar Carbodiimide (HNCNH): A New Astronomical Detection from the GBT PRIMOS Survey via Maser Emission Features", *The Astrophysical Journal Letters* **758** (2): L33–L38, arXiv:1209.1590, Bibcode:2012Ap J...758L..33M,doi:10.1088/2041-8205/758/2/L33

[125] Ohishi, Masatoshi; et al. (1996), "Detection of a New Interstellar Molecular Ion, H_2COH^+ (Protonated Formaldehyde)", *Astrophysical Journal* **471** (1): L61–4, Bibcode:1996ApJ...471L..61O, doi:10.1086/310325, PMID 11541244

[126] Cernicharo, J.; et al. (2007), "Astronomical detection of C_4H<sup–, the second interstellar anion", *Astronomy and Astrophysics* **61** (2): L37–L40, Bibcode:2007A&A...467L..37C, doi:10.1051/0004-6361:20077415

[127] Walmsley, C. M.; Winnewisser, G.; Toelle, F. (1990), "Cyanoacetylene and cyanodiacetylene in interstellar clouds", *Astronomy and Astrophysics* **81** (1–2): 245–250, Bibcode:1980A&A....81..245W

[128] Kawaguchi, Kentarou; et al. (1992), "Detection of isocyanoacetylene HCCNC in TMC-1", *Astrophysical Journal* **386** (2): L51–L53, Bibcode:1992ApJ...386L..51K, doi:10.1086/186290

[129] Turner,B.E.;et al. (1975), "Microwave detection of interstellar cyanamide",*Astrophysical Journal***201**: L149–L152,Bibcode: doi:10.1086/181963

[130] Agúndez, M.; et al. (2015), "Probing non-polar interstellar molecules through their protonated form: Detection of protonated cyanogen (NCCNH+)", *Astronomy and Astrophysics* **579**: L10, Bibcode:2015A&A...579L..10A, doi:10.1051/0004-6361/201526650

[131] Remijan, Anthony J.; et al. (2008), "Detection of interstellar cyanoformaldehyde (CNCHO)", *Astrophysical Journal* **675** (2): L85–L88, Bibcode:2008ApJ...675L..85R, doi:10.1086/533529

[132] Goldhaber, D. M.; Betz, A. L. (1984), "Silane in IRC +10216", *Astrophysical Journal, Part 2 - Letters to the Editor* **279**: –L55–L58, Bibcode:1984ApJ...279L..55G, doi:10.1086/184255

[133] Hollis, J. M.; et al. (2006), "Detection of Acetamide (CH_3CONH_2): The Largest Interstellar Molecule with a Peptide Bond", *Astrophysical Journal* **643** (1): L25–L28, Bibcode:2006ApJ...643L..25H, doi:10.1086/505110

[134] Zaleski, D. P.; et al. (2013), "Detection of E-Cyanomethanimine toward Sagittarius B2(N) in the Green Bank Telescope PRIMOS Survey", *Astrophysical Journal Letters* **765**: L109, arXiv:1302.0909, Bibcode:2013ApJ...765L..10Z, doi:10.1088/2041-8205/765/1/L10

[135] Betz, A. L. (1981), "Ethylene in IRC +10216",*Astrophysical Journal, Part 2 - Letters to the Editor***244**: –L105,Bibcode:1981ApJ doi:10.1086/183490

[136] Remijan, Anthony J.; et al. (2005), "Interstellar Isomers: The Importance of Bonding Energy Differences", *Astrophysical Journal* **632** (1): 333–339, arXiv:astro-ph/0506502, Bibcode:2005ApJ...632..333R, doi:10.1086/432908

[137] "Complex Organic Molecules Discovered in Infant Star System". *NRAO* (Astrobiology Web). 8 April 2015. Retrieved 2015-04-09.

[138] Cernicharo, José; et al. (1997), "Infrared Space Observatory's Discovery of C_4H_2, C_6H_2, and Benzene in CRL 618", *Astrophysical Journal Letters* **546** (2): L123–L126, Bibcode:2001ApJ...546L.123C, doi:10.1086/318871

[139] Guelin, M.; Neininger, N.; Cernicharo, J. (1998), "Astronomical detection of the cyanobutadiynyl radical C_5N", *Astronomy and Astrophysics* **335**: L1–L4, arXiv:astro-ph/9805105, Bibcode:1998A&A...335L...1G

[140] Irvine, W. M.; et al. (1988), "A new interstellar polyatomic molecule - Detection of propynal in the cold cloud TMC-1", *Astrophysical Journal, Part 2 - Letters* **335**: L89–L93, Bibcode:1988ApJ...335L..89I, doi:10.1086/185346

[141] Agúndez, M.; et al. (2014), "New molecules in IRC +10216: confirmation of C_5S and tentative identification of MgCCH, NCCP, and SiH_3CN", *Astronomy and Astrophysics* **570**: A45, doi:10.1051/0004-6361

[142] *Scientists Toast the Discovery of Vinyl Alcohol in Interstellar Space*, National Radio Astronomy Observatory, October 1, 2001, retrieved 2006-12-20

[143] Dickens, J. E.; et al. (1997), "Detection of Interstellar Ethylene Oxide (c-C2H4O)", *The Astrophysical Journal* **489** (2): 753–757, Bibcode:1997ApJ...489..753D, doi:10.1086/304821, PMID 11541726

[144] Kaifu, N.; Takagi, K.; Kojima, T. (1975), "Excitation of interstellar methylamine", *Astrophysical Journal* **198**: L85–L88, Bibcode:1975ApJ...198L..85K, doi:10.1086/181818

[145] McCarthy, M. C.; et al. (2006), "Laboratory and Astronomical Identification of the Negative Molecular Ion C_6H^-", *Astrophysical Journal* **652** (2): L141–L144, Bibcode:2006ApJ...652L.141M, doi:10.1086/510238

[146] Halfven, D. T.; et al. (2015), "INTERSTELLAR DETECTION OF METHYL ISOCYANATE CH_3NCO IN Sgr B2(N): A LINK FROM MOLECULAR CLOUDS TO COMETS", *Astrophysical Journal* **812**: L5

[147] Mehringer, David M.; et al. (1997), "Detection and Confirmation of Interstellar Acetic Acid", *Astrophysical Journal Letters* **480**: L71, Bibcode:1997ApJ...480L..71M, doi:10.1086/310612

[148] Lovas, F. J.; et al. (2006), "Hyperfine Structure Identification of Interstellar Cyanoallene toward TMC-1", *Astrophysical Journal Letters* **637** (1): L37–L40, Bibcode:2006ApJ...637L..37L, doi:10.1086/500431

[149] Sincell, Mark (June 27, 2006), *The Sweet Signal of Sugar in Space*, American Association for the Advancement of Science, retrieved 2006-12-20

[150] Loomis, R. A.; et al. (2013), "The Detection of Interstellar Ethanimine CH_3CHNH) from Observations Taken during the GBT PRIMOS Survey", *Astrophysical Journal Letters* **765**: L9, arXiv:1302.1121, Bibcode:2013ApJ...765L...9L, doi:10.1088/2041-8205/765/1/L9

[151] Guelin, M.; et al. (1997), "Detection of a new linear carbon chain radical: C_7H", *Astronomy and Astrophysics* **317**: L37–L40, Bibcode:1997A&A...317L...1G

[152] Belloche, A.; et al. (2008), "Detection of amino acetonitrile in Sgr B2(N)", *Astronomy & Astrophysics* **482**: 179–196, arXiv:080 Bibcode:2008A&A...482..179B, doi:10.1051/0004-6361:20079203

[153] Remijan, Anthony J.; et al. (2014), "OBSERVATIONAL RESULTS OF A MULTI-TELESCOPE CAMPAIGN IN SEARCH OF INTERSTELLAR UREA [$(NH2)_2CO$]", *Astrophysical Journal* **783** (2): 77, arXiv:1401.4483, Bibcode:2014ApJ...783...77R, doi:10.1088/0004-637X/783/2/77

[154] Remijan, Anthony J.; et al. (2006), "Methyltriacetylene (CH_3C_6H) toward TMC-1: The Largest Detected Symmetric Top", *Astrophysical Journal* **643** (1): L37–L40, Bibcode:2006ApJ...643L..37R, doi:10.1086/504918

[155] Snyder, L. E.; et al. (1974), "Radio Detection of Interstellar Dimethyl Ether", *Astrophysical Journal* **191**: L79–L82, Bibcode:197 doi:10.1086/181554

[156] Zuckerman, B.; et al. (1975), "Detection of interstellar trans-ethyl alcohol", *Astrophysical Journal* **196** (2): L99–L102, Bibcode:1975ApJ...196L..99Z, doi:10.1086/181753

[157] Cernicharo, J.; Guelin, M. (1996), "Discovery of the C_8H radical", *Astronomy and Astrophysics* **309**: L26–L30, Bibcode:1996A

[158] Brünken, S.; et al. (2007), "Detection of the Carbon Chain Negative Ion C_8H^- in TMC-1", *Astrophysical Journal* **664** (1): L43–L46, Bibcode:2007ApJ...664L..43B, doi:10.1086/520703

[159] Bell, M. B.; et al. (1997), "Detection of $HC_{11}N$ in the Cold Dust Cloud TMC-1", *Astrophysical Journal Letters* **483** (1): L61–L64, arXiv:astro-ph/9704233, Bibcode:1997ApJ...483L..61B, doi:10.1086/310732

[160] Kroto, H. W.; et al. (1978), "The detection of cyanohexatriyne, H ($C\equiv C)_3CN$, in Heiles's cloud 2", *The Astrophysical Journal* **219**: L133–L137, Bibcode:1978ApJ...219L.133K, doi:10.1086/182623

[161] Marcelino, N.; et al. (2007), "Discovery of Interstellar Propylene (CH_2CHCH_3): Missing Links in Interstellar Gas-Phase Chemistry", *Astrophysical Journal* **665** (2): L127–L130, arXiv:0707.1308, Bibcode:2007ApJ...665L.127M, doi:10.1086/5213

[162] Kolesniková, L.; et al. (2014), "Spectroscopic Characterization and Detection of Ethyl Mercaptan in Orion", *Astrophysical Journal Letters* **784** (1): L7, arXiv:1401.7810, Bibcode:2014ApJ...784L...7K, doi:10.1088/2041-8205/784/1/L7

[163] Snyder, Lewis E.; et al. (2002),"Confirmation of Interstellar Acetone",*The Astrophysical Journal*578(1): 245–255,Bibcode doi:10.1086/342273

[164] Hollis, J. M.; et al. (2002),"Interstellar Antifreeze: Ethylene Glycol",*Astrophysical Journal*571(1): L59–L62,Bibcode:2002 doi:10.1086/341148, retrieved 2010-07-18

[165] Hollis, J. M. (2005), "Complex Molecules and the GBT: Is Isomerism the Key?" (PDF), *Complex Molecules and the GBT: Is Isomerism the Key?*, Proceedings of the IAU Symposium 231, Astrochemistry throughout the Universe, Asilomar, CA, pp. 119–127

[166] Eyre, Michael (26 September 2014). "Complex organic molecule found in interstellar space". *BBC News*. Retrieved 2014-09-26.

[167] Belloche, Arnaud; Garrod, Robin T.; Müller, Holger S. P.; Menten, Karl M. (26 September 2014). "Detection of a branched alkyl molecule in the interstellar medium: iso-propyl cyanide". *Science* **345** (6204): 1584–1587. arXiv:1410.2607. Bibcode:2014Sci...345.1584B.doi:10.1126/science.1256678. Retrieved2014-09-26.

[168] Belloche, A.; et al. (May 2009), "Increased complexity in interstellar chemistry: Detection and chemical modeling of ethyl formate and n-propyl cyanide in Sgr B2(N)", *Astronomy and Astrophysics* **499** (1): 215–232, arXiv:0902.4694, Bibcode:2009A&A...499..215B,doi:10.1051/0004-6361/200811550

[169] Tercero, B.; et al. (2013), "Discovery of Methyl Acetate and Gauche Ethyl Formate in Orion", *Astrophysical Journal Letters* **770**: L13, arXiv:1305.1135, Bibcode:2013ApJ...770L..13T, doi:10.1088/2041-8205/770/1/L13

[170] Cami, Jan; et al. (July 22, 2010), "Detection of C_{60} and C_{70} in a Young Planetary Nebula", *Science* **329** (5996): 1180–2, Bibcode:2010Sci...329.1180C, doi:10.1126/science.1192035, PMID 20651118

[171] Foing, B. H.; Ehrenfreund, P. (1994), "Detection of two interstellar absorption bands coincident with spectral features of C60+", *Nature* **369** (6478): 296–298, Bibcode:1994Natur.369..296F, doi:10.1038/369296a0.

[172] Berné, Olivier; Mulas, Giacomo; Joblin, Christine (2013), "Interstellar $C_{60}{}^{+}$",*Astronomy & Astrophysics***550**: L4,arXiv:1211.7 Bibcode:2013A&A...550L...4B, doi:10.1051/0004-6361/201220730

[173] Lacour, S.; et al. (2005), "Deuterated molecular hydrogen in the Galactic ISM. New observations along seven translucent sightlines", *Astronomy and Astrophysics* **430** (3): 967–977, arXiv:astro-ph/0410033, Bibcode:2005A&A...430..967L, doi:10.1051/0004-6361:20041589

[174] Ceccarelli, Cecilia (2002), "Millimeter and infrared observations of deuterated molecules", *Planetary and Space Science* **50** (12–13): 1267–1273, Bibcode:2002P&SS...50.1267C, doi:10.1016/S0032-0633(02)00093-4

[175] Green, Sheldon (1989), "Collisional excitation of interstellar molecules - Deuterated water, HDO", *Astrophysical Journal Supplement Series* **70**: 813–831, Bibcode:1989ApJS...70..813G, doi:10.1086/191358

[176] Butner, H. M.; et al. (2007), "Discovery of interstellar heavy water",*Astrophysical Journal***659**(2): L137–L140,Bibcode:2007 doi:10.1086/517883

[177] Turner, B. E.; Zuckerman, B. (1978), "Observations of strongly deuterated molecules - Implications for interstellar chemistry", *Astrophysical Journal, Part 2 - Letters to the Editor* **225**: L75–L79, Bibcode:1978ApJ...225L..75T, doi:10.1086/182797

[178] Lis, D. C.; et al. (2002), "Detection of Triply Deuterated Ammonia in the Barnard 1 Cloud", *Astrophysical Journal* **571** (1): L55–L58, Bibcode:2002ApJ...571L..55L, doi:10.1086/341132.

[179] Hatchell, J. (2003), "High NH_2D/NH_3 ratios in protostellar cores", *Astronomy and Astrophysics* **403** (2): L25–L28, arXiv:astro-ph/0302564, Bibcode:2003A&A...403L..25H, doi:10.1051/0004-6361:20030297.

[180] Turner, B. E. (1990), "Detection of doubly deuterated interstellar formaldehyde (D2CO) - an indicator of active grain surface chemistry", *Astrophysical Journal, Part 2 - Letters* **362**: L29–L33, Bibcode:1990ApJ...362L..29T, doi:10.1086/185840.

[181] Cernicharo, J.; et al. (2013), "Detection of the Ammonium ion in space",*Astrophysical Journal Letters***771**: L10,arXiv:1306.3, Bibcode:2013ApJ...771L..10C, doi:10.1088/2041-8205/771/1/L10

[182] Doménech, J. L.; et al. (2013), "Improved Determininination of the 1_0-0_0 Rotational Frequency of NH_3D^+ from the High-Resolution Spectrum of the ν_4 Infrared Band", *Astrophysical Journal Letters* **771**: L11, arXiv:1306.3792, Bibcode:2013ApJ...771L..11D,doi:10.1088/2041-8205/771/1/L10

[183] Gerin, M.; et al. (1992), "Interstellar detection of deuterated methyl acetylene", *Astronomy and Astrophysics* **253** (2): L29–L32, Bibcode:1992A&A...253L..29G.

[184] Markwick, A. J.; Charnley, S. B.; Butner, H. M.; Millar, T. J. (2005), "Interstellar CH3CCD", *The Astrophysical Journal* **627** (2): L117–L120, Bibcode:2005ApJ...627L.117M, doi:10.1086/432415.

[185] Agúndez, M.; et al. (2008-06-04), "Tentative detection of phosphine in IRC +10216", *Astronomy & Astrophysics* **485** (3): L33, arXiv:0805.4297, Bibcode:2008A&A...485L..33A, doi:10.1051/0004-6361:200810193

[186] Gupta, H.; et al. (2013), "Laboratory Measurements and Tentative Astronomical Identification of H_2NCO^+", *Astrophysical Journal Letters* **778**: L1, Bibcode:2013ApJ...778L...1G, doi:10.1088/2041-8205/778/1/L1

[187] Kuan, Y. J.; et al. (2003), "Interstellar Glycine", *Astrophysical Journal* **593** (2): 848–867, Bibcode:2003ApJ...593..848K, doi:10.1086/375637.

[188] Widicus Weaver, S. L.; Blake, G. A. (2005), "1,3-Dihydroxyacetone in Sagittarius B2(N-LMH): The First Interstellar Ketose", *Astrophysical Journal Letters* **624** (1): L33–L36, Bibcode:2005ApJ...624L..33W, doi:10.1086/430407

[189] Fuchs, G. W.; et al. (2005), "Trans-Ethyl Methyl Ether in Space: A new Look at a Complex Molecule in Selected Hot Core Regions", *Astronomy & Astrophysics* **444** (2): 521–530, arXiv:astro-ph/0508395, Bibcode:2005A&A...444..521F, doi:10.1051/0004-6361:20053599,retrieved2010-07-18

[190] Iglesias-Groth, S.; et al. (2008-09-20), "Evidence for the Naphthalene Cation in a Region of the Interstellar Medium with Anomalous Microwave Emission", *The Astrophysical Journal Letters* **685**: L55–L58, arXiv:0809.0778, Bibcode:2008ApJ...685L..55I,doi:10.1086/592349-This spectral assignment has not been independently confirmed,and is described by the authors as "ten-tative" (page L58).

[191] García-Hernández, D. A.; et al. (2011), "The Formation of Fullerenes: Clues from New C_{60}, C_{70}, and (Possible) Planar C_{24} Detections in Magellanic Cloud Planetary Nebulae", *Astrophysical Journal Letters* **737** (2): L30, arXiv:1107.2595, Bibcode:2011ApJ...737L..30G, doi:10.1088/2041-8205/737/2/L30, retrieved 2011-08-12.

[192] Iglesias-Groth, S.; et al. (May 2010), "A search for interstellar anthracene toward the Perseus anomalous microwave emission region", *Monthly Notices of the Royal Astronomical Society* **407** (4): 2157–2165, arXiv:1005.4388, Bibcode:2010MNRAS.407.2157I,doi:10.1111/j.1365-2966.2010.17075.x

12.7 External links

- Woon, David E. (October 1, 2010). "Interstellar and Circumstellar Molecules". Retrieved 2010-10-04.
- "Molecules in Space". Universität zu Köln. August 2010. Retrieved 2010-10-04.
- Dworkin, Jason P. (February 1, 2007). "Interstellar Molecules". NASA's Cosmic Ice Lab. Retrieved 2010-12-23.
- Wootten, Al (November 2005). "The 129 reported interstellar and circumstellar molecules". National Radio Astronomy Observatory. Retrieved 2007-02-13.
- Lovas, F. J.; Dragoset, R. A. (February 2004). "NIST Recommended Rest Frequencies for Observed Interstellar Molecular Microwave Transitions, 2002 Revision". National Institute of Standards and Technology. Retrieved 2007-02-13.

Chapter 13

Extraterrestrial life

For other uses, see Astrobiology.

Extraterrestrial life[n 1] is life that does not originate from Earth. It is also called **alien life**, or, if it is a sentient and/or relatively complex individual, an "extraterrestrial" or "alien" (or, to avoid confusion with the legal sense of "alien", a "**space alien**"). These as-yet-hypothetical life forms range from simple bacteria-like organisms to beings with civilizations far more advanced than humanity. Although many scientists expect extraterrestrial life to exist, so far no unambiguous evidence for its existence exists.[1][2]

The science of extraterrestrial life is known as exobiology. The science of astrobiology also considers life on Earth as well, and in the broader astronomical context. Meteorites that have fallen to Earth have sometimes been examined for signs of microscopic extraterrestrial life. In 2015, "remains of biotic life" were found in 4.1 billion-year-old rocks in Western Australia, when the young Earth was about 400 million years old.[3][4] According to one of the researchers, "If life arose relatively quickly on Earth ... then it could be common in the universe."[3]

Since the mid-20th century, there has been an ongoing search for signs of extraterrestrial intelligence, from radios used to detect possible extraterrestrial signals, to telescopes used to search for potentially habitable extrasolar planets.[5] It has also played a major role in works of science fiction. Over the years, science fiction works, especially Hollywood's involvement, has increased the public's interest in the possibility of extraterrestrial life. Some encourage aggressive methods to try to get in contact with life in outer space, whereas others argue that it might be dangerous to actively call attention to Earth.[6][7]

13.1 Background

Alien life, such as microorganisms, has been hypothesized to exist in the Solar System and throughout the universe. This hypothesis relies on the vast size and consistent physical laws of the observable universe. According to this argument, made by scientists such as Carl Sagan and Stephen Hawking,[8] it would be improbable for life *not* to exist somewhere other than Earth.[9][10] This argument is embodied in the Copernican principle, which states that Earth does not occupy a unique position in the Universe, and the mediocrity principle, which states that there is nothing special about life on Earth.[11] The chemistry of life may have begun shortly after the Big Bang, 13.8 billion years ago, during a habitable epoch when the universe was only 10–17 million years old.[12][13][14] Life may have emerged independently at many places throughout the universe. Alternatively, life may have formed less frequently, then spread—by meteoroids, for example—between habitable planets in a process called panspermia.[15][16] In any case, complex organic molecules may have formed in the protoplanetary disk of dust grains surrounding the Sun before the formation of Earth.[17] According to these studies, this process may occur outside Earth on several planets and moons of the Solar System and on planets of other stars.[17]

Since the 1950s, scientists have argued the idea that "habitable zones" around stars are the most likely places to find life. Numerous discoveries in these zones since 2007 have generated estimations of frequencies of Earth-like planets —in terms of composition— numbering in the many billions[18] though as of 2013, only a small number of planets have been discovered in these zones.[19] Nonetheless, on November 4, 2013, astronomers reported, based on *Kepler* space mission data, that there could be as many as 40 billion Earth-sized planets orbiting in the habitable zones of Sun-like stars and red dwarfs in the Milky Way,[20][21] 11 billion of which may be orbiting Sun-like stars.[22] The nearest such planet may be 12 light-years away, according to the scientists.[20][21] Astrobiologists have also considered a "follow the energy" view of potential habitats.[23][24]

13.2 Possible basis

13.2.1 Biochemistry

Main articles: Biochemistry, Hypothetical types of biochemistry and Water and life

It is often hypothesized that life forms elsewhere in the universe would, like life on Earth, be based on carbon chemistry and rely on liquid water. Life forms based on ammonia (rather than water) have been suggested, though this solvent appears less suitable than water. It is also conceivable that there are forms of life whose solvent is a liquid hydrocarbon, such as methane, ethane or propane.[25]

About 29 chemical elements are thought to play an active positive role in living organisms on Earth.[26] About 95% of this living matter is built upon only six elements: carbon, hydrogen, nitrogen, oxygen, phosphorus and sulfur. These six elements form the basic building blocks of virtually all life on Earth, whereas most of the remaining elements are found only in trace amounts.[27] The unique characteristics of carbon made it unlikely that any other element could replace carbon, even on another planet, to generate the biochemistry necessary for life. The carbon atom has the unique ability to make four strong chemical bonds with other atoms, including other carbon atoms. These covalent bonds have a direction in space, so that carbon atoms can form the skeletons of complex 3-dimensional structures with definite architectures such as nucleic acids and proteins. Carbon forms more compounds than all other elements combined. The great versatility of the carbon atom makes it the element most likely to provide the bases—even exotic ones—to the chemical composition of life on other planets.[28]

Life on Earth requires water as its solvent in which biochemical reactions take place. Sufficient quantities of carbon and the other elements along with water, may enable the formation of living organisms on other planets with a chemical make-up and temperature range similar to that of Earth.[29] Terrestrial planets such as Earth are formed in a process that allows for the possibility of having compositions similar to Earth's.[30] The combination of carbon, hydrogen and oxygen in the chemical form of carbohydrates (e.g. sugar) can be a source of chemical energy on which life depends, and can provide structural elements for life. Plants derive energy through the conversion of light energy into chemical energy via photosynthesis. Life, as currently recognized, requires carbon in both reduced (methane derivatives) and partially oxidized (carbon oxides) states. Nitrogen is needed as a reduced ammonia derivative in all proteins, sulfur as a derivative of hydrogen sulfide in some necessary proteins, and phosphorus oxidized to phosphates in genetic material and in energy transfer.

13.3 Planetary habitability in the Solar System

See also: Planetary habitability and Natural satellite habitability

Some bodies in the Solar System have been suggested as having the potential for an environment that could host extraterrestrial life, particularly those with possible subsurface oceans.[31] Should life be discovered elsewhere in the Solar System, astrobiologists suggest that it will more likely be in the form of extremophile microorganisms.

Some speculate that Mars may possess niche subsurface environments where microbial life might exist.[32][33][34] A subsurface marine environment on Jupiter's moon Europa might be the most likely habitat in the Solar System, outside Earth, for extremophile microorganisms.[35][36][37]

The panspermia hypothesis proposes that life elsewhere in the Solar System may have a common origin. If extraterrestrial life was found on another body in the Solar System, it could have originated from Earth just as life on Earth may have been seeded from elsewhere (exogenesis). The first known mention of the term 'panspermia' was in the writings of the 5th century BC Greek philosopher Anaxagoras.[38] In the nineteenth century it was again revived in modern form by several scientists, including Jöns Jacob Berzelius (1834),[39] Kelvin (1871),[40] Hermann von Helmholtz (1879)[41] and, somewhat later, by Svante Arrhenius (1903).[42] Sir Fred Hoyle (1915–2001) and Chandra Wickramasinghe (born 1939) are important proponents of the hypothesis who further contended that lifeforms continue to enter Earth's atmosphere, and may be responsible for epidemic outbreaks, new diseases, and the genetic novelty necessary for macroevolution.[43]

Directed panspermia concerns the deliberate transport of microorganisms in space, sent to Earth to start life here, or sent from Earth to seed new stellar systems with life. The Nobel prize winner Francis Crick, along with Leslie Orgel proposed that seeds of life may have been purposely spread by an advanced extraterrestrial civilization,[44] but considering an early "RNA world" Crick noted later that life may have originated on Earth.[45]

13.3.1 Mars

Main article: Life on Mars

Life on Mars has been long speculated. Liquid water is widely thought to have existed on Mars in the past, and now can occasionally be found as low-volume liquid brines in shallow Martian soil.[46] The origin of the potential biosignature of methane observed in Mars atmosphere is unexplained, although abiotic hypotheses have also been proposed.[47] By July 2008, laboratory tests aboard NASA's *Phoenix* Mars lander identified water in a surface soil sample. Photographs from the Mars Global Surveyor from 2006 showed evidence of recent (i.e. within 10 years) flows of a liquid on Mars's frigid surface.[48] There is evidence that Mars had a warmer and wetter past: dried-up river beds, polar ice caps, volcanos, and minerals that form in the presence of water have all been found. Nevertheless, present conditions on Mars subsurface may support life.[49][50] Evidence obtained by the *Curiosity* rover studying Aeolis Palus, Gale Crater in 2013, strongly suggest an ancient freshwater lake that could have been a hospitable environment for microbial life.[51][52]

Current studies on Mars by the *Curiosity* and *Opportunity* rovers are now searching for evidence of ancient life, including a biosphere based on autotrophic, chemotrophic and/or chemolithoautotrophic microorganisms, as well as ancient water, including fluvio-lacustrine environments (plains related to ancient rivers or lakes) that may have been habitable.[53][54][55][56] The search for evidence of habitability, taphonomy (related to fossils), and organic carbon on Mars is now a primary NASA objective.[53]

13.3.2 Ceres

Ceres, the only dwarf planet in the asteroid belt, was confirmed by the Herschel Space Observatory to have a thin water vapor atmosphere.[57][58] Frost on the surface may also have been detected in the form of bright spots.[59][60][61] The presence of water on Ceres has led to speculation that life may be possible there.[62][63][64]

13.3.3 Jupiter system

Jupiter

Carl Sagan and others in the 1960s and 1970s computed conditions for hypothetical microorganisms living in the atmosphere of Jupiter,[65] however, the intense radiation and other conditions do not appear to permit encapsulation and molecular biochemistry, so life there is thought unlikely.[66] In contrast, some of Jupiter's moons may have habitats capable of sustaining life. Scientists have indications that heated subsurface oceans of liquid water may exist deep under the crusts of the three outer Galilean moons —Europa,[35][67][68] Ganymede,[69][70][71][72][73] and Callisto.[74][75][76] The EJSM/Laplace mission is planned to determine the habitability of these environments.

Europa

Main article: Life on Europa

Jupiter's moon Europa has been subject to speculation about the existence of life due to the strong possibility of a liquid water ocean beneath its ice surface.[35][37] Hydrothermal vents on the bottom of the ocean, if they exist, may warm the ice and could be capable of supporting multicellular microorganisms.[78] It is also possible that Europa could support aerobic macrofauna using oxygen created by cosmic rays impacting its surface ice.[79]

The case for life on Europa was greatly enhanced in 2011 when it was discovered that vast lakes exist within Europa's thick, icy shell. Scientists found that ice shelves surrounding the lakes appear to be collapsing into them, thereby providing a mechanism through which life-forming chemicals created in sunlit areas on Europa's surface could be transferred to its interior.[80][81]

On December 11, 2013, NASA reported the detection of "clay-like minerals" (specifically, phyllosilicates), often associated with organic materials, on the icy crust of Europa.[82] The presence of the minerals may have been the result of a collision with an asteroid or comet according to the scientists.[82] The *Europa Multiple-Flyby Mission*, which would assess the habitability of Europa, is planned for launch in 2025.[83][84] Europa's subsurface ocean is considered the best target for the discovery of life.[35][37]

13.3.4 Saturn system

Titan and Enceladus have been speculated to host possible habitats supportive of life.

Enceladus

Enceladus, a moon of Saturn, has some of the conditions for life, including geothermal activity and water vapor, as well as possible under-ice oceans heated by tidal effects.[85][86] The *Cassini–Huygens* probe detected carbon, hydrogen, nitrogen and oxygen—all key elements for supporting life—during its 2005 flyby through one of Enceladus's geysers spewing ice and gas. The temperature and density of the plumes indicate a warmer, watery source beneath the surface.[47]

Titan

Main article: Life on Titan

Titan, the largest moon of Saturn, is the only known moon in the Solar System with a significant atmosphere. Data from the *Cassini–Huygens* mission refuted the hypothesis of a global hydrocarbon ocean, but later demonstrated the existence of liquid hydrocarbon lakes in the polar regions—the first stable bodies of surface liquid discovered outside Earth.[87][88][89] Analysis of data from the mission has uncovered aspects of atmospheric chemistry near the surface that are consistent with—but do not prove—the hypothesis that organisms there if present, could be consuming hydrogen, acetylene and ethane, and producing methane.[90][91][92]

13.3.5 Small Solar System bodies

Small Solar System bodies have also been speculated to host habitats for extremophiles. Fred Hoyle and Chandra Wickramasinghe have proposed that microbial life might exist on comets and asteroids.[93][94][95][96]

13.4 Scientific search

The scientific search for extraterrestrial life is being carried out both directly and indirectly.

ar planets

r planets
ary systems
ch for extrasolar planets that may be conducive to life, narrowing the search to terrestrial planets e of their star.[128][129] Since 1992, nearly 2000 exoplanets have been discovered (1977 planets ary systems including 490 multiple planetary systems as of 1 November 2015).[130] The extrasolar planets so far discovered range in size from that of terrestrial planets similar to Earth's size to that of gas giants larger than Jupiter.[130] The number of observed exoplanets is expected to increase greatly in the coming years.[131]

The *Kepler space telescope* has also detected a few thousand[132][133] candidate planets,[134][135] of which about 11% may be false positives.[136] There is at least one planet on average per star.[137]

About 1 in 5 Sun-like stars[lower-alpha 1] have an "Earth-sized"[lower-alpha 2] planet in the habitable zone,[lower-alpha 3] with the nearest expected to be within 12 light-years distance from Earth.[138][139] Assuming 200 billion stars in the Milky Way,[lower-alpha 4] that would be 11 billion potentially habitable Earth-sized planets in the Milky Way, rising to 40 billion if red dwarfs are included.[22] The rogue planets in the Milky Way possibly number in the trillions.[140]

The nearest known exoplanet, if confirmed, would be Alpha Centauri Bb, located 4.37 light-years from Earth in the southern constellation of Centaurus.[141] As of March 2014, the least massive planet known is PSR B1257+12 A, which is about twice the mass of the Moon. The most massive planet listed on the NASA Exoplanet Archive is DENIS-P J082303.1-491201 b,[142][143] about 29 times the mass of Jupiter, although according to most definitions of a planet, it is too massive to be a planet and may be a brown dwarf instead. Almost all of the planets detected so far are within the Milky Way, but there have also been a few possible detections of extragalactic planets. The study of planetary habitability also considers a wide range of other factors in determining the suitability of a planet for hosting life.[5]

13.5 The Drake equation

Main article: Drake equation

In 1961, University of California, Santa Cruz, astronomer and astrophysicist Frank Drake devised the Drake equation as a way to stimulate scientific dialogue at a meeting on the search for extraterrestrial intelligence (SETI).[144] The Drake equation is a probabilistic argument used to estimate the number of active, communicative extraterrestrial civilizations in the Milky Way galaxy. The equation is best understood not as an equation in the strictly mathematical sense, but to summarize all the various concepts which scientists must contemplate when considering the question of life elsewhere.[145] The Drake equation is:

$$N = R_* \cdot f_p \cdot n_e \cdot f_\ell \cdot f_i \cdot f_c \cdot L$$

where:

> N = the number of civilizations in our galaxy with which radio-communication might be possible (i.e. which are on our current past light cone);

and

> $R*$ = the average rate of star formation in our galaxy
>
> fp = the fraction of those stars that have planets
>
> ne = the average number of planets that can potentially support life per star that has planets
>
> fl = the fraction of planets that could support life that actually develop life at some point
>
> fi = the fraction of planets with life that actually go on to develop intelligent life (civilizations)

fc = the fraction of civilizations that develop a technology that releases detectable signs of their existence into space

L = the length of time for which such civilizations release detectable signals into space[146]

The Drake equation has proved controversial since several of its factors are extremely uncertain and are entirely based on conjecture. Thus the equation cannot be used to draw firm conclusions of any kind.[147] This has led critics to label the equation a guesstimate, or even meaningless.

Based on observations from the Hubble Space Telescope, there are at least 125 billion galaxies in the observable universe. It is estimated that at least ten percent of all Sun-like stars have a system of planets,[148] i.e. there are 6.25×10^{18} stars with planets orbiting them in the observable universe. Even if we assume that only one out of a billion of these stars have planets supporting life, there would be some 6.25×10^{9} (billion) life-supporting planetary systems in the observable universe.

The apparent contradiction between high estimates of the probability of the existence of extraterrestrial civilizations and the lack of evidence for such civilizations, is known as the Fermi paradox.[149]

13.6 Cultural impact

13.6.1 Cosmic pluralism

Main article: Cosmic pluralism

Cosmic pluralism, the plurality of worlds, or simply pluralism, describes the philosophical belief in numerous "worlds" in addition to Earth, which might harbor extraterrestrial life. Before the development of the heliocentric theory and a recognition that our Sun is just one of many stars,[150] the notion of pluralism was largely mythological and philosophical.[151][152] With the scientific and Copernican revolutions, and later, during the later, during the Enlightenment, cosmic pluralism became a mainstream notion, supported by the likes of Bernard le Bovier de Fontenelle in his 1686 work*Entretiens sur la pluralité des mondes*.[154] Pluralism was also championed by philosophers such as John Locke, Giordano Bruno and astronomers such as William Herschel. The astronomer Camille Flammarion promoted the notion of cosmic pluralism in his 1862 book *La pluralité des mondes habités*.[155] None of these notions of pluralism were based on any specific observation or scientific information.

13.6.2 Early modern period

There was a dramatic shift in thinking initiated by the invention of the telescope and the Copernican assault on geocentric cosmology. Once it became clear that Earth was merely one planet amongst countless bodies in the universe, the theory of extraterrestrial life started to become a topic in the scientific community. The best known early-modern proponent of such ideas was the Italian philosopher Giordano Bruno, who argued in the 16th century for an infinite universe in which every star is surrounded by its own planetary system. Bruno wrote that other worlds "have no less virtue nor a nature different to that of our earth" and, like Earth, "contain animals and inhabitants".[156]

In the early 17th century, the Czech astronomer Anton Maria Schyrleus of Rheita mused that "if Jupiter has (...) inhabitants (...) they must be larger and more beautiful than the inhabitants of Earth, in proportion to the [characteristics] of the two spheres".[157]

In Baroque literature such as *The Other World: The Societies and Governments of the Moon* by Cyrano de Bergerac, extraterrestrial societies are presented as humoristic or ironic parodies of earthly society. The didactic poet Henry More took up the classical theme of the Greek Democritus in "Democritus Platonissans, or an Essay Upon the Infinity of Worlds" (1647). In "The Creation: a Philosophical Poem in Seven Books" (1712), Sir Richard Blackmore observed: "We may pronounce each orb sustains a race / Of living things adapted to the place". With the new relative viewpoint that the Copernican revolution had wrought, he suggested "our world's sunne / Becomes a starre elsewhere". Fontanelle's

"Conversations on the Plurality of Worlds" (translated into English in 1686) offered similar excursions on the possibility of extraterrestrial life, expanding, rather than denying, the creative sphere of a Maker.

The possibility of extraterrestrials remained a widespread speculation as scientific discovery accelerated. William Herschel, the discoverer of Uranus, was one of many 18th–19th-century astronomers who believed that the Solar System is populated by alien life. Other luminaries of the period who championed "cosmic pluralism" included Immanuel Kant and Benjamin Franklin. At the height of the Enlightenment, even the Sun and Moon were considered candidates for extraterrestrial inhabitants.

13.6.3 19th century

Speculation about life on Mars increased in the late 19th century, following telescopic observation of apparent Martian canal—which soon, however, turned out to be optical illusions.[158] Despite this, in 1895, American astronomer Percival Lowell published his book *Mars,* followed by *Mars and its Canals* in 1906, proposing that the canals were the work of a long-gone civilization.[159] This idea led British writer H. G. Wells to write the novel *The War of the Worlds* in 1897, telling of an invasion by aliens from Mars who were fleeing the planet's desiccation.

Spectroscopic analysis of Mars's atmosphere began in earnest in 1894, when U.S. astronomer William Wallace Campbell showed that neither water nor oxygen was present in the Martian atmosphere.[160] By 1909 better telescopes and the best perihelic opposition of Mars since 1877 conclusively put an end to the canal hypothesis.

The science fiction genre, although not so named during the time, develops during the late 19th century. Jules Verne's *Around the Moon* (1870) features a discussion of the possibility of life on the Moon, but with the conclusion that it is barren. Stories involving extraterrestrials are found in e.g. Garrett P. Serviss's *Edison's Conquest of Mars* (1898), an unauthorized sequel to *The War of the Worlds* by H. G. Wells was published in 1897 which stands at the beginning of the popular idea of the "Martian invasion" of Earth prominent in 20th-century pop culture.

13.6.4 20th century

See also: Space exploration

Most unidentified flying objects or UFO sightings[161] can be readily explained as sightings of Earth-based aircraft, known astronomical objects, or as hoaxes.[162] Nonetheless, a certain fraction of the public believe that UFOs might actually be of extraterrestrial origin, and, indeed, the notion has had influence on popular culture.

The possibility of extraterrestrial life on the Moon was ruled out in the 1960s, and during the 1970s it became clear that most of the other bodies of the Solar System do not harbor highly developed life, although the question of primitive life on bodies in the Solar System remains an open question.

13.6.5 Recent history

The failure so far of the SETI program to detect an intelligent radio signal after decades of effort has at least partially dimmed the prevailing optimism of the beginning of the space age. Notwithstanding, belief in extraterrestrial beings continues to be voiced in pseudoscience, conspiracy theories, and in popular folklore, notably "Area 51" and legends. It has become a pop culture trope given less-than-serious treatment in popular entertainment.

In the words of SETI's Frank Drake, "All we know for sure is that the sky is not littered with powerful microwave transmitters".[163] Drake noted that it is entirely possible that advanced technology results in communication being carried out in some way other than conventional radio transmission. At the same time, the data returned by space probes, and giant strides in detection methods, have allowed science to begin delineating habitability criteria on other worlds, and to confirm that at least other planets are plentiful, though aliens remain a question mark. The Wow! signal, detected in 1977 by a SETI project, remains a subject of speculative debate.

In 2000, geologist and paleontologist Peter Ward and astrobiologist Donald Brownlee published a book entitled *Rare Earth: Why Complex Life is Uncommon in the Universe*.[164] In it, they discussed the Rare Earth hypothesis, in which they claim that Earth-like life is rare in the universe, whereas microbial life is common. Ward and Brownlee are open to the idea of evolution on other planets that is not based on essential Earth-like characteristics (such as DNA and carbon).

Theoretical physicist Stephen Hawking in 2010 warned that humans should not try to contact alien life forms. He warned that aliens might pillage Earth for resources. "If aliens visit us, the outcome would be much as when Columbus landed in America, which didn't turn out well for the Native Americans", he said.[165] Jared Diamond has expressed similar concerns.[166]

In November 2011, the White House released an official response to two petitions asking the U.S. government to acknowledge formally that aliens have visited Earth and to disclose any intentional withholding of government interactions with extraterrestrial beings. According to the response, "The U.S. government has no evidence that any life exists outside our planet, or that an extraterrestrial presence has contacted or engaged any member of the human race."[167][168] Also, according to the response, there is "no credible information to suggest that any evidence is being hidden from the public's eye."[167][168] The response noted "odds are pretty high" that there may be life on other planets but "the odds of us making contact with any of them—especially any intelligent ones—are extremely small, given the distances involved."[167][168]

In 2013, the exoplanet Kepler-62f was discovered, along with Kepler-62e and Kepler-62c. A related special issue of the journal Science, published earlier, described the discovery of the exoplanets.[169]

On 17 April 2014, the discovery of the Earth-size exoplanet Kepler-186f, 500 light-years from Earth, was publicly announced;[170] it is the first Earth-size planet to be discovered in the habitable zone and it has been hypothesized that there may be liquid water on its surface.

On 13 February 2015, scientists (including Geoffrey Marcy, Seth Shostak, Frank Drake and David Brin) at a convention of the American Association for the Advancement of Science, discussed Active SETI and whether transmitting a message to possible intelligent extraterrestrials in the Cosmos was a good idea;[171][172] one result was a statement, signed by many, that a "worldwide scientific, political and humanitarian discussion must occur before any message is sent".[173]

On 20 July 2015, Stephen Hawking, British physicist, and Yuri Milner, Russian billionaire, along with the SETI Institute, announced a well-funded effort, called the Breakthrough Initiatives, to expand efforts to search for extraterrestrial life. The group contracted the services of the 100-meter Robert C. Byrd Green Bank Telescope in West Virginia in the United States and the 64-meter Parkes Telescope in New South Wales, Australia.[174]

13.7 See also

Searches for extraterrestrial life

- Allen Telescope Array
- Astrobiology
- Communication with extraterrestrial intelligence
- Nexus for Exoplanet System Science
- SETI

Subjects

- Extraterrestrial liquid water
- Planetary protection
- Potential cultural impact of extraterrestrial contact

Hypotheses

- Aurelia and Blue Moon

- Fermi paradox
- Kardashev scale
- Metalaw
- Rare Earth hypothesis
- Sentience quotient
- Zoo hypothesis

13.8 Notes

[1] Where "extraterrestrial" is derived from the Latin *extra* ("beyond", "not of") and *terrestris* ("of Earth", "belonging to Earth").

[1] For the purpose of this 1 in 5 statistic, "Sun-like" means G-type star. Data for Sun-like stars wasn't available so this statistic is an extrapolation from data about K-type stars

[2] For the purpose of this 1 in 5 statistic, Earth-sized means 1–2 Earth radii

[3] For the purpose of this 1 in 5 statistic, "habitable zone" means the region with 0.25 to 4 times Earth's stellar flux (corresponding to 0.5–2 AU for the Sun).

[4] About 1/4 of stars are GK Sun-like stars. The number of stars in the galaxy is not accurately known, but assuming 200 billion stars in total, the Milky Way would have about 50 billion Sun-like (GK) stars, of which about 1 in 5 (22%) or 11 billion would be Earth-sized in the habitable zone. Including red dwarfs would increase this to 40 billion.

13.9 References

[1] Davies, Paul (18 November 2013). "Are We Alone in the Universe?". *New York Times*. Retrieved 20 November 2013.

[2] Pickrell, John (4 September 2006). "Top 10: Controversial pieces of evidence for extraterrestrial life". *New Scientist*. Retrieved 2011-02-18.

[3] Borenstein, Seth (19 October 2015). "Hints of life on what was thought to be desolate early Earth". *Excite* (Yonkers, NY: Mindspark Interactive Network). Associated Press. Retrieved 2015-10-20.

[4] Bell, Elizabeth A.; Boehnike, Patrick; Harrison, T. Mark; et al. (19 October 2015). "Potentially biogenic carbon preserved in a 4.1 billion-year-old zircon" (PDF). *Proc. Natl. Acad. Sci. U.S.A.* (Washington, D.C.: National Academy of Sciences). doi:10.1073/pnas.1517557112. ISSN 1091-6490. Retrieved 2015-10-20. Early edition, published online before print.

[5] Overbye, Dennis (January 6, 2015). "As Ranks of Goldilocks Planets Grow, Astronomers Consider What's Next". *New York Times*. Retrieved January 6, 2015.

[6] BBC News – Scientists in US are urged to seek contact with aliens

[7] Baum, Seth; Haqq-Misra, Jacob; Domagal-Goldman, Shawn. "Would Contact with Extraterrestrials Benefit or Harm Humanity? A Scenario Analysis", *Acta Astronautica, 2011, 68 (11–12):2014–2129*, April 22, 2011, accessed August 18, 2011.

[8] Weaver, Rheyanne. "Ruminations on other worlds". *State Press*. Retrieved 10 March 2014.

[9] Brad Steiger, John White, eds. (1986). *Other Worlds, Other Universes*. Health Research Books. p. 3. ISBN 0-7873-1291-6.

[10] Filkin, David; Hawking, Stephen W. (1998). *Stephen Hawking's universe: the cosmos explained*. Art of Mentoring Series. Basic Books. p. 194. ISBN 0-465-08198-3.

[11] Rauchfuss, Horst (2008). *Chemical Evolution and the Origin of Life*. T. N. Mitchell. Springer. ISBN 3-540-78822-0

[12] Loeb, Abraham (October 2014). "The Habitable Epoch of the Early Universe". *International Journal of Astrobiology* **13** (04): 337–339. arXiv:1312.0613. Bibcode:2014IJAsB..13..337L. doi:10.1017/S1473550414000196. Retrieved 15 December 2014.

[13] Loeb, Abraham (2 December 2013). "The Habitable Epoch of the Early Universe" (PDF). *Arxiv*. arXiv:1312.0613v3. Retrieved 15 December 2014.

[14] Dreifus, Claudia (2 December 2014). "Much-Discussed Views That Go Way Back – Avi Loeb Ponders the Early Universe, Nature and Life". *New York Times*. Retrieved 3 December 2014.

[15] Rampelotto, P.H. (2010). "Panspermia: A Promising Field Of Research" (PDF). Astrobiology Science Conference. Retrieved 3 December 2014.

[16] Gonzalez, Guillermo; Richards, Jay Wesley (2004). *The privileged planet: how our place in the cosmos is designed for discovery*. Regnery Publishing. pp. 343–345. ISBN 0-89526-065-4.

[17] Moskowitz, Clara (29 March 2012). "Life's Building Blocks May Have Formed in Dust Around Young Sun". Space.com. Retrieved 30 March 2012.

[18] Choi, Charles Q. (21 March 2011). "New Estimate for Alien Earths: 2 Billion in Our Galaxy Alone". Space.com. Retrieved 2011-04-24.

[19] Torres, Abel Mendez (April 26, 2013). "Ten potentially habitable exoplanets now". *Habitable Exoplanets Catalog*. University of Puerto Rico. Retrieved April 29, 2013.

[20] Overbye, Dennis (November 4, 2013). "Far-Off Planets Like the Earth Dot the Galaxy". *New York Times*. Retrieved November 5, 2013.

[21] Petigura, Eric A.; Howard, Andrew W.; Marcy, Geoffrey W. (October 31, 2013). "Prevalence of Earth-size planets orbiting Sun-like stars". *Proceedings of the National Academy of Sciences of the United States of America*. arXiv:1311.6806. Bibcode:2013PNAS..11019273P. doi:10.1073/pnas.1319909110. Retrieved November 5, 2013.

[22] Khan, Amina (November 4, 2013). "Milky Way may host billions of Earth-size planets". *Los Angeles Times*. Retrieved November 5, 2013.

[23] Hoehler, Tori M.; Amend, Jan P.; Shock, Everett L. (2007). "A "Follow the Energy" Approach for Astrobiology". *Astrobiology* **7** (6): 819–823. Bibcode:2007AsBio...7..819H. doi:10.1089/ast.2007.0207. ISSN 1531-1074.

[24] Jones, Eriita G.; Lineweaver, Charles H. (2010). "To What Extent Does Terrestrial Life "Follow The Water"?". *Astrobiology* **10** (3): 349–361. Bibcode:2010AsBio..10..349J. doi:10.1089/ast.2009.0428. ISSN 1531-1074.

[25] Committee on the Limits of Organic Life in Planetary Systems, Committee on the Origins and Evolution of Life, National Research Council; The Limits of Organic Life in Planetary Systems; The National Academies Press, 2007; p 74

[26] Ultratrace minerals. Authors: Nielsen, Forrest H. USDA, ARS Source: Modern nutrition in health and disease / editors, Maurice E. Shils ... et al.. Baltimore : Williams & Wilkins, c1999., p. 283-303. Issue Date: 1999 URI:

[27] Mix, Lucas John (2009). *Life in space: astrobiology for everyone*. Harvard University Press. p. 76. ISBN 0-674-03321-3. Retrieved 2011-08-08.

[28] Norman H. Horowitz. To Utopia and Back: The Search for Life in the Solar System 1986 W.H. Freeman & Co, NY ISBN 0-7167-1765-4 ISBN 0-7167-1766-2

[29] Pace, Norman R. (January 20, 2001). "The universal nature of biochemistry". *Proceedings of the National Academy of Sciences of the United States of America* **98** (3): 805–808. Bibcode:2001PNAS...98..805P. doi:10.1073/pnas.98.3.805. PMC 33372. PMID 11158550.

[30] Bond, Jade C.; O'Brien, David P.; Lauretta, Dante S. (June 2010). "The Compositional Diversity of Extrasolar Terrestrial Planets. I. In Situ Simulations". *The Astrophysical Journal* **715** (2): 1050–1070. arXiv:1004.0971. Bibcode:2010ApJ...715.1050B. doi:10.1088/0004-637X/715/2/1050.

[31] Dyches, Preston; Chou, Felcia (7 April 2015). "The Solar System and Beyond is Awash in Water". *NASA*. Retrieved 8 April 2015.

[32] Summons, Roger E.; Amend, Jan P.; Bish, David; Buick, Roger; Cody, George D.; Des Marais, David J.; Dromart, Gilles; Eigenbrode, Jennifer L.; et al. (2011). "Preservation of Martian Organic and Environmental Records: Final Report of the Mars Biosignature Working Group". *Astrobiology* **11** (2): 157–81. Bibcode:2011AsBio..11..157S. doi:10.1089/ast.2010.0506. PMID 21417945. There is general consensus that extant microbial life on Mars would probably exist (if at all) in the subsurface and at low abundance.

[33] Michalski, Joseph R.; Cuadros, Javier; Niles, Paul B.; Parnell, John; Deanne Rogers, A.; Wright, Shawn P. (2013). "Groundwater activity on Mars and implications for a deep biosphere". *Nature Geoscience* **6** (2): 133–8. Bibcode:2013NatGe...6..133M. doi:10.1038/ngeo1706.

[34] "Habitability and Biology: What are the Properties of Life?". *Phoenix Mars Mission*. The University of Arizona. Retrieved 2013-06-06. If any life exists on Mars today, scientists believe it is most likely to be in pockets of liquid water beneath the Martian surface.

[35] Tritt, Charles S. (2002). "Possibility of Life on Europa". Milwaukee School of Engineering. Archived from the original on 9 June 2007. Retrieved 10 August 2007.

[36] Jeffrey S. Kargel, Jonathan Z. Kaye, James W. Head, III; et al. (2000). "Europa's Crust and Ocean: Origin, Composition, and the Prospects for Life" (PDF). *Icarus* (Planetary Sciences Group, Brown University) **148** (1): 226–265. Bibcode:2000Icar..148..226K. doi:10.1006/icar.2000.6471.

[37] Schulze-Makuch, Dirk; and Irwin, Louis N. (2001). "Alternative Energy Sources Could Support Life on Europa" (PDF). *Departments of Geological and Biological Sciences, University of Texas at El Paso*. Archived from the original (PDF) on 3 July 2006. Retrieved 21 December 2007.

[38] Margaret O'Leary (2008) Anaxagoras and the Origin of Panspermia Theory, iUniverse publishing Group, # ISBN 978-0-595-49596-2

[39] Berzelius (1799–1848), J. J. "Analysis of the Alais meteorite and implications about life in other worlds".

[40] Thomson (Lord Kelvin), W. (1871). "Inaugural Address to the British Association Edinburgh. "We must regard it as probably to the highest degree that there are countless seed-bearing meteoritic stones moving through space."". *Nature* **4** (92): 261–278 [262]. Bibcode:1871Natur...4..261.. doi:10.1038/004261a0.

[41] Darwin's contribution to the development of the Panspermia theory. By Demets R. *Astrobiology*. 2012 October; 12(10) pages: 946–50. doi: 10.1089/ast.2011.0790

[42] Arrhenius, S., *Worlds in the Making: The Evolution of the Universe*. New York, Harper & Row, 1908.

[43] Fred Hoyle, Chandra Wickramasinghe and John Watson, *Viruses from Space and Related Matters*, University College Cardiff Press, 1986.

[44] Crick, F. H.; Orgel, L. E. (1973). "Directed Panspermia". *Icarus* **19**: 341–348. Bibcode:1979JBIS...32..419M. doi:10.1016/0019-1035(73)90110-3+.

[45] "Anticipating an RNA world. Some past speculations on the origin of life: where are they today?" by L. E. Orgel and F. H. C. Crick in *FASEB J.* (1993) Volume 7 pages 238–239.

[46] Ojha, L.; Wilhelm, M. B.; Murchie, S. L.; McEwen, A. S.; Wray, J. J.; Hanley, J.; Massé, M.; Chojnacki, M. (2015). "Spectral evidence for hydrated salts in recurring slope lineae on Mars". *Nature Geoscience*. doi:10.1038/ngeo2546.

[47] "Top 10 Places To Find Alien Life : Discovery News". News.discovery.com. 2010-06-08. Retrieved 2012-06-13.

[48] "Water 'flowed recently' on Mars". *BBC News*. 2006-12-06. Retrieved 2010-05-02.

[49] Baldwin, Emily (26 April 2012). "Lichen survives harsh Mars environment". Skymania News. Retrieved 27 April 2012.

[50] de Vera, J.-P.; Kohler, Ulrich (26 April 2012). "The adaptation potential of extremophiles to Martian surface conditions and its implication for the habitability of Mars" (PDF). European Geosciences Union. Retrieved 27 April 2012.

[51] Chang, Kenneth (December 9, 2013). "On Mars, an Ancient Lake and Perhaps Life". *New York Times*. Retrieved December 9, 2013.

[52] Various (December 9, 2013). "Science – Special Collection – Curiosity Rover on Mars". *Science*. Retrieved December 9, 2013.

[53] Grotzinger, John P. (January 24, 2014). "Introduction to Special Issue – Habitability, Taphonomy, and the Search for Organic Carbon on Mars". *Science* **343** (6169): 386–387. Bibcode:2014Sci...343..386G. doi:10.1126/science.1249944. Retrieved January 24, 2014.

[54] Various (January 24, 2014). "Special Issue – Table of Contents – Exploring Martian Habitability". *Science* **343** (6169): 345–452. Retrieved 24 January 2014.

[55] Various (January 24, 2014). "Special Collection – Curiosity – Exploring Martian Habitability". *Science*. Retrieved January 24, 2014.

[56] Grotzinger, J.P.; et al. (January 24, 2014). "A Habitable Fluvio-Lacustrine Environment at Yellowknife Bay, Gale Crater, Mars". *Science* **343** (6169). Bibcode:2014Sci...343A.386G. doi:10.1126/science.1242777. Retrieved January 24, 2014.

[57] Küppers, M.; O'Rourke, L.; Bockelée-Morvan, D.; Zakharov, V.; Lee, S.; Von Allmen, P.; Carry, B.; Teyssier, D.; Marston, A.; Müller, T.; Crovisier, J.; Barucci, M. A.; Moreno, R. (2014-01-23). "Localized sources of water vapour on the dwarf planet (1) Ceres". *Nature* **505** (7484): 525–527. Bibcode:2014Natur.505..525K. doi:10.1038/nature12918. ISSN 0028-0836. PMID 24451541.

[58] Campins, H.; Comfort, C. M. (23 January 2014). "Solar system: Evaporating asteroid". *Nature* **505** (7484): 487–488. Bibcode:2014Natur.505..487C. doi:10.1038/505487a. PMID 24451536.

[59] A'Hearn, Michael F.; Feldman, Paul D. (1992). "Water vaporization on Ceres". *Icarus* **98** (1): 54–60. Bibcode:1992Icar...98...54A. doi:10.1016/0019-1035(92)90206-M.

[60] A. Duffy – Cosmos – What on Ceres are those bright spots?

[61] Rivkin, Andrew (21 July 2015). "Dawn at Ceres: A haze in Occator crater?". *The Planetary Society*. Retrieved 2015-07-24.

[62] O'Neill, Ian (5 March 2009). "Life on Ceres: Could the Dwarf Planet be the Root of Panspermia". *Universe Today*. Retrieved 30 January 2012.

[63] Catling, David C. (2013). *Astrobiology: A Very Short Introduction*. Oxford: Oxford University Press. p. 99. ISBN 0-19-958645-4.

[64] Boyle, Alan (22 January 2014). "Is There Life on Ceres? Dwarf Planet Spews Water Vapor". NBC. Retrieved 10 February 2015.

[65] Ponnamperuma, Cyril; Molton, Peter (January 1973). "The prospect of life on Jupiter". *Space Life Sciences* **4** (1): 32–44. Bibcode:1973SLSci...4...32P. doi:10.1007/BF02626340.

[66] Irwin, Louis Neal; Schulze-Makuch, Dirk (June 2001). "Assessing the Plausibility of Life on Other Worlds". *Astrobiology* **1** (2): 143–160. Bibcode:2001AsBio...1..143I. doi:10.1089/153110701753198918. PMID 12467118.

[67] Dyches, Preston; Brown, Dwayne (12 May 2015). "NASA Research Reveals Europa's Mystery Dark Material Could Be Sea Salt". *NASA*. Retrieved 12 May 2015.

[68] Jeffrey S. Kargel, Jonathan Z. Kaye, James W. Head, III; et al. (2000). "Europa's Crust and Ocean: Origin, Composition, and the Prospects for Life" (PDF). *Icarus* (Planetary Sciences Group, Brown University) **148** (1): 226–265. Bibcode:2000Icar..148..226K. doi:10.1006/icar.2000.6471.

[69] Staff (March 12, 2015). "NASA's Hubble Observations Suggest Underground Ocean on Jupiter's Largest Moon". *NASA News*. Retrieved 2015-03-15.

[70] "Jupiter moon Ganymede could have ocean with more water than Earth – NASA". *Russia Today (RT)*. 13 March 2015. Retrieved 2015-03-13.

[71] Clavin, Whitney (1 May 2014). "Ganymede May Harbor 'Club Sandwich' of Oceans and Ice". *NASA* (Jet Propulsion Laboratory). Retrieved 2014-05-01.

[72] Vance, Steve; Bouffard, Mathieu; Choukroun, Mathieu; Sotina, Christophe (12 April 2014). "Ganymede's internal structure including thermodynamics of magnesium sulfate oceans in contact with ice". *Planetary and Space Science*. Bibcode:2014P&SS...96...62V. doi:10.1016/j.pss.2014.03.011. Retrieved 2014-05-02.

[73] Staff (1 May 2014). "Video (00:51) – Jupiter's 'Club Sandwich' Moon". *NASA*. Retrieved 2014-05-02.

[74] Chang, Kenneth (March 12, 2015). "Suddenly, It Seems, Water Is Everywhere in Solar System". *New York Times*. Retrieved March 12, 2015.

[75] Kuskov, O.L.; Kronrod, V.A. (2005). "Internal structure of Europa and Callisto". *Icarus* **177** (2): 550–369. Bibcode:2005Icar..177..550K. doi:10.1016/j.icarus.2005.04.014.

[76] Showman, Adam P.; Malhotra, Renu (1999). "The Galilean Satellites" (PDF). *Science* **286** (5437): 77–84. doi:10.1126/science.286.5437.77. PMID 10506564.

[77] Possibility of Life on Europa. University of Victoria, Department of Physics And Astronomy. Canada

[78] Friedman, Louis (December 14, 2005). "Projects: Europa Mission Campaign". The Planetary Society. Retrieved 2011-08-08

[79] Nancy Atkinson (2009). "Europa Capable of Supporting Life, Scientist Says". Universe Today. Retrieved 2011-08-18.

[80] Phil Plait, "Huge lakes of water may exist under Europa's ice", "Bad Astronomy Blog"

[81] "SCIENTISTS FIND EVIDENCE FOR "GREAT LAKE" ON EUROPA AND POTENTIAL NEW HABITAT FOR LIFE"

[82] Cook, Jia-Rui c. (December 11, 2013). "Clay-Like Minerals Found on Icy Crust of Europa". *NASA*. Retrieved December 11, 2013.

[83] Wall, Mike (5 March 2014). "NASA hopes to launch ambitious mission to icy Jupiter moon". *Space.com*. Retrieved 2014-04-15.

[84] Clark, Stephen (14 March 2014). "Economics, water plumes to drive Europa mission study". *Spaceflight Now*. Retrieved 2014-04-15.

[85] Coustenis, A.; et al. (March 2009). "TandEM: Titan and Enceladus mission". *Experimental Astronomy* **23** (3): 893–946. Bibcode:2009ExA....23..893C. doi:10.1007/s10686-008-9103-z.

[86] Lovett, Richard A. (31 May 2011). "Enceladus named sweetest spot for alien life". *Nature* (Nature). doi:10.1038/news.2011.337. Retrieved 2011-06-03.

[87] SPACE.com – Scientists Reconsider Habitability of Saturn's Moon

[88] SPACE.com – Lakes Found on Saturn's Moon Titan

[89] "Lakes on Titan, Full-Res: PIA08630". 2006-07-24.

[90] "What is Consuming Hydrogen and Acetylene on Titan?". NASA/JPL. 2010. Archived from the original on 6 June 2010. Retrieved 2010-06-06.

[91] Darrell F. Strobel (2010). "Molecular hydrogen in Titan's atmosphere: Implications of the measured tropospheric and thermospheric mole fractions". *Icarus* **208** (2): 878. Bibcode:2010Icar..208..878S. doi:10.1016/j.icarus.2010.03.003.

[92] McKay, C. P.; Smith, H. D. (2005). "Possibilities for methanogenic life in liquid methane on the surface of Titan". *Icarus* **178** (1): 274–276. Bibcode:2005Icar..178..274M. doi:10.1016/j.icarus.2005.05.018.

[93] Hoyle, Fred, *Evolution from Space*, Omni Lecture, Royal Institution, London, 12 January 1982; *Evolution from Space* (1982) pp. 27–28 ISBN 0-89490-083-8; *Evolution from Space: A Theory of Cosmic Creationism* (1984) ISBN 0-671-49263-2

[94] Hoyle, Fred (1985). *Living Comets*. Cardiff: University College, Cardiff Press.

[95] Wickramasinghe, Chandra (June 2011). "Viva Panspermia". *The Observatory*.

[96] Wesson, P (2010). "Panspermia, Past and Present: Astrophysical and Biophysical Conditions for the Dissemination of Life in Space". *Sp. Sci.Rev.* 1–4 **156**: 239–252. arXiv:1011.0101. Bibcode:2010SSRv..156..239W. doi:10.1007/s11214-010-9671-x.

[97] Crenson, Matt (6 August 2006). "Experts: Little Evidence of Life on Mars". Associated Press (on discovery.com). Archived from the original on 16 April 2011. Retrieved 8 March 2011.

[98] McKay DS, Gibson EK, ThomasKeprta KL, Vali H, Romanek CS, Clemett SJ, Chillier XDF, Maechling CR, Zare RN (1996). "Search for past life on Mars: Possible relic biogenic activity in Martian meteorite ALH84001". *Science* **273** (5277): 924–930. Bibcode:1996Sci...273..924M. doi:10.1126/science.273.5277.924. PMID 8688069.

[99] McKay DS, Thomas-Keprta KL, Clemett, SJ, Gibson, EK Jr, Spencer L, Wentworth SJ (2009). Hoover, Richard B; Levin, Gilbert V; Rozanov, Alexei Y; Retherford, Kurt D, eds. "Life on Mars: new evidence from martian meteorites". *Proc. SPIE.* Proceedings of SPIE **7441** (1): 744102. doi:10.1117/12.832317. Retrieved 8 March 2011.

[100] Webster, Guy (27 February 2014). "NASA Scientists Find Evidence of Water in Meteorite, Reviving Debate Over Life on Mars". *NASA.* Retrieved 27 February 2014.

[101] White, Lauren M.; Gibson, Everett K.; Thomnas-Keprta, Kathie L.; Clemett, Simon J.; McKay, David (19 February 2014). "Putative Indigenous Carbon-Bearing Alteration Features in Martian Meteorite Yamato 000593". *Astrobiology* **14** (2): 170–181. Bibcode:2014AsBio..14..170W. doi:10.1089/ast.2011.0733. Retrieved 27 February 2014.

[102] Gannon, Megan (28 February 2014). "Mars Meteorite with Odd 'Tunnels' & 'Spheres' Revives Debate Over Ancient Martian Life". *Space.com.* Retrieved 28 February 2014.

[103] Chambers, Paul (1999). *Life on Mars; The Complete Story.* London: Blandford. ISBN 0-7137-2747-0.

[104] Klein, Harold P.; Levin, Gilbert V.; Levin, Gilbert V.; Oyama, Vance I.; Lederberg, Joshua; Rich, Alexander; Hubbard, Jerry S.; Hobby, George L.; Straat, Patricia A.; Berdahl, Bonnie J.; Carle, Glenn C.; Brown, Frederick S.; Johnson, Richard D. (1976-10-01). "The Viking Biological Investigation: Preliminary Results". *Science* **194** (4260): 99–105. Bibcode:1976Sci...194...99K. doi:10.1126/science.194.4260.99. PMID 17793090. Retrieved 2008-08-15.

[105] Beegle, Luther W.; Wilson, Michael G.; Abilleira, Fernando; Jordan, James F.; Wilson, Gregory R. (August 2007). "A Concept for NASA's Mars 2016 Astrobiology Field Laboratory". *Astrobiology* **7** (4): 545–577. Bibcode:2007AsBio...7..545B. doi:10.1089/ast.2007.0153. PMID 17723090. Retrieved 2009-07-20.

[106] "ExoMars rover". ESA. Retrieved 2014-04-14.

[107] Berger, Brian (2005). "Exclusive: NASA Researchers Claim Evidence of Present Life on Mars".

[108] "NASA denies Mars life reports". spacetoday.net. 2005.

[109] Spotts, Peter N. (2005-02-28). "Sea boosts hope of finding signs of life on Mars". The Christian Science Monitor. Retrieved 2006-12-18.

[110] Chow, Dennis (22 July 2011). "NASA's Next Mars Rover to Land at Huge Gale Crater". Space.com. Retrieved 2011-07-22.

[111] Amos, Jonathan (22 July 2011). "Mars rover aims for deep crater". *BBC News.* Retrieved 2011-07-22.

[112] Glaser, Linda (January 27, 2015). "Introducing: The Carl Sagan Institute". Retrieved 2015-05-11.

[113] "Carl Sagan Institute - Research". May 2015. Retrieved 2015-05-11.

[114] Cofield, Calla (30 March 2015). "Catalog of Earth Microbes Could Help Find Alien Life". *Space.com.* Retrieved 2015-05-11.

[115] Callahan, M.P.; Smith, K.E.; Cleaves, H.J.; Ruzica, J.; Stern, J.C.; Glavin, D.P.; House, C.H.; Dworkin, J.P. (11 August 2011). "Carbonaceous meteorites contain a wide range of extraterrestrial nucleobases". PNAS. doi:10.1073/pnas.1106493108. Retrieved 2011-08-15.

[116] Steigerwald, John (8 August 2011). "NASA Researchers: DNA Building Blocks Can Be Made in Space". NASA. Retrieved 2011-08-10.

[117] ScienceDaily Staff (9 August 2011). "DNA Building Blocks Can Be Made in Space, NASA Evidence Suggests". ScienceDaily. Retrieved 2011-08-09.

[118] Chow, Denise (26 October 2011). "Discovery: Cosmic Dust Contains Organic Matter from Stars". Space.com. Retrieved 2011-10-26.

[119] ScienceDaily Staff (26 October 2011). "Astronomers Discover Complex Organic Matter Exists Throughout the Universe". ScienceDaily. Retrieved 2011-10-27.

[120] Kwok, Sun; Zhang, Yong (26 October 2011). "Mixed aromatic–aliphatic organic nanoparticles as carriers of unidentified infrared emission features". *Nature* **479** (7371): 80–3. Bibcode:2011Natur.479...80K. doi:10.1038/nature10542. PMID 22031328.

[121] Than, Ker (August 29, 2012). "Sugar Found In Space". *National Geographic.* Retrieved August 31, 2012.

[122] Staff (August 29, 2012). "Sweet! Astronomers spot sugar molecule near star". Associated Press. Retrieved August 31, 2012.

[123] Jørgensen, J. K.; Favre, C.; Bisschop, S.; Bourke, T.; Dishoeck, E.; Schmalzl, M. (2012). "Detection of the simplest sugar, glycolaldehyde, in a solar-type protostar with ALMA" (PDF). eprint.

[124] Schenkel, Peter (May 2006). "SETI Requires a Skeptical Reappraisal". Skeptical Inquirer. Retrieved June 28, 2009.

[125] Moldwin, Mark (November 2004). "Why SETI is science and UFOlogy is not". Skeptical Inquirer.

[126] "The Search for Extraterrestrial Intelligence (SETI) in the Optical Spectrum". The Columbus Optical SETI Observatory.

[127] Whitmire, Daniel P.; Wright, David P. (April 1980). "Nuclear waste spectrum as evidence of technological extraterrestrial civilizations". *Icarus* **42** (1): 149–156. Bibcode:1980Icar...42..149W. doi:10.1016/0019-1035(80)90253-5.

[128] "Discovery of OGLE 2005-BLG-390Lb, the first cool rocky/icy exoplanet". *IAP.fr*. 25 January 2006.

[129] SPACE.com – Major Discovery: New Planet Could Harbor Water and Life

[130] Schneider, Jean (10 September 2011). "Interactive Extra-solar Planets Catalog". *The Extrasolar Planets Encyclopaedia*. Retrieved 2012-01-30.

[131] Mike Wall, "NASA's Kepler Observatory to continue hunt for strange new worlds", "The Christian Science Monitor"

[132] "NASA – Kepler". Retrieved 4 November 2013.

[133] Harrington, J. D.; Johnson, M. (4 November 2013). "NASA Kepler Results Usher in a New Era of Astronomy".

[134] Tenenbaum, P.; Jenkins, J. M.; Seader, S.; Burke, C. J.; Christiansen, J. L.; Rowe, J. F.; Caldwell, D. A.; Clarke, B. D.; Li, J.; Quintana, E. V.; Smith, J. C.; Thompson, S. E.; Twicken, J. D.; Borucki, W. J.; Batalha, N. M.; Cote, M. T.; Haas, M. R.; Hunter, R. C.; Sanderfer, D. T.; Girouard, F. R.; Hall, J. R.; Ibrahim, K.; Klaus, T. C.; McCauliff, S. D.; Middour, C. K.; Sabale, A.; Uddin, A. K.; Wohler, B.; Barclay, T.; Still, M. (2013). "Detection of Potential Transit Signals in the First 12 Quarters of *Kepler* Mission Data". *The Astrophysical Journal Supplement Series* **206**: 5. arXiv:1212.2915. Bibcode:2013ApJS..206....5T. doi:10.1088/0067-0049/206/1/5.

[135] "My God, it's full of planets! They should have sent a poet." (Press release). Planetary Habitability Laboratory, University of Puerto Rico at Arecibo. 3 January 2012.

[136] Santerne, A.; Díaz, R. F.; Almenara, J.-M.; Lethuillier, A.; Deleuil, M.; Moutou, C. (2013). "Astrophysical false positives in exoplanet transit surveys: Why do we need bright stars?". arXiv:1310.2133 [astro-ph.EP].

[137] Cassan, A.; et al. (January 11, 2012). "One or more bound planets per Milky Way star from microlensing observations". *Nature* **481** (7380): 167–169. arXiv:1202.0903. Bibcode:2012Natur.481..167C. doi:10.1038/nature10684. PMID 22237108.

[138] Sanders, R. (4 November 2013). "Astronomers answer key question: How common are habitable planets?". *newscenter.berkeley.edu*.

[139] Petigura, E. A.; Howard, A. W.; Marcy, G. W. (2013). "Prevalence of Earth-size planets orbiting Sun-like stars". *Proceedings of the National Academy of Sciences***110**(48): 19273. arXiv:1311.6806. Bibcode:2013PNAS..11019273P.doi:10.1073/pnas.1319

[140] Strigari, L. E.; Barnabè, M.; Marshall, P. J.; Blandford, R. D. (2012). "Nomads of the Galaxy". *Monthly Notices of the Royal Astronomical Society* **423** (2): 1856–1865. arXiv:1201.2687. Bibcode:2012MNRAS.423.1856S. doi:10.1111/j.1365-2966.2012.21009.x. estimates 700 objects >10^{-6} solar masses (roughly the mass of Mars) per main-sequence star between 0.08 and 1 Solar mass, of which there are billions in the Milky Way.

[141] Dumusque, X.; Pepe, F.; Lovis, C.; Ségransan, D.; Sahlmann, J.; Benz, W.; Bouchy, F.; Mayor, M.; et al. (17 October 2012). "An Earth mass planet orbiting Alpha Centauri B" (PDF). *Nature* **490** (7423): 207–11. Bibcode:2012Natur.491..207D. doi:10.1038/nature11572. PMID 23075844. Retrieved 17 October 2012.

[142] "DENIS-P J082303.1-491201 b". *Caltech*. Retrieved 8 March 2014.

[143] Sahlmann, J.; Lazorenko, P. F.; Ségransan, D.; Martín, E. L.; Queloz, D.; Mayor, M.; Udry, S. (August 2013). "Astrometric orbit of a low-mass companion to an ultracool dwarf". *Harvard University* **556**: 133. arXiv:1306.3225. Bibcode:2013A&A...55 .doi:10.1051/0004-6361/201321871.

[144] "Chapter 3 — Philosophy: "Solving the Drake Equation". SETI League. December 2002. Retrieved July 24, 2015.

[145] Burchell, M.J. (2006). "W(h)ither the Drake equation?".*International Journal of Astrobiology***5**(3): 243–250. Bibcode:2006IJ doi:10.1017/S1473550406003107.

[146] Aguirre, L. (1 July 2008). "The Drake Equation". *Nova ScienceNow*. PBS. Retrieved 2010-03-07.

[147] Jack Cohen and Ian Stewart (2002). *Evolving the Alien*. John Wiley and Sons, Inc., Hoboken, NJ. Chapter 6, *What does a Martian look like?*

[148] Marcy, G.; Butler, R.; Fischer, D.; et al. (2005). "Observed Properties of Exoplanets: Masses, Orbits and Metallicities". *Progress of Theoretical Physics Supplement***158**: 24–42. arXiv:astro-ph/0505003. Bibcode:2005PThPS.158...24M.doi:10.114

[149] Overbye, Dennis (August 3, 2015). "The Flip Side of Optimism About Life on Other Planets". *New York Times*. Retrieved October 29, 2015.

[150] *Who discovered that the Sun was a Star?* Stanford Solar Center.

[151] Michael J. Crowe (1999). *The Extraterrestrial Life Debate, 1750–1900*. Courier Dover Publications. ISBN 0-486-40675-X.

[152] Wiker, Benjamin D. (November 4, 2002). "Alien Ideas: Christianity and the Search for Extraterrestrial Life". *Crisis Magazine*. Archived from the original on February 10, 2003.

[153] Irwin, Robert (2003). *The Arabian Nights: A Companion*. Tauris Parke Paperbacks. p. 204 & 209. ISBN 1-86064-983-1.

[154] Conversations on the Plurality of Worlds— Bernard le Bovier de Fontenelle

[155] Flammarion, (Nicolas) Camille (1842–1925)— The Internet Encyclopedia of Science

[156] "Giordano Bruno: On the Infinite Universe and Worlds (De l'Infinito Universo et Mondi) Introductory Epistle: Argument of the Third Dialogue". Retrieved 4 October 2014.

[157] "Rheita.htm". cosmovisions.com.

[158] Evans, J. E. and Maunder, E. W. (1903) "Experiments as to the Actuality of the 'Canals' observed on Mars", MNRAS, **63** (1903) 488

[159] *Is Mars habitable? A critical examination of Professor Percival Lowell's book "Mars and its canals."*, an alternative explanation, by Alfred Russel Wallace, F.R.S., etc. *London, Macmillan and co., 1907.*

[160] Chambers, Paul (1999). *Life on Mars; The Complete Story*. London: Blandford. ISBN 0-7137-2747-0.

[161] Cross, Anne (2004). "The Flexibility of Scientific Rhetoric: A Case Study of UFO Researchers". *Qualitiative Sociology* **27** (1): 3–34. doi:10.1023/B:QUAS.0000015542.28438.41.

[162] Ailleris, Philippe (January–February 2011). "The lure of local SETI: Fifty years of field experiments". *Acta Astronautica* **68** (1–2): 2–15. Bibcode:2011AcAau..68....2A. doi:10.1016/j.actaastro.2009.12.011.

[163] "LECTURE 4: MODERN THOUGHTS ON EXTRATERRESTRIAL LIFE". *The University of Antarctica*. Retrieved 2015-07-25.

[164] Amazon.com: Rare Earth: Why Complex Life is Uncommon in the Universe: Books: Peter Ward, Donald Brownlee

[165] "Hawking warns over alien beings". *BBC News*. 2010-04-25. Retrieved 2010-05-02.

[166] Diamond, Jared. "The Third Chimpanzee", Harper Perennial, 2006, Chapter 12.

[167] Larson, Phil (5 November 2011). "Searching for ET, But No Evidence Yet". White House. Retrieved 2011-11-06.

[168] Atkinson, Nancy (5 November 2011). "No Alien Visits or UFO Coverups, White House Says". UniverseToday. Retrieved 2011-11-06.

[169] Staff (May 3, 2013). "Special Issue: Exoplanets". *Science*. Retrieved May 18, 2013.

[170] Chang, Kenneth (17 April 2014). "Scientists Find an 'Earth Twin', or Maybe a Cousin". *New York Times*.

[171] Borenstein, Seth (of AP News) (13 February 2015). "Should We Call the Cosmos Seeking ET? Or Is That Risky?". *New York Times*. Retrieved 14 February 2015.

[172] Ghosh, Pallab (12 February 2015). "Scientist: 'Try to contact aliens'". *BBC News*. Retrieved 12 February 2015.

[173] Various (13 February 2015). "Statement – Regarding Messaging To Extraterrestrial Intelligence (METI) / Active Searches For Extraterrestrial Intelligence (Active SETI)". *University of California, Berkeley*. Retrieved 14 February 2015.

[174] Katz, Gregory (20 July 2015). "Searching for ET: Hawking to look for extraterrestrial life". *AP News*. Retrieved 20 July 2015.

13.10 Further reading

- Baird, John C. (1987). *The Inner Limits of Outer Space: A Psychologist Critiques Our Efforts to Communicate With Extraterrestrial Beings*. Hanover: University Press of New England. ISBN 0-87451-406-1.
- Cohen, Jack; Stewart, Ian (2002). *Evolving the Alien: The Science of Extraterrestrial Life*. Ebury Press. ISBN 0-09-187927-2.
- Crowe, Michael J. (1986). *The Extraterrestrial Life Debate, 1750–1900*. Cambridge. ISBN 0-521-26305-0.
- Crowe, Michael J. (2008). *The extraterrestrial life debate Antiquity to 1915: A Source Book*. University of Notre Dame Press. ISBN 0-268-02368-9.
- Dick, Steven J. (1984). *Plurality of Worlds: The Extraterrestrial Life Debate from Democratis to Kant*. Cambridge.
- Dick, Steven J. (1996). *The Biological Universe: The Twentieth Century Extraterrestrial Life Debate and the Limits of Science*. Cambridge. ISBN 0-521-34326-7.
- Dick, Steven J. (2001). *Life on Other Worlds: The 20th Century Extraterrestrial Life Debate*. Cambridge. ISBN 0-521-79912-0.
- Dick, Steven J.; Strick, James E. (2004). *The Living Universe: NASA And the Development of Astrobiology*. Rutgers. ISBN 0-8135-3447-X.
- Fasan, Ernst (1970). *Relations with alien intelligences – the scientific basis of metalaw*. Berlin: Berlin Verlag.
- Goldsmith, Donald (1997). *The Hunt for Life on Mars*. New York: A Dutton Book. ISBN 0-525-94336-6.
- Grinspoon, David (2003). *Lonely Planets: The Natural Philosophy of Alien Life*. HarperCollins. ISBN 0-06-018540-6.
- Lemnick, Michael T. (1998). *Other Worlds: The Search for Life in the Universe*. New York: A Touchstone Book.
- Michaud, Michael (2006). *Contact with Alien Civilizations – Our Hopes and Fears about Encountering Extraterrestrials*. Berlin: Springer. ISBN 0-387-28598-9.
- Pickover, Cliff (2003). *The Science of Aliens*. New York: Basic Books. ISBN 0-465-07315-8.
- Roth, Christopher F. (2005). Debbora Battaglia, ed. *Ufology as Anthropology: Race, Extraterrestrials, and the Occult. E.T. Culture: Anthropology in Outerspaces* (Durham, NC: Duke University Press).
- Sagan, Carl; Shklovskii, I. S. (1966). *Intelligent Life in the Universe*. Random House.
- Sagan, Carl (1973). *Communication with Extraterrestrial Intelligence*. MIT Press. ISBN 0-262-19106-7.
- Ward, Peter D. (2005). *Life as we do not know it-the NASA search for (and synthesis of) alien life*. New York: Viking. ISBN 0-670-03458-4.
- Tumminia, Diana G. (2007). *Alien Worlds – Social and Religious Dimensions of Extraterrestrial Contact*. Syracuse: Syracuse University Press. ISBN 978-0-8156-0858-5.

13.11 External links

- "Is it true that there could be intelligent life out there?". physics.org. Retrieved 2 November 2012.
- Minerals and the Origins of Life (Robert Hazen, NASA) (video, 60m, April 2014).
- "Search for Life in the Universe" (NASA) (video, 87m, July 14, 2014).

Some major international efforts to search for extraterrestrial life. Clockwise from top left:
1 The search for extrasolar planets
(image: Kepler *telescope)*
2 Listening for extraterrestrial signals
indicating intelligence (image: Allen array)
3 Robotic exploration of the Solar System
(image: Curiosity *rover on Mars)*

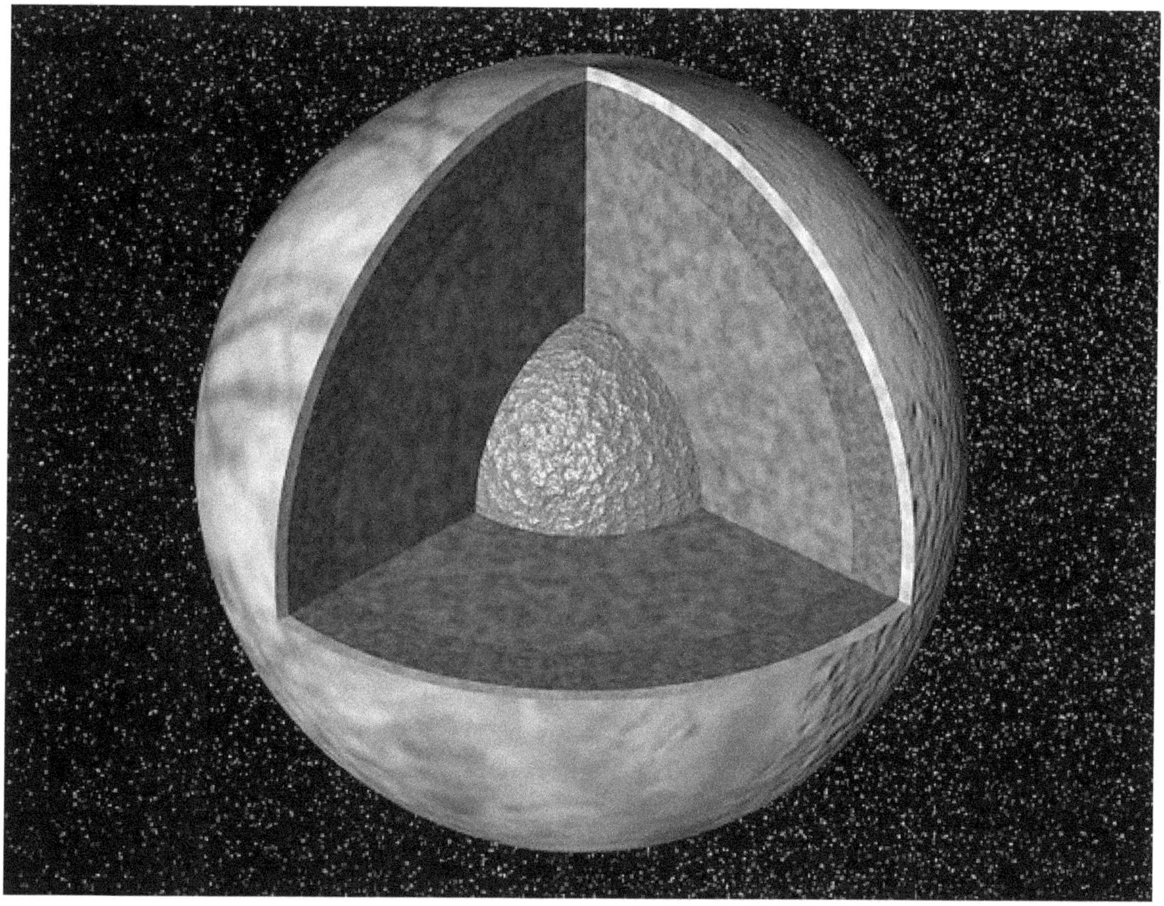

Subsurface oceans such as the one pictured of Europa could possibly harbor life.[77]

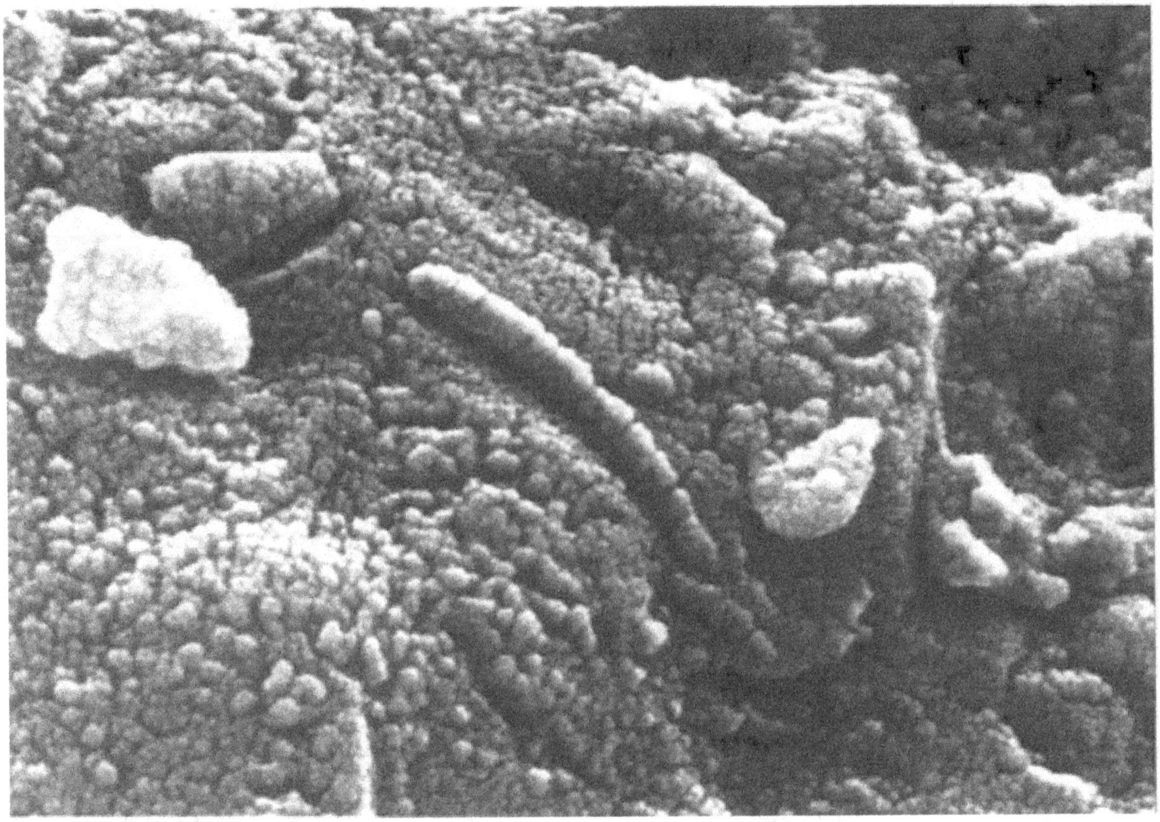

Electron micrograph of martian meteorite ALH84001 showing structures that some scientists think could be fossilized bacteria-like life forms.

Artist's Impression of Gliese 581 c, the first terrestrial extrasolar planet discovered within its star's habitable zone.

Artist's impression of the Kepler telescope in space.

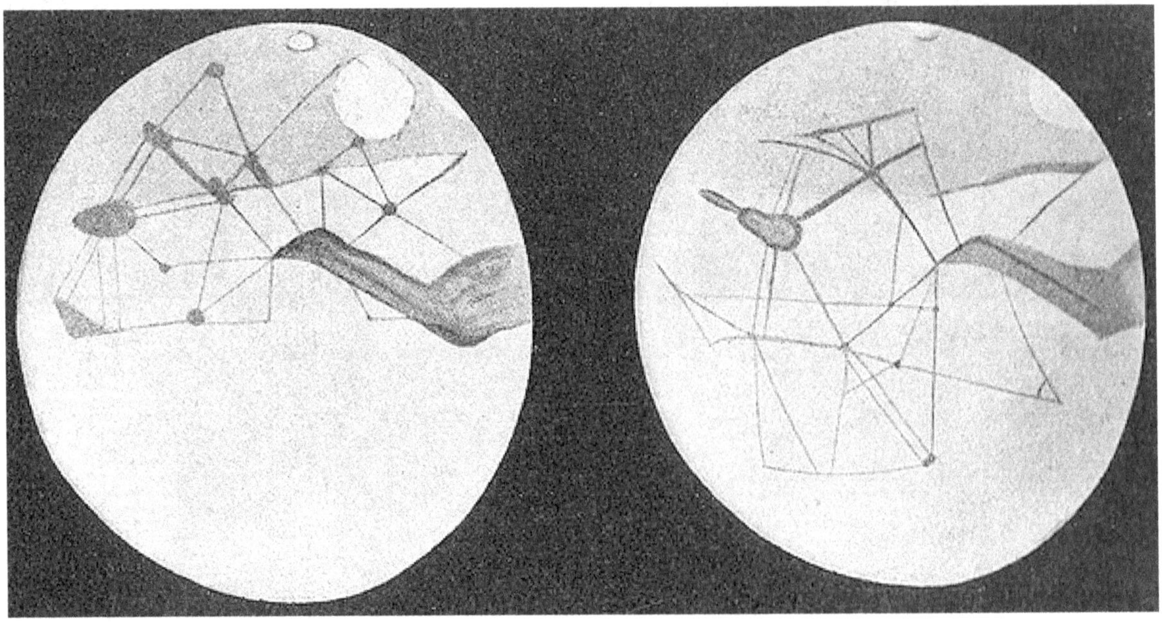

Artificial Martian channels, depicted by Percival Lowell

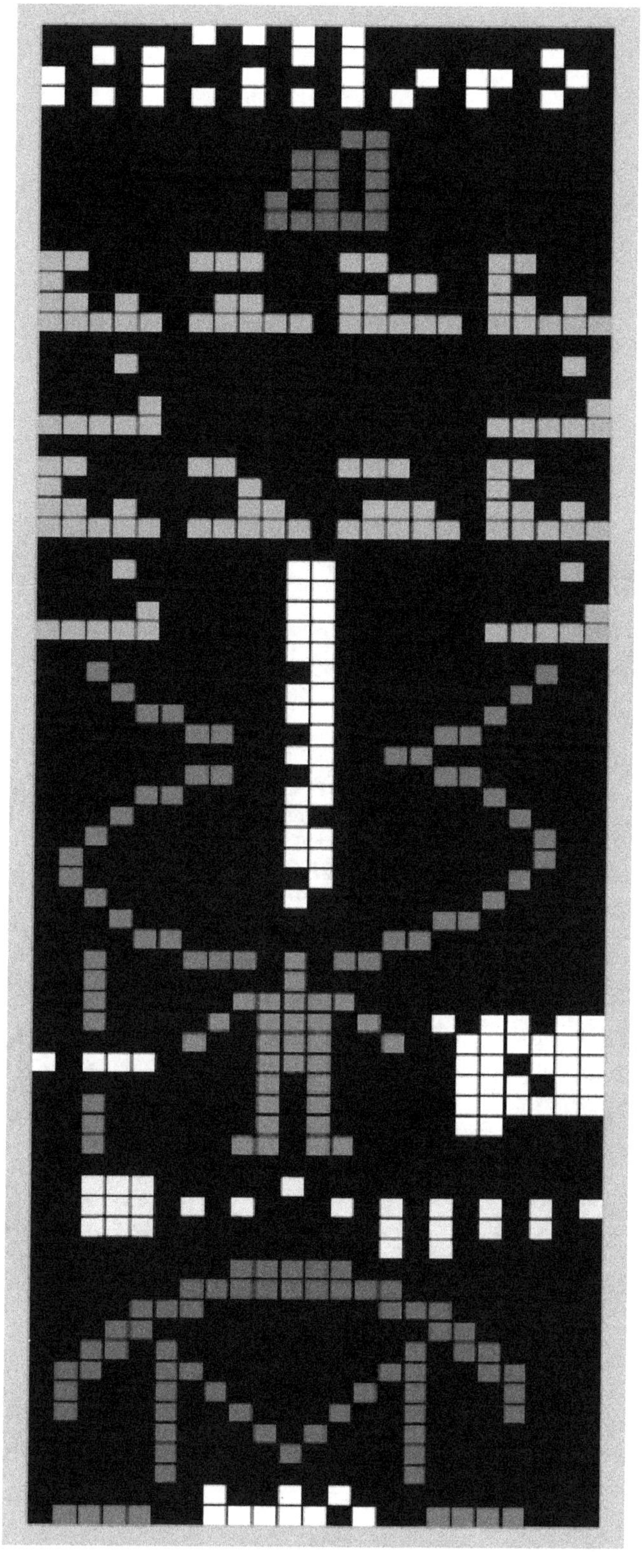

The Arecibo message is a digital message sent to globular star cluster M13, and is a well-known symbol of human attempts to contact extraterrestrials.

Chapter 14

Extremophile

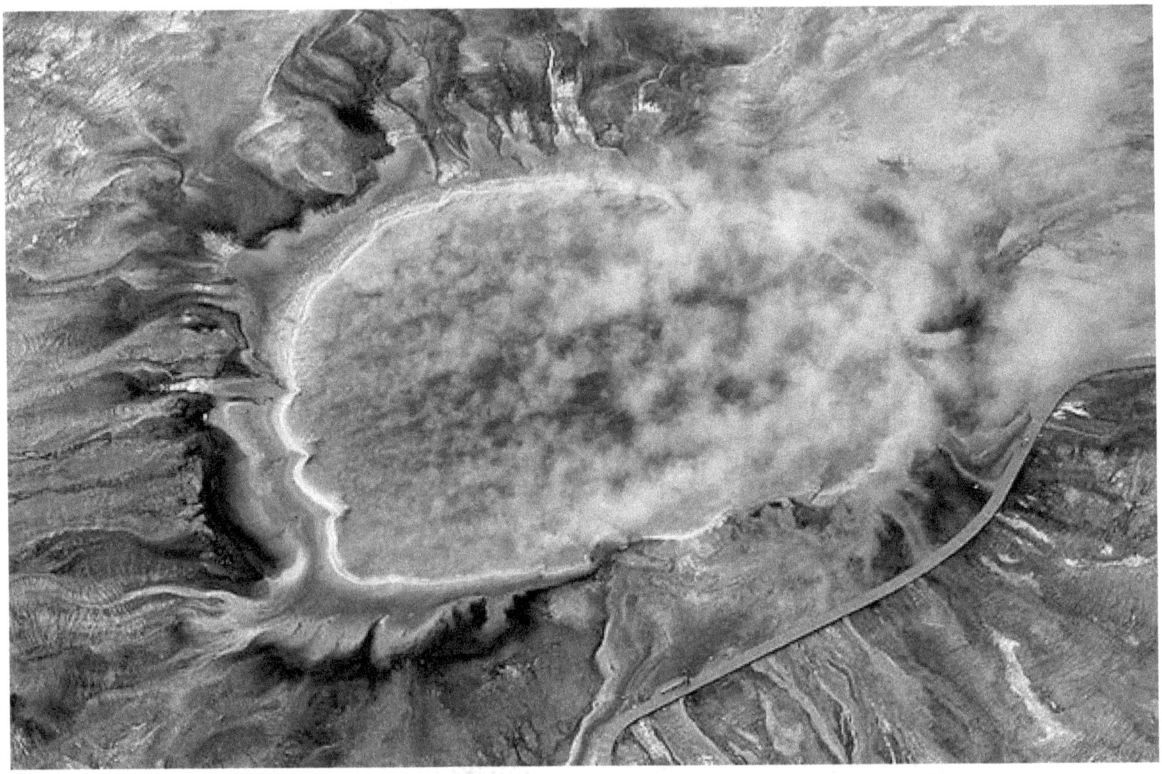

Thermophiles, a type of extremophile, produce some of the bright colors of Grand Prismatic Spring, Yellowstone National Park

An **extremophile** (from Latin *extremus* meaning "extreme" and Greek *philiā* (φιλία) meaning "love") is an organism that thrives in physically or geochemically extreme conditions that are detrimental to most life on Earth.[1][2] In contrast, organisms that live in more moderate environments may be termed mesophiles or neutrophiles.

14.1 Characteristics

In the 1980s and 1990s, biologists found that microbial life has an amazing flexibility for surviving in extreme environments — niches that are extraordinarily hot, or acidic, for example — that would be completely inhospitable to complex organisms. Some scientists even concluded that life may have begun on Earth in hydrothermal vents far under the ocean's surface.[3] According to astrophysicist Dr. Steinn Sigurdsson, "There are viable bacterial spores that have been found that are 40 million years old on Earth — and we know they're very hardened to radiation."[4] On 6 February 2013, scientists reported that bacteria were found living in the cold and dark in a lake buried a half-mile deep under the ice in Antarctica.[5] On 17 March 2013, researchers reported data that suggested microbial life forms thrive in the Mariana Trench, the deepest spot on the Earth.[6][7] Other researchers reported related studies that microbes thrive inside rocks up to 1900 feet below the sea floor under 8500 feet of ocean off the coast of the northwestern United States.[6][8] According to one of the researchers,"You can find microbes everywhere — they're extremely adaptable to conditions, and survive wherever they are."[6]

14.2 Morphology

Most known extremophiles are microbes. The domain Archaea contains renowned examples, but extremophiles are present in numerous and diverse genetic lineages of bacteria and archaeans. Furthermore, it is erroneous to use the term extremophile to encompass all archaeans, as some are mesophilic. Neither are all extremophiles unicellular; protostome animals found in similar environments include the Pompeii worm, the psychrophilic Grylloblattidae (insects) and Antarctic krill (a crustacean). Many would also classify tardigrades (water bears) as extremophiles but while tardigrades can survive in extreme environments, they are not considered extremophiles because they are not adapted to live in these conditions. Their chances of dying increase the longer they are exposed to the extreme environment.

14.3 Classifications

There are many classes of extremophiles that range all around the globe, each corresponding to the way its environmental niche differs from mesophilic conditions. These classifications are not exclusive. Many extremophiles fall under multiple categories and termed as polyextremophiles. For example, organisms living inside hot rocks deep under Earth's surface are thermophilic and barophilic such as *Thermococcus barophilus*.[9] A polyextremophile living at the summit of a mountain in the Atacama Desert might be a radioresistant xerophile, a psychrophile, and an oligotroph. Polyextremophiles are well known for their ability to tolerate both high and low pH levels.

14.3.1 Terms

Acidophile An organism with optimal growth at pH levels of 3 or below

Alkaliphile An organism with optimal growth at pH levels of 9 or above

Anaerobe An organism that does not require oxygen for growth such as *Spinoloricus Cinzia*. Two sub-types exist: facultative anaerobe and obligate anaerobe. A facultative anaerobe can tolerate anaerobic and aerobic conditions; however, an obligate anaerobe would die in presence of even trace levels of oxygen.

Cryptoendolith An organism that lives in microscopic spaces within rocks, such as pores between aggregate grains; these may also be called Endolith, a term that also includes organisms populating fissures, aquifers, and faults filled with groundwater in the deep subsurface.

Halophile An organism requiring at least 0.2M concentrations of salt (NaCl) for growth[10]

Hyperthermophile An organism that can thrive at temperatures between 80–122 °C, such as those found in hydrothermal systems

Hypolith An organism that lives underneath rocks in cold deserts

Lithoautotroph An organism (usually bacteria) whose sole source of carbon is carbon dioxide and exergonic inorganic oxidation (chemolithotrophs) such as *Nitrosomonas europaea*; these organisms are capable of deriving energy from reduced mineral compounds like pyrites, and are active in geochemical cycling and the weathering of parent bedrock to form soil

Metallotolerant capable of tolerating high levels of dissolved heavy metals in solution, such as copper, cadmium, arsenic, and zinc; examples include *Ferroplasma sp., Cupriavidus metallidurans* and GFAJ-1.[11][12][13]

Oligotroph An organism capable of growth in nutritionally limited environments

Osmophile An organism capable of growth in environments with a high sugar concentration

Piezophile (Also referred to as barophile). An organism that lives optimally at high pressures such as those deep in the ocean or underground;[14] common in the deep terrestrial subsurface, as well as in oceanic trenches

Polyextremophile A **polyextremophile** (faux Ancient Latin/Greek for 'affection for many extremes') is an organism that qualifies as an extremophile under more than one category.

Psychrophile/Cryophile An organism capable of survival, growth or reproduction at temperatures of -15 °C or lower for extended periods; common in cold soils, permafrost, polar ice, cold ocean water, and in or under alpine snowpack

Radioresistant Organisms resistant to high levels of ionizing radiation, most commonly ultraviolet radiation, but also including organisms capable of resisting nuclear radiation

Thermophile An organism that can thrive at temperatures between 45–122 °C

Thermoacidophile Combination of thermophile and acidophile that prefer temperatures of 70–80 °C and pH between 2 and 3

Xerophile An organism that can grow in extremely dry, desiccating conditions; this type is exemplified by the soil microbes of the Atacama Desert

14.4 In astrobiology

Astrobiology is the field concerned with forming theories, such as panspermia, about the distribution, nature, and future of life in the universe. In it, microbial ecologists, astronomers, planetary scientists, geochemists, philosophers, and explorers cooperate constructively to guide the search for life on other planets. Astrobiologists are particularly interested in studying extremophiles, as many organisms of this type are capable of surviving in environments similar to those known to exist on other planets. For example, Mars may have regions in its deep subsurface permafrost that could harbor endolith communities. The subsurface water ocean of Jupiter's moon Europa may harbor life, especially at hypothesized hydrothermal vents at the ocean floor.

Recent research carried out on extremophiles in Japan involved a variety of bacteria including *Escherichia coli* and *Paracoccus denitrificans* being subject to conditions of extreme gravity. The bacteria were cultivated while being rotated in an ultracentrifuge at high speeds corresponding to 403,627 g (i.e. 403,627 times the gravity experienced on Earth). *Paracoccus denitrificans* was one of the bacteria which displayed not only survival but also robust cellular growth under these conditions of hyperacceleration which are usually found only in cosmic environments, such as on very massive stars or in the shock waves of supernovas. Analysis showed that the small size of prokaryotic cells is essential for successful growth under hypergravity. The research has implications on the feasibility of panspermia.[15][16]

On 26 April 2012, scientists reported that lichen survived and showed remarkable results on the adaptation capacity of photosynthetic activity within the simulation time of 34 days under Martian conditions in the Mars Simulation Laboratory (MSL) maintained by the German Aerospace Center (DLR).[17][18]

On 29 April 2013, scientists at Rensselaer Polytechnic Institute, funded by NASA, reported that, during spaceflight on the International Space Station, microbes seem to adapt to the space environment in ways "not observed on Earth" and in ways that "can lead to increases in growth and virulence".[19]

On 19 May 2014, scientists announced that numerous microbes, like *Tersicoccus phoenicis*, may be resistant to methods usually used in spacecraft assembly clean rooms. It's not currently known if such resistant microbes could have withstood space travel and are present on the *Curiosity* rover now on the planet Mars.[20]

On 20 August 2014, scientists confirmed the existence of microorganisms living half a mile below the ice of Antarctica.[21]

14.5 Examples

New sub-types of -philes are identified frequently and the sub-category list for extremophiles is always growing. For example, microbial life lives in the liquid asphalt lake, Pitch Lake. Research indicates that extremophiles inhabit the asphalt lake in populations ranging between 10^6 to 10^7 cells/gram.[23][24] Likewise, until recently boron tolerance was unknown but a strong borophile was discovered in bacteria. With the recent isolation of *Bacillus boroniphilus*, borophiles came into discussion.[25] Studying these borophiles may help illuminate the mechanisms of both boron toxicity and boron deficiency.

14.6 Industrial uses

The thermoalkaliphilic catalase, which initiates the breakdown of hydrogen peroxide into oxygen and water, was isolated from an organism, *Thermus brockianus*, found in Yellowstone National Park by Idaho National Laboratory researchers. The catalase operates over a temperature range from 30 °C to over 94 °C and a pH range from 6-10. This catalase is extremely stable compared to other catalases at high temperatures and pH. In a comparative study, the *T. brockianus* catalase exhibited a half life of 15 days at 80 °C and pH 10 while a catalase derived from *Aspergillus niger* had a half life of 15 seconds under the same conditions. The catalase will have applications for removal of hydrogen peroxide in industrial processes such as pulp and paper bleaching, textile bleaching, food pasteurization, and surface decontamination of food packaging.[26]

DNA modifying enzymes such as *Taq* DNA polymerase and some *Bacillus* enzymes used in clinical diagnostics and starch liquefaction are produced commercially by several biotechnology companies.[27]

14.7 DNA transfer

Over 65 prokaryotic species are known to be naturally competent for genetic transformation, the ability to transfer DNA from one cell to another cell followed by integration of the donor DNA into the recipient cell's chromosome.[28] Several extremophiles are able to carry out species-specific DNA transfer, as described below. However, it is not yet clear how common such a capability is among extremophiles.

The bacterium *Deinococcus radiodurans* is one of the most radioresistant organisms known. This bacterium can also survive cold, dehydration, vacuum and acid and is thus known as a polyextremophile. *D. radiodurans* is competent to perform genetic transformation.[29] Recipient cells are able to repair DNA damage in donor transforming DNA that had been UV irradiated as efficiently as they repair cellular DNA when the cells themselves are irradiated. The extreme thermophilic bacterium *Thermus thermophilus* and other related *Thermus* species are also capable of genetic transformation.[30]

Halobacterium volcanii, an extreme halophilic (saline tolerant) archaeon, is capable of natural genetic transformation. Cytoplasmic bridges are formed between cells that appear to be used for DNA transfer from one cell to another in either direction.[31]

Sulfolobus solfataricus and *Sulfolobus acidocaldarius* are hyperthermophilic archaea. Exposure of these organisms to the DNA damaging agents UV irradiation, bleomycin or mitomycin C induces species-specific cellular aggregation.[32][33] UV-induced cellular aggregation of *S. acidocaldarius* mediates chromosomal marker exchange with high frequency.[33] Recombination rates exceed those of uninduced cultures by up to three orders of magnitude. Frols et al.[32] and Ajon et al.[33] hypothesized that cellular aggregation enhances species-specific DNA transfer between *Sulfolobus* cells in order to repair damaged DNA by means of homologous recombination. Van Wolferen et al.[34] noted that this DNA exchange process may be crucial under DNA damaging conditions such as high temperatures. It has also been suggested that DNA transfer in *Sulfolobus* may be an early form of sexual interaction similar to the more well-studied bacterial transformation systems that involve species-specific DNA transfer leading to homologous recombinational repair of DNA damage[35] (and see Transformation (genetics)).

Extracellular membrane vesicles (MVs) might be involved in DNA transfer between different hyperthermophilic archaeal species.[36] It has been shown that both plasmids[37] and viral genomes[36] can be transferred via MVs. Notably, a horizontal plasmid transfer has been documented between hyperthermophilic *Thermococcus* and *Methanocaldococcus* species, respectively belonging to the orders *Thermococcales* and *Methanococcales*.[38]

14.8 See also

- *Deinococcus radiodurans*
- Extremotroph
- List of microorganisms tested in outer space
- Tardigrade
- Thermophile

14.9 References

[1] Rampelotto, P. H. (2010). "Resistance of microorganisms to extreme environmental conditions and its contribution to Astrobiology". *Sustainability* **2** (6): 1602–1623. Bibcode:2010Sust....2.1602R. doi:10.3390/su2061602.

[2] Rothschild, L.J.; Mancinelli, R.L. (2001). "Life in extreme environments". *Nature* **409** (6823): 1092–1101. Bibcode:2001Natu doi:10.1038/35059215. PMID 11234023.

[3] "Mars Exploration - Press kit" (PDF). NASA. June 2003. Retrieved 14 July 2009.

[4] BBC Staff (23 August 2011). "Impacts 'more likely' to have spread life from Earth". BBC. Retrieved 24 August 2011.

[5] Gorman, James (6 February 2013). "Bacteria Found Deep Under Antarctic Ice, Scientists Say". *New York Times*. Retrieved 6 February 2013.

[6] Choi, Charles Q. (17 March 2013). "Microbes Thrive in Deepest Spot on Earth". LiveScience. Retrieved 17 March 2013.

[7] Glud, Ronnie; Wenzhöfer, Frank; Middleboe, Mathias; Oguri, Kazumasa; Turnewitsch, Robert; Canfield, Donald E.; Kitazato, Hiroshi (17 March 2013). "High rates of microbial carbon turnover in sediments in the deepest oceanic trench on Earth". *Nature Geoscience* **6** (4): 284–288. Bibcode:2013NatGe...6..284G. doi:10.1038/ngeo1773. Retrieved 17 March 2013.

[8] Oskin, Becky (14 March 2013). "Intraterrestrials: Life Thrives in Ocean Floor". LiveScience. Retrieved 17 March 2013.

[9] Thermococcus barophilus sp. nov., a new barophilic and hyperthermophilic archaeon isolated under high hydrostatic pressure from a deep-sea hydrothermal vent. *IJSEM*, p. 351-359, 49, 1999.

[10] Cavicchioli, R. & Thomas, T. 2000. Extremophiles. In: J. Lederberg. (ed.) Encyclopedia of Microbiology, Second Edition, Vol. 2, pp. 317–337. Academic Press, San Diego.

[11] "Studies refute arsenic bug claim". *BBC News*. 9 July 2012. Retrieved 10 July 2012.

[12] Erb, Tobias J.; Kiefer, Patrick; Hattendorf, Bodo; Günther, Detlef; Vorholt, Julia A. (8 July 2012). "GFAJ-1 Is an Arsenate-Resistant, Phosphate-Dependent Organism". *Science* **337** (6093): 467–70. Bibcode:2012Sci...337..467E. doi:10.1126/science. PMID22773139. Retrieved10 July2012.

[13] Reaves, Marshall Louis; Sinha, Sunita; Rabinowitz, Joshua D.; Kruglyak, Leonid; Redfield, Rosemary J. (8 July 2012). "Absence of Detectable Arsenate in DNA from Arsenate-Grown GFAJ-1 Cells". *Science* **337** (6093): 470–3. arXiv:1201.6643. Bibcode: 2012Sci...337..470R.doi:10.1126/science.1219861. PMC3845625. PMID22773140. Retrieved10July2012.

[14] Dworkin, Martin; Falkow, Stanley (13 July 2006). *The Prokaryotes: Vol. 1: Symbiotic Associations, Biotechnology, Applied Microbiology*. Springer. p. 94. ISBN 978-0-387-25476-0.

[15] Than, Ker (25 April 2011). "Bacteria Grow Under 400,000 Times Earth's Gravity". *National Geographic- Daily News*. National Geographic Society. Retrieved 28 April 2011.

[16] Deguchi, Shigeru; Hirokazu Shimoshige, Mikiko Tsudome, Sada-atsu Mukai, Robert W. Corkery, Susumu Ito, and Koki Horikoshi; Tsudome, M.; Mukai, S.-a.; Corkery, R. W.; Ito, S.; Horikoshi, K. (2011). "Microbial growth at hyperaccelerations up to 403,627 xg". *Proceedings of the National Academy of Sciences* **108** (19): 7997–8002. Bibcode:2011PNAS..108.7997D. doi:10.1073/pnas.1018027108. Retrieved 28 April 2011.

[17] Baldwin, Emily (26 April 2012). "Lichen survives harsh Mars environment". Skymania News. Retrieved 27 April 2012.

[18] de Vera, J.-P.; Kohler, Ulrich (26 April 2012). "The adaptation potential of extremophiles to Martian surface conditions and its implication for the habitability of Mars" (PDF). European Geosciences Union. Retrieved 27 April 2012.

[19] Kim W; et al. (29 April 2013). "Spaceflight Promotes Biofilm Formation by Pseudomonas aeruginosa". *Plos One* **8** (4): e6237. Bibcode:2013PLoSO...862437K. doi:10.1371/journal.pone.0062437. Retrieved 5 July 2013.

[20] Madhusoodanan, Jyoti (19 May 2014). "Microbial stowaways to Mars identified".*Nature (journal)*. doi:10.1038/nature.2014. Retrieved 23 May 2014.

[21] Fox, Douglas (20 August 2014). "Lakes under the ice: Antarctica's secret garden". *Nature (journal)* **512** (7514): 244–246. Bibcode:2014Natur.512..244F. doi:10.1038/512244a. Retrieved 21 August 2014.

[22] Mack, Eric (20 August 2014). "Life Confirmed Under Antarctic Ice; Is Space Next?". *Forbes*. Retrieved 21 August 2014.

[23] Microbial Life Found in Hydrocarbon Lake. *the physics arXiv blog* 15 April 2010.

[24] Schulze-Makuch, Haque, Antonio, Ali, Hosein, Song, Yang, Zaikova, Beckles, Guinan, Lehto, Hallam. Microbial Life in a Liquid Asphalt Desert.

[25] Ahmed, Iftikhar; Yokota, Akira; Fujiwara, Toru (2006). "A novel highly boron tolerant bacterium, Bacillus boroniphilus sp. nov., isolated from soil, that requires boron for its growth". *Extremophiles* **11** (2): 217–224. doi:10.1007/s00792-006-0027-0. PMID 17072687.

[26] "Bioenergy and Industrial Microbiology". *Idaho National Laboratory*. U.S. Department of Energy. Retrieved 3 February 2014.

[27] Anitori, RP (editor) (2012). *Extremophiles: Microbiology and Biotechnology*. Caister Academic Press. ISBN 978-1-904455-98-1.

[28] Johnsborg, O; Eldholm, V; Håvarstein, LS. (2007). "Natural genetic transformation: prevalence, mechanisms and function". *Res Microbiol* **158** (10): 767–78. doi:10.1016/j.resmic.2007.09.004. PMID 17997281.

[29] Moseley, BE; Setlow, JK. (1968). "Transformation in Micrococcus radiodurans and the ultraviolet sensitivity of its transforming DNA". *Proc Natl Acad Sci U S A* **61** (1): 176–83. Bibcode:1968PNAS...61..176M. doi:10.1073/pnas.61.1.176. PMID 5303325.

[30] Koyama, Y; Hoshino, T; Tomizuka, N; Furukawa, K. (1986). "Genetic transformation of the extreme thermophile Thermus thermophilus and of other Thermus spp". *J Bacteriol* **166** (1): 338–40. PMID 3957870.

[31] Rosenshine, I; Tchelet, R; Mevarech, M. (1989). "The mechanism of DNA transfer in the mating system of an archaebacterium". *Science* **245** (4924): 1387–9. Bibcode:1989Sci...245.1387R. doi:10.1126/science.2818746. PMID 2818746.

[32] Fröls, S; Ajon, M; Wagner, M; Teichmann, D; Zolghadr, B; Folea, M; Boekema, EJ; Driessen, AJ; Schleper, C; et al. (2008). "UV-inducible cellular aggregation of the hyperthermophilic archaeon Sulfolobus solfataricus is mediated by pili formation". *Mol Microbiol* **70** (4): 938–52. doi:10.1111/j.1365-2958.2008.06459.x. PMID 18990182.

[33] Ajon, M; Fröls, S; van Wolferen, M; Stoecker, K; Teichmann, D; Driessen, AJ; Grogan, DW; Albers, SV; Schleper, C.; et al. (2011). "UV-inducible DNA exchange in hyperthermophilic archaea mediated by type IV pili". *Mol Microbiol* **82** (4): 807–17. doi:10.1111/j.1365-2958.2011.07861.x. PMID 21999488.

[34] Van Wolferen, M; Ajon, M; Driessen, AJ; Albers, SV. (2013). "How hyperthermophiles adapt to change their lives: DNA exchange in extreme conditions". *Extremophiles* **17** (4): 545–63. doi:10.1007/s00792-013-0552-6. PMID 23712907.

[35] Bernstein H and Bernstein C (2013). Evolutionary Origin and Adaptive Function of Meiosis, Meiosis, Dr. Carol Bernstein (Ed.), ISBN 978-953-51-1197-9, InTech,http://www.intechopen.com/books/meiosis/evolutionary-origin-and-adaptive-function-of

[36] Gaudin M, Krupovic M, Marguet E, Gauliard E, Cvirkaite-Krupovic V, Le Cam E, Oberto J, Forterre P; Krupovic; Marguet; Gauliard; Cvirkaite-Krupovic; Le Cam; Oberto; Forterre (2014). "Extracellular membrane vesicles harbouring viral genomes". *Environ Microbiol* **16** (4): 1167–75. doi:10.1111/1462-2920.12235. PMID 24034793.

[37] Gaudin M, Gauliard E, Schouten S, Houel-Renault L, Lenormand P, Marguet E, Forterre P.; Gauliard; Schouten; Houel-Renault; Lenormand; Marguet; Forterre (2013). "Hyperthermophilic archaea produce membrane vesicles that can transfer DNA". *Environ Microbiol Rep* **5** (1): 109–16. doi:10.1111/j.1758-2229.2012.00348.x. PMID 23757139.

[38] Krupovic M, Gonnet M, Hania WB, Forterre P, Erauso G; Gonnet; Hania; Forterre; Erauso (2013). "Insights into dynamics of mobile genetic elements in hyperthermophilic environments from five new Thermococcus plasmids". *PLoS ONE* **8** (1): e49044. Bibcode:2013PLoSO...849044K. doi:10.1371/journal.pone.0049044. PMC 3543421. PMID 23326305.

14.10 Further reading

- Wilson, Z. E. and Brimble, M. A. (January 2009). "Molecules derived from the extremes of life". *Nat. Prod. Rep.* **26** (1): 44–71. doi:10.1039/b800164m. PMID 19374122.
- Rossi M; et al. (July 2003). "Extremophiles 2002". *J Bacteriol.* **185** (13): 3683–9. doi:10.1128/JB.185.13.3683-3689.2003. PMC 161588. PMID 12813059.
- C.Michael Hogan (2010). "Extremophile". *Encyclopedia of Earth, National Council of Science & the Environment, eds. E,Monosson & C.Cleveland.*
- Joseph Seckbach, et al.: *Polyextremophiles: life under multiple forms of stress.* Springer, Dordrecht 2013, ISBN 978-94-007-6488-0.

14.11 External links

- Extreme Environments - Science Education Resource Center
- Extremophile Research
- Eukaryotes in extreme environments
- The Research Center of Extremophiles
- DaveDarling's Encyclopedia of Astrobiology, Astronomy, and Spaceflight
- The International Society for Extremophiles
- Idaho National Laboratory
- Polyextremophile on David Darling's *Encyclopedia of Astrobiology, Astronomy, and Spaceflight*

Chapter 15

Exobiology Radiation Assembly

EURECA facility deployment in 1992

Exobiology Radiation Assembly (ERA) was an experiment that investigated the biological effects of space radiation, on board the European Retrievable Carrier (EURECA), an unmanned 4.5 tonne satellite with a payload of 15 experiments.[1] It was an astrobiology mission developed by the European Space Agency (ESA).

It was launched 31 July 1992 by STS-46 - *Space Shuttle Atlantis*, and put into an orbit at an altitude of 508 km. EURECA was retrieved on 1 July 1993 by STS-57- *Space Shuttle Endeavour* and returned to Earth for further analysis.

15.1 Objectives

The general goal of the experiment was to study the response of dehydrated and metabolically dormant microorganisms (spores of *Bacillus subtilis*, cells of *Deinococcus radiodurans*, conidial spores of *Aspergillus species*) and cellular constituents (plasmid DNA, proteins, purple membranes, amino acids, urea) to the extremely dehydrating conditions of outer space, in some cases in combination with irradiation by solar UV light.[2]

15.2 Results

The Exobiology Radiation Assembly (ERA) provided information on the exposure of invertebrates, microorganisms and organic molecules to long-term exposure to outer space conditions, such as ultraviolet (UV) radiation, cosmic radiation and vacuum.

Spores of different strains of *Bacillus subtilis* and the *Escherichia coli* plasmid pUC19 were exposed to selected conditions of space (space vacuum and/or defined wavebands and intensities of solar ultraviolet radiation). After the approximately 11 months lasting mission, their responses were studied in terms of survival, mutagenesis in the *his* (*B. subtilis*) or *lac* locus (pUC19), induction of DNA strand breaks, efficiency of DNA repair systems, and the role of external protective agents. The data were compared with those of a simultaneously running ground control experiment:[2][3]

- The survival of spores treated with the vacuum of space, however shielded against solar radiation, is substantially increased, if they are exposed in multilayers and/or in the presence of glucose as protective.
- All spores in "artificial meteorites", i.e. embedded in clays or simulated Martian soil, are killed.
- Vacuum treatment leads to an increase of mutation frequency in spores, but not in plasmid DNA.
- Extraterrestrial solar ultraviolet radiation is mutagenic, induces strand breaks in the DNA and reduces survival substantially.
- Action spectroscopy confirms results of previous space experiments of a synergistic action of space vacuum and solar UV radiation with DNA being the critical target.
- The decrease in viability of the microorganisms could be correlated with the increase in DNA damage.
- The purple membranes, amino acids and urea were not measurably affected by the dehydrating condition of open space, if sheltered from solar radiation. Plasmid DNA, however, suffered a significant amount of strand breaks under these conditions.[2]

15.3 See also

- Bion
- BIOPAN
- Biosatellite program
- EXPOSE
- List of microorganisms tested in outer space
- O/OREOS
- OREOcube

15.4 References

[1] "Exobiology and Radiation Assembly (ERA)". *ESA*. NASA. 1992. Retrieved 2013-07-22.

[2] Dose, K.; Bieger-Dose, A.; Dillmann, R.; Gill, M.; Kerz, O.; Klein, A.; Meinert, H.; Nawroth, T.; et al. (1995). "ERA-experiment 'space biochemistry'". *Advances in Space Research* **16** (8): 119–29. Bibcode:1995AdSpR..16..119D. doi:10.1016/0273-1177(95)00280-R. PMID 11542696.

[3] Horneck, G; Eschweiler, U; Reitz, G; Wehner, J; Willimek, R; Strauch, K (1995). "Biological responses to space: Results of the experiment 'Exobiological Unit' of ERA on EURECA I". *Advances in Space Research* **16** (8): 105–18. Bibcode:1995AdSpR..16..105H. doi:10.1016/0273-1177(95)00279-N. PMID 11542695.

Chapter 16

BIOPAN

BIOPAN is a multi-user research program by the European Space Agency (ESA) designed to investigate the effect of the space environment on biological material.[1][2] The experiments in BIOPAN are exposed to solar and cosmic radiation, the space vacuum and weightlessness, or a selection thereof. Optionally, the experiment temperature can be stabilized. BIOPAN hosts astrobiology, radiobiology and materials science experiments.

The BIOPAN facility is installed on the external surface of Russian Foton descent capsules protruding from the thermal blanket that envelops the satellite.

16.1 Design and features

The BIOPAN program started in the early nineties with an ESA contract for the a joint development by Kayser-Threde and Kayser Italia. It was based on the heritage of a low-tech Russian exposure container called KNA (*Kontejner Nauchnoj Apparatury*). The BIOPAN facilities are installed on the external surface of Foton descent capsules. It has a motor-driven hinged lid, which opens 180° in Earth orbit to expose the experiment samples to the harsh space environment. For re-entry, the closed facility is protected with an Ablative heat shield.

The BIOPAN facilities are equipped with thermometers, UV sensors, a radiometer, a pressure sensor and an active radiation dosimeter. Data acquired by the sensors is stored by BIOPAN throughout each mission and can be accessed after flight.[1] The possibility of overheating during atmospheric re-entry was acknowledged early during the development, therefore, a quite massive heat shield was designed for it. While the total weight of BIOPAN is close to 27 kg, including the experiments, the heat shield is responsible for 12 kg of that figure.[3]

The BIOPAN electronics consists of the following units: signal acquisition board, microcontroller board with its flight software, memory board and EGSE.[1]

16.2 Missions

The missions flown so far are:[3][4]

16.3 See also

- Bion
- Biosatellite program
- EXPOSE

Foton-12 capsule on display

- List of microorganisms tested in outer space
- O/OREOS
- OREOcube
- Tanpopo

16.4 References

[1] "BIOPAN Pan for exposure to space environment". *Kayser Italia*. 2013. Retrieved 2013-07-17.

[2] Charles S. Cockell, Karen Olsson-Francis (23 October 2009). "Experimental methods for studying microbial survival in extraterrestrial environments" (PDF). *Journal of Microbiological Methods* **80** (1): 1–13. doi:10.1016/j.mimet.2009.10.004. PMID 19854226. Retrieved 2013-07-31.

[3] Gerda Horneck and Petra Rettberg (2007). *Complete Course in Astrobiology*. Germany: WILEY-VCH. ISBN 978-3-527-40660-9. Retrieved 2013-07-17.

[4] Surviving the Final Frontier. 25 November 2002.

[5] "Lichen survives in space". *European Space Agency*. 8 November 2005. Retrieved 2013-07-17.

[6] Pascale Ehrenfreunda; Richard Ruiterkampa; Zan Peters; Bernard Foing; Farid Salama; Zita Martins. (March 2007). "The ORGANICS experiment on BIOPAN V: UV and space exposure of aromatic compounds". *Planetary and Space Science* **55** (4): 383–400. Bibcode:2007P&SS...55..383E. doi:10.1016/j.pss.2006.07.001. Retrieved 2013-07-17.

[7] "Foton-M3 experiments return to Earth". *European Space Agency*. 26 September 2007. Retrieved 2013-07-17.

[8] Ledford, Heidi (8 September 2008). "Spacesuits optional for 'water bears'". *Nature* (Nature Publishing Group). doi:10.1038/new Retrieved 13 August 2012. s.2008.1087.

[9] "ESA-space experiments: from BIOPAN 6 experiment "Lithopanspermia" to EXPOSE".

[10] The BIOPAN experiment MARSTOX II of the FOTON M-3 mission July 2008.

Chapter 17

Long Duration Exposure Facility

NASA's **Long Duration Exposure Facility**, or **LDEF**, was a school bus-sized cylindrical facility designed to provide long-term experimental data on the outer space environment and its effects on space systems, materials, operations and selected spore's survival.[1][2] It was placed in low Earth orbit by Space Shuttle *Challenger* in April 1984. The original plan called for the LDEF to be retrieved in March 1985, but after a series of delays it was eventually returned to Earth by *Columbia* in January 1990.[2]

It successfully carried science and technology experiments for about 5.7 years, that have revealed a broad and detailed collection of space environmental data. LDEF's 69 months in space provided scientific data on the long-term effects of space exposure on materials, components and systems that has benefited NASA spacecraft designers to this day.[3]

17.1 History

Researchers identified the potential of the planned Space Shuttle to deliver a payload to space, leave it there for a long-term exposure to the harsh outer space environment, and on a separate mission retrieve the payload and return it to Earth for analysis. The LDEF concept evolved from a spacecraft proposed by NASA's Langley Research Center in 1970 to study the meteoroid environment, the Meteoroid and Exposure Module (MEM).[1] The project was approved in 1974 and LDEF was built at NASA's Langley Research Center.[3]

17.2 Launch

The STS-41-C crew of *Challenger* deployed LDEF on April 7, 1984. Attitude control of LDEF was achieved with gravity gradient and inertial distribution to maintain three-axis stability in orbit. Therefore, propulsion or other attitude control systems were not required, making LDEF free of acceleration forces and contaminants from jet firings.[3]

17.3 Experiments

The LDEF facility was designed to glean information vital to the development of the future space station and other spacecraft, especially the reactions of various space building materials to radiation, extreme temperature changes and collisions with space matter.

Engineers originally intended that the first mission would last about one year, and that several long-duration exposure missions would use the same frame. The exposure facility was actually used for a single 5.7-year mission. Fifty-seven science and technology experiments – involving government and university investigators from the United States, Canada, Denmark, France, Germany, Ireland, the Netherlands, Switzerland, and the United Kingdom – flew on the LDEF mission.[3] A total of 57 experiments were conducted on the LDEF.[2] Interstellar gases also would be trapped in an attempt to find clues into the formation of the Milky Way and the evolution of heavier elements.[3] Some examples are investigation exposure effects on:

- materials, coatings, and thermal systems
- power and spacecraft propulsion
- optical fibers and pure crystals for use in electronics
- electronics and optics
- survival of tomato seeds and bacterial spores[3]

17.3.1 EXOSTACK

In the German experiment EXOSTACK, 30% of *Bacillus subtilis* spores survived the nearly 6 years exposure to outer space when embedded in salt crystals, whereas 80% survived in the presence of glucose, which stabilize the structure of the cellular macromolecules, especially during vacuum-induced dehydration.[4][5]

If shielded against solar UV, spores of *B. subtilis* were capable of surviving in space for up to 6 years, especially if embedded in clay or meteorite powder (artificial meteorites). The data may support the likelihood of interplanetary transfer of microorganisms within meteorites, the so-called lithopanspermia hypothesis.[5]

17.4 Retrieval

At LDEF's launch, retrieval was scheduled for March 19, 1985, eleven months after deployment.[3] Schedules slipped, postponing the retrieval mission first to 1986, then indefinitely due to the *Challenger* disaster. It was finally recovered by *Columbia* on mission STS-32 on January 12, 1990.[6] *Columbia* approached LDEF in such a way as to minimize possible contamination to LDEF from thruster exhaust.[7] While LDEF was still attached to the RMS arm, an extensive 4.5 hour survey photographed each individual experiment tray, as well as larger areas.[7]

Columbia landed at Edwards Air Force Base on January 20, 1990.[3] With LDEF still in its bay, *Columbia* was ferried back on the Shuttle Carrier Aircraft to the Kennedy Space Center on January 26. Special efforts were taken to ensure protection against contamination of the payload bay during the ferry flight.[3]

Between January 30 and 31, LDEF was removed from *Columbia*'s payload bay in KSC's Orbiter Processing Facility, placed in a special payload canister, and transported to the Operations and Checkout Building. On February 1, 1990, LDEF was transported in the LDEF Assembly and Transportation System to the Spacecraft Assembly and Encapsulation Facility - 2, where the LDEF project team led deintegration activities.[7]

Columbia arrives at Kennedy Space Center with LDEF still in its payload bay.

LDEF after retrieval.

LDEF is removed from *Columbia*'s payload bay

17.5 See also

- Materials International Space Station Experiment
- Mir Environmental Effects Payload

17.6 References

[1] "The Long Duration Exposure Facility". *NASA*. Langley Research Center. Retrieved 2013-07-29.

[2] Allen, Carlton. "Long Duration Exposure Facility (LDEF)". *NASA*. Retrieved 2014-01-22.

[3] Grinter, Kay (8 January 2010). "Retrieval of LDEF provided resolution, better data" (PDF). *Spaceport News* (NASA). p. 7. Retrieved 2014-01-22.

[4] Paul Clancy (Jun 23, 2005). *Looking for Life, Searching the Solar System*. Cambridge University Press.

[5] Horneck, Gerda; David M. Klaus; Rocco L. Mancinelli (March 2010). "Space Microbiology". *Microbiology and Molecular Biology Reviews* **74** (1): 121–156. doi:10.1128/mmbr.00016-09. Retrieved 2013-07-29.

[6] "LDEF Archive". Langley Research Center. Retrieved July 16, 2010.

[7] Kramer, Herbert J. "LDEF (Long Duration Exposure Facility)". *NASA*. Earth Observation Portal. Retrieved 2014-01-22.

17.7 External links

- NASA Langley LDEF site
- *The Long Duration Exposure Facility (LDEF), Mission 1 Experiments*, 1984. NASA SP-473
- Photographic Survey of the LDEF Mission report

Chapter 18

Rosetta (spacecraft)

Rosetta is a space probe built by the European Space Agency launched on 2 March 2004. Along with *Philae*, its lander module, *Rosetta* is performing a detailed study of comet 67P/Churyumov–Gerasimenko (67P).[6][7] On 6 August 2014, the spacecraft reached the comet and performed a series of manoeuvres to be captured in its orbit. On 12 November, the lander module performed the first successful landing on a comet.[8] As of 2015, the mission continues to return data from the spacecraft in orbit and from the lander in the comet's surface. During its journey to the comet, the spacecraft flew by Mars and the asteroids 21 Lutetia and 2867 Šteins.[9][10][11]

The probe is named after the Rosetta Stone, a stele of Egyptian origin featuring a decree in three scripts. The lander is named after the Philae obelisk, which bears a bilingual Greek and Egyptian hieroglyphic inscription. A comparison of its hieroglyphs with those on the Rosetta Stone catalysed the deciphering of the Egyptian writing system. Similarly, it is hoped that these spacecraft will result in better understanding of comets and the early Solar System.[12][13] In a more direct analogy to its namesake, the *Rosetta* spacecraft also carries a micro-etched nickel alloy Rosetta disc donated by the Long Now Foundation inscribed with 13,000 pages of text in 1200 languages.[14]

18.1 Mission overview

Rosetta was launched on 2 March 2004 from the Guiana Space Centre in French Guiana on an Ariane 5 rocket and reached Comet Churyumov–Gerasimenko on 6 August 2014,[15] becoming the first spacecraft to orbit a comet.[16][17][18] (Previous missions had conducted successful flybys of seven other comets).[19] It is one of ESA's Horizon 2000 cornerstone missions.[20] The spacecraft consists of the *Rosetta* orbiter, which features 12 instruments, and the *Philae* lander, with nine additional instruments.[21] The *Rosetta* mission will orbit Comet Churyumov–Gerasimenko for 17 months and is designed to complete the most detailed study of a comet ever attempted. The spacecraft is controlled from the European Space Operations Centre (ESOC), in Darmstadt, Germany.[22] The planning for the operation of the scientific payload, together with the data retrieval, calibration, archiving and distribution, is performed from the European Space Astronomy Centre (ESAC), in Villanueva de la Cañada, near Madrid, Spain.[23] It has been estimated that in the decade preceding 2014, some 2,000 people assisted in the mission in some capacity.[24]

In 2007, *Rosetta* made a Mars gravity assist (flyby) on its way to Comet Churyumov–Gerasimenko.[25] The spacecraft also performed two asteroid flybys.[26] The craft completed its flyby of asteroid 2867 Šteins in September 2008 and of 21 Lutetia in July 2010.[27] Later, on 20 January 2014, *Rosetta* was taken out of a 31-month hibernation mode as it approached Comet Churyumov–Gerasimenko.[28][29]

Rosetta's *Philae* lander successfully made the first soft landing on a comet nucleus when it touched down on Comet Churyumov–Gerasimenko on 12 November 2014.[30][31][32] Astrophysicist Elizabeth Pearson said that although the future of the lander *Philae* is uncertain, *Rosetta* is the workhorse of the mission and its work will carry on.[33]

Comet Churyumov–Gerasimenko in September 2014 as imaged by Rosetta

18.2 History

18.2.1 Background

During the 1986 approach of Halley's Comet, international space probes were sent to explore the comet, most prominent among them being ESA's *Giotto*. After the probes returned valuable scientific information, it became obvious that follow-ons were needed that would shed more light on cometary composition and answer new questions.

Both ESA and NASA started cooperatively developing new probes. The NASA project was the Comet Rendezvous Asteroid Flyby (CRAF) mission. The ESA project was the follow-on Comet Nucleus Sample Return (CNSR) mission. Both missions were to share the Mariner Mark II spacecraft design, thus minimising costs. In 1992, after NASA cancelled CRAF due to budgetary limitations, ESA decided to develop a CRAF-style project on its own. By 1993 it was evident that the ambitious sample return mission was infeasible with the existing ESA budget, so the mission was redesigned and subsequently approved by the ESA,[24] with the final flight plan resembling the cancelled CRAF mission: an asteroid flyby followed by a comet rendezvous with in-situ examination, including a lander. After the spacecraft launch, Gerhard Schwehm was named mission manager; he retired in March 2014.[24]

18.2.2 Mission firsts

The *Rosetta* mission planned to achieve many historic firsts.[34]

On its way to comet 67P, *Rosetta* passed through the main asteroid belt, and made the first European close encounter with several of these primitive objects. *Rosetta* was the first spacecraft to fly close to Jupiter's orbit using solar cells as its main power source.

Rosetta is the first spacecraft to orbit a comet nucleus,[35] and is the first spacecraft to fly alongside a comet as it heads towards the inner Solar System. It is planned to be the first spacecraft to examine at close proximity how a frozen comet is transformed by the warmth of the Sun. Shortly after its arrival at 67P, the *Rosetta* orbiter dispatched the *Philae* lander for the first controlled touchdown on a comet nucleus. The robotic lander's instruments obtained the first images from a comet's surface and made the first in-situ analysis of its composition.

18.2.3 Design and construction

The *Rosetta* bus is a 2.8 × 2.1 × 2.0 m (9.2 × 6.9 × 6.6 ft) central frame and aluminium honeycomb platform. Its total mass is approximately 2,900 kg (6,400 lb), which includes the 100 kg (220 lb) *Philae* lander and 165 kg (364 lb) of science instruments. The Payload Support Module is mounted on top of the spacecraft and houses the scientific instruments, while the Bus Support Module is on the bottom and contains spacecraft support subsystems. Heaters placed around the spacecraft keep its systems warm while it is distant from the Sun. *Rosetta*'s communications suite includes a 2.2 m (7.2 ft) steerable high-gain parabolic dish antenna, a 0.8 m (2.6 ft) fixed-position medium-gain antenna, and two omnidirectional low-gain antennas.[36]

Electrical power for the spacecraft comes from two solar arrays totalling 64 square metres (690 sq ft).[37] Each solar array is subdivided into five solar panels, with each panel being 2.25 × 2.736 m (7.38 × 8.98 ft). The individual solar cells are made of silicon, 200 μm thick, and 61.95 × 37.75 mm (2.44 × 1.49 in).[38] The solar arrays generate a maximum of approximately 1,500 watts at perihelion,[38] a minimum of 400 watts in hibernation mode at 5.2 AU, and 850 watts when comet operations begin at 3.4 AU.[36] Spacecraft power is controlled by a redundant Terma power module also used in the *Mars Express* spacecraft,[39][40] and is stored in four 10-A·h NiCd batteries supplying 28 volts to the bus.[36]

Main propulsion comprises 24 paired bipropellant 10 N thrusters,[37] with four pairs of thrusters being used for delta-*v* burns. The spacecraft carried 1,719.1 kg (3,790 lb) of propellant at launch: 659.6 kg (1,454 lb) of monomethylhydrazine fuel and 1,059.5 kg (2,336 lb) of dinitrogen tetroxide oxidiser, contained in two 1,108-litre (244 imp gal; 293 US gal) grade 5 titanium alloy tanks and providing delta-*v* of at least 2,300 metres per second (7,500 ft/s) over the course of the mission. Propellant pressurisation is provided by two 68-litre (15 imp gal; 18 US gal) high-pressure helium tanks.[41]

Rosetta was built in a clean room according to COSPAR rules, but "sterilisation [was] generally not crucial since comets are usually regarded as objects where you can find prebiotic molecules, that is, molecules that are precursors of life, but not living microorganisms", according to Gerhard Schwehm, *Rosetta*'s project scientist.[42] The total cost of the mission is about €1.3 billion (US$1.8 billion).[43]

18.2.4 Launch

Rosetta was set to be launched on 12 January 2003 to rendezvous with the comet 46P/Wirtanen in 2011.

This plan was abandoned after the failure of an Ariane 5 carrier rocket during Hot Bird 7's launch on 11 December 2002, grounding it until the cause of the failure could be determined. A new plan was formed to target the comet Churyumov–Gerasimenko, with a revised launch date of 26 February 2004 and comet rendezvous in 2014. The larger mass and the resulting increased impact velocity made modification of the landing gear necessary.[44] After two scrubbed launch attempts, *Rosetta* was launched on 2 March 2004 at 7:17 GMT from the Guiana Space Centre in French Guiana. Aside from the changes made to launch time and target, the mission profile remained almost identical.

18.2.5 Deep space manoeuvres

To achieve the required velocity to rendezvous with 67P, *Rosetta* used gravity assist manoeuvres to accelerate throughout the inner Solar System. The comet's orbit was known before *Rosetta*'s launch, from ground-based measurements, to an accuracy of approximately 100 km (62 mi). Information gathered by the onboard cameras beginning at a distance of 24

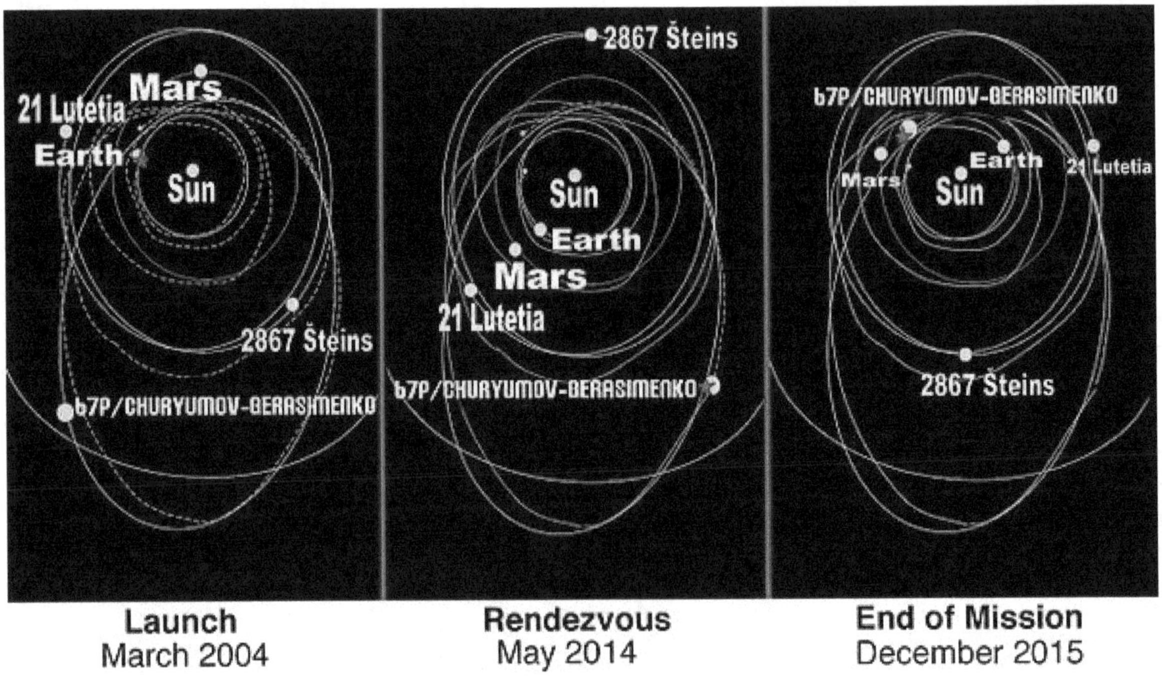

Trajectory of the Rosetta *space probe*

million kilometres (15,000,000 mi) were processed at ESA's Operation Centre to refine the position of the comet in its orbit to a few kilometres.

The first Earth flyby was on 4 March 2005.

On 25 February 2007, the craft was scheduled for a low-altitude flyby of Mars, to correct the trajectory. This was not without risk, as the estimated altitude of the flyby was a mere 250 kilometres (160 mi). During that encounter, the solar panels could not be used since the craft was in the planet's shadow, where it would not receive any solar light for 15 minutes, causing a dangerous shortage of power. The craft was therefore put into standby mode, with no possibility to communicate, flying on batteries that were originally not designed for this task.[45] This Mars manoeuvre was therefore nicknamed "The Billion Euro Gamble".[46] The flyby was successful, with *Rosetta* even returning detailed images of the surface and atmosphere of the planet, and the mission continued as planned.[9][25]

The second Earth flyby was on 13 November 2007 at a distance of 5,700 km (3,500 mi).[47][48] In observations made on 7 and 8 November, *Rosetta* was briefly mistaken for a near-Earth asteroid about 20 m (66 ft) in diameter by an astronomer of the Catalina Sky Survey and was given the provisional designation 2007 VN_{84}.[49] Calculations showed that it would pass very close to Earth, which led to speculation that it could impact Earth.[50] However, astronomer Denis Denisenko recognised that the trajectory matched that of *Rosetta*, which the Minor Planet Center confirmed in an editorial release on 9 November.[51][52]

The spacecraft performed a close flyby of asteroid 2867 Šteins on 5 September 2008. Its onboard cameras were used to fine-tune the trajectory, achieving a minimum separation of less than 800 km (500 mi). Onboard instruments measured the asteroid from 4 August to 10 September. Maximum relative speed between the two objects during the flyby was 8.6 km/s (19,000 mph; 31,000 km/h).[53]

Rosetta's third and final flyby of Earth happened on 12 November 2009.[54]

On 10 July 2010, *Rosetta* flew by 21 Lutetia, a large main-belt asteroid, at a minimum distance of 3,168±7.5 km (1,969±4.7 mi) at a velocity of 15 kilometres per second (9.3 mi/s).[11] The flyby provided images of up to 60 metres (200 ft) per pixel resolution and covered about 50% of the surface, mostly in the northern hemisphere.[27][55] The 462 images were obtained in 21 narrow- and broad-band filters extending from 0.24 to 1 μm.[27] Lutetia was also observed by the visible–near-infrared imaging spectrometer VIRTIS, and measurements of the magnetic field and plasma environment were taken as well.[27][55]

Rosetta's *signal received at ESOC in Darmstadt, Germany, on 20 January 2014*

In May 2014, *Rosetta* began a series of eight burns. These reduced the relative velocity between the spacecraft and 67P from 775 m/s (2,540 ft/s) to 7.9 m/s (26 ft/s).[15]

18.2.6 Orbit around 67P

In August 2014, *Rosetta* rendezvoused with the comet 67P/Churyumov–Gerasimenko (67P) and commenced a series of manoeuvres that took it on two successive triangular paths, averaging 100 and 50 kilometres (62 and 31 mi) from the nucleus, whose segments are hyperbolic escape trajectories alternating with thruster burns.[16][17] After closing to within about 30 km (19 mi) from the comet on 10 September, the spacecraft entered actual orbit about it.[16][17][18]

The surface layout of 67P was unknown before *Rosetta*'s arrival. The orbiter mapped the comet in anticipation of detaching its lander.[56] By 25 August 2014, five potential landing sites had been determined.[57] On 15 September 2014, ESA announced Site J, named *Agilkia* in honour of Agilkia Island by an ESA public contest and located on the "head" of the comet,[58] as the lander's destination.[59]

18.2.7 *Philae* lander

Main article: Philae (spacecraft)

Philae detached from *Rosetta* on 12 November 2014 at 08:35 UTC, and approached 67P at a relative speed of about 1 m/s (3.6 km/h; 2.2 mph).[60] It initially landed on 67P at 15:33 UTC, but bounced twice, coming to rest at 17:33 UTC.[8][61] Confirmation of contact with 67P reached Earth at 16:03 UTC.[62]

On contact with the surface, two harpoons were to be fired into the comet to prevent the lander from bouncing off as the comet's escape velocity is only around 1 m/s (3.6 km/h; 2.2 mph).[63] Analysis of telemetry indicated that the surface

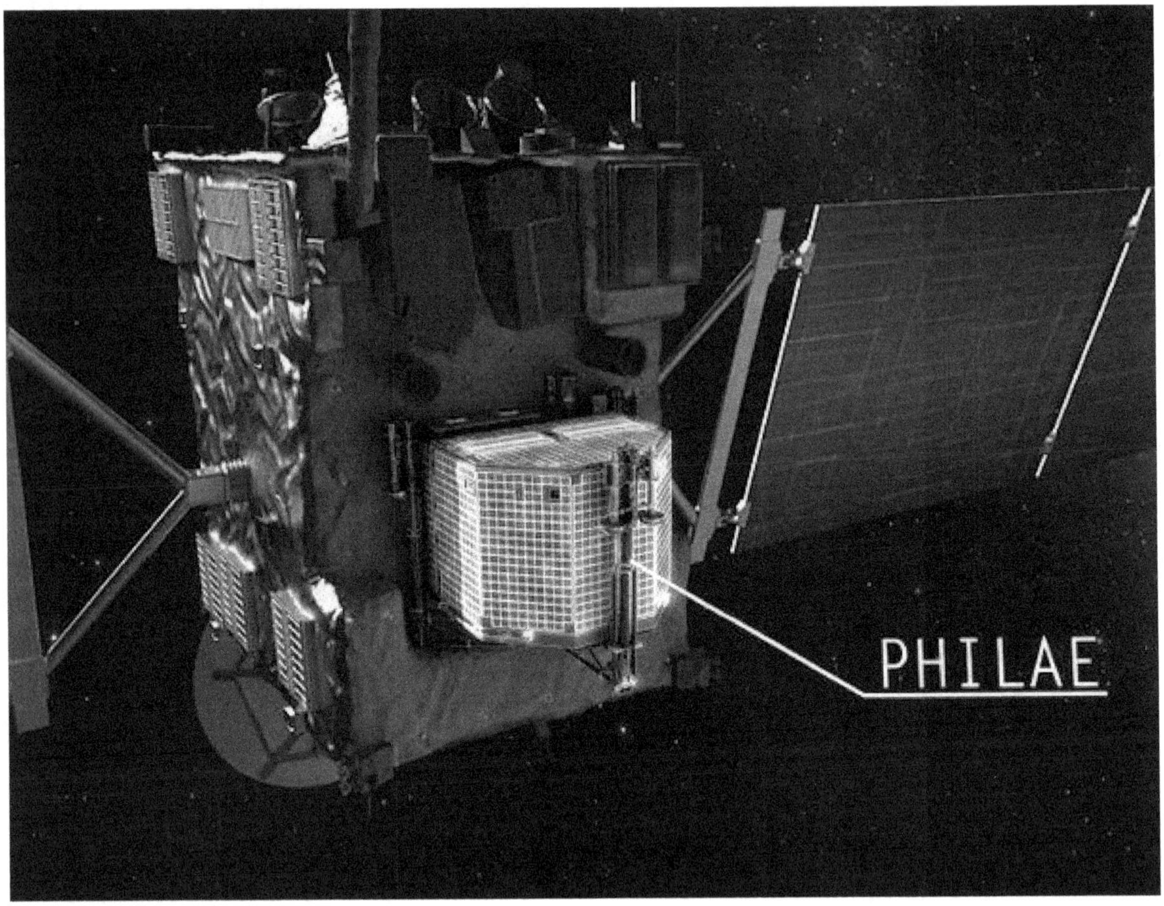

Rosetta *and* Philae

at the initial touchdown site is relatively soft, covered with a layer of granular material about 0.82 feet (0.25 meters) deep,[64] and that the harpoons had not fired upon landing. After landing on the comet, the *Philae* had been scheduled to commence its science mission, which included:

- Characterisation of the nucleus
- Determination of the chemical compounds present, including amino acid enantiomers[65]
- Study of comet activities and developments over time

Philae landed oddly, likely in the shadow of a nearby cliff or crater wall and canted at an angle of around 30 degrees. This made it unable to adequately collect solar power, and it lost contact with *Rosetta* when its batteries ran out after two days, well before much of the planned science objectives could be attempted.[66] Contact was briefly and intermittently reestablished several months later at various times between June 13 and July 9, before contact was lost once again.

18.2.8 Results

One of the first discoveries was that the magnetic field of 67P oscillated at 40–50 millihertz. Scientists modified the signal by speeding it up 10,000 times so that people could hear a rendition of it. While a natural phenomenon, it has been described as a "song",[67] and has been compared to *Continuum* for harpsichord by György Ligeti.[68] However, results from *Philae*'s landing show that the comet's nucleus has no magnetic field, and that the field originally detected by *Rosetta* likely resulted from the influence of solar winds.[69][70]

The isotopic signature of water vapour from comet 67P, as determined by the *Rosetta* spacecraft, is substantially different from that found on Earth. That is, the ratio of deuterium to hydrogen in the water from the comet was determined to be three times that found for terrestrial water. This makes it very unlikely that water found on Earth came from comets such as comet 67P, according to the scientists.[71][72][73] On 22 January 2015, NASA reported that, between June and August 2014, the rate at which water vapor was released by the comet increased up to tenfold.[74]

On 2 June 2015, NASA reported that the ALICE spectrograph on *Rosetta* determined that electrons within 1 km (0.62 mi) above the comet nucleus — produced from photoionization of water molecules by solar radiation, and not photons from the Sun as thought earlier — are responsible for the degradation of water and carbon dioxide molecules released from the comet nucleus into its coma.[75][76]

18.3 Instruments

For the instruments on *Philae*, see Philae (spacecraft)#Instruments.

18.3.1 Nucleus

The investigation of the nucleus is done by three optical spectrometers, one microwave radio antenna and one radar:

- **ALICE** (an ultraviolet imaging spectrograph). The ultraviolet spectrograph will search for and quantify the noble gas content in the comet nucleus, from which the temperature during the comet creation could be estimated. The detection is done by an array of potassium bromide and caesium iodide photocathodes. The 3.1 kg (6.8 lb) instrument uses 2.9 watts and was produced in the USA, and an improved version is used in the New Horizons spacecraft. It operates in the extreme and far ultraviolet spectrum, between 700 and 2,050 ångströms (70 and 205 nm).[77][78]
- **OSIRIS** (Optical, Spectroscopic, and Infrared Remote Imaging System). The camera system has a narrow-angle lens (700 mm) and a wide-angle lens (140 mm), with a 2048×2048 pixel CCD chip. The instrument was constructed in Germany.[79]
- **VIRTIS** (Visible and Infrared Thermal Imaging Spectrometer). The Visible and IR spectrometer is able to make pictures of the nucleus in the IR and also search for IR spectra of molecules in the coma. The detection is done by a mercury cadmium telluride array for IR and with a CCD chip for the visible wavelength range. The instrument was produced in Italy, and improved versions were used for Dawn and Venus Express.[80]
- **MIRO** (Microwave Instrument for the Rosetta Orbiter). The abundance and temperature of volatile substances like water, ammonia and carbon dioxide can be detected by MIRO via their microwave emissions. The 30 cm (12 in) radio antenna was constructed in Germany, while the rest of the 18.5 kg (41 lb) instrument was provided by the USA.
- **CONSERT** (Comet Nucleus Sounding Experiment by Radiowave Transmission). The CONSERT experiment will provide information about the deep interior of the comet using a radar. The radar will perform tomography of the nucleus by measuring electromagnetic wave propagation between the *Philae* lander and the *Rosetta* orbiter through the comet nucleus. This allows it to determine the comet's internal structure and deduce information on its composition. The electronics were developed by France and both antennas were constructed in Germany.[81]
- **RSI** (Radio Science Investigation). RSI makes use of the probe's communication system for physical investigation of the nucleus and the inner coma of the comet.[82]

18.3.2 Gas and particles

- **ROSINA** (Rosetta Orbiter Spectrometer for Ion and Neutral Analysis). The instrument consists of a double-focus magnetic mass spectrometer DFMS and a reflectron type time of flight mass spectrometer RTOF. The DFMS has a high resolution (can resolve N_2 from CO) for molecules up to 300 amu. The RTOF is highly sensitive for neutral molecules and for ions.[83] ROSINA was developed at the University of Bern in Switzerland.

- **MIDAS** (Micro-Imaging Dust Analysis System). The high-resolution atomic force microscope will investigate several physical aspects of the dust particles which are deposited on a silicon plate.[84]

- **COSIMA** (Cometary Secondary Ion Mass Analyser). COSIMA analyses the composition of dust particles by secondary ion mass spectrometry, using indium ions. It can detect ions up to a mass of 6500 amu.[85]

- **GIADA** (Grain Impact Analyser and Dust Accumulator). GIADA will analyse the dust environment of the comet coma measuring the optical cross section, momentum, speed and mass of each grain entering inside the instrument.[86][87]

18.3.3 Solar wind interaction

- **RPC** (Rosetta Plasma Consortium).[88][89]

18.4 Search for organic compounds

Previous observations have shown that comets contain complex organic compounds.[90][91][92][93] These are the elements that make up nucleic acids and amino acids, essential ingredients for life as we know it. Comets are thought to have delivered a vast quantity of water to Earth, and they may have also seeded Earth with organic molecules.[94] *Rosetta* and *Philae* will also search for organic molecules, nucleic acids (the building blocks of DNA and RNA) and amino acids (the building blocks of proteins) by sampling and analysing the comet's nucleus and coma cloud of gas and dust,[94] helping assess the contribution comets made to the beginnings of life on Earth.[90] Before succumbing to falling power levels, *Philae*'s COSAC instrument was able to detect organic molecules in the comet's atmosphere, and may be able to continue its investigation if it can establish a stable contact with *Rosetta*.[95]

Amino acids

Upon landing on the comet, *Philae* will also test some hypotheses as to why essential amino acids are almost all "left-handed", which refers to how the atoms arrange in orientation in relation to the carbon core of the molecule.[96] Most asymmetrical molecules are oriented in approximately equal numbers of left- and right-handed configurations (chirality), and the primarily left-handed structure of essential amino acids used by living organisms is unique. One hypothesis that will be tested was proposed in 1983 by William A. Bonner and Edward Rubenstein, Stanford University professors emeritus of chemistry and medicine respectively. They conjectured that when spiralling radiation is generated from a supernova, the circular polarisation of that radiation could then destroy one type of "handed" molecules. The supernova could wipe out one type of molecules while also flinging the other surviving molecules into space, where they could eventually end up on a planet.[97]

18.4.1 Preliminary results

The VIRTIS spectrometer on board the *Rosetta* spacecraft has provided evidence of nonvolatile organic macromolecular compounds everywhere on the surface of comet 67P with little to no water ice visible.[98] Preliminary analyses strongly suggest the carbon is present as polyaromatic organic solids mixed with sulfides and iron-nickel alloys.[99] In turn, the *Philae* lander's COSAC instrument detected organic molecules in the comet's atmosphere as it descended to its surface.[100][101] In contrast, the COSIMA instrument on-board *Rosetta* collected and analyzed comet grains from the near-nucleus region and the inner coma, and has not yet found carbon-bearing compounds.[102]

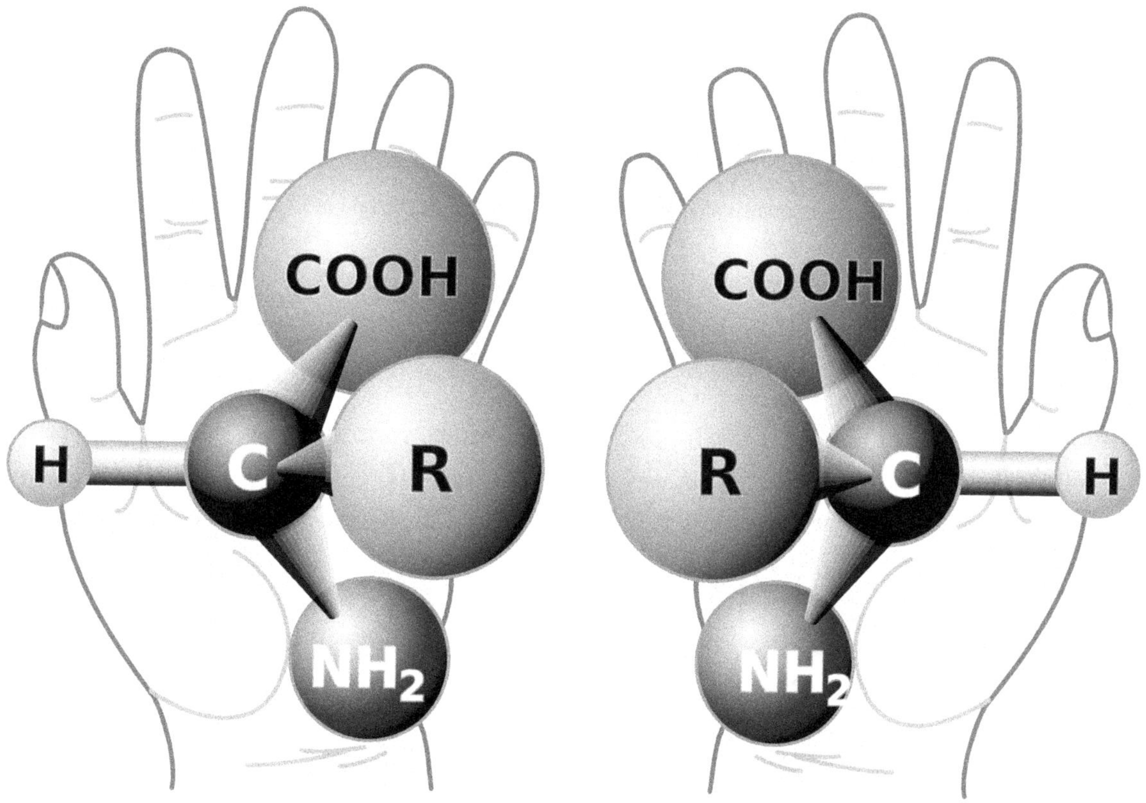

Two enantiomers of a generic amino acid. The mission will study why one chirality of some amino acids seems to be dominant in the universe.

18.5 Reaction control system problems

In 2006, *Rosetta* suffered a leak in its reaction control system (RCS).[103] The system, which consists of 24 bipropellant 10-newton thrusters,[15] is responsible for fine tuning the trajectory of *Rosetta* throughout its journey. The RCS will operate at a lower pressure than designed due to the leak. This may cause the propellants to mix incompletely and so burn 'dirtier' and less efficiently, though ESA engineers are confident that they have sufficient fuel reserves to allow successful completion of the mission.[104]

Rosetta's reaction wheels are showing higher than expected vibration, though testing revealed the system can be operated more efficiently resulting in less wear on the wheels. Before hibernation, two of the spacecraft's four reaction wheels began exhibiting “noise”. Engineers turned on three of the wheels after the spacecraft awoke, including one of the bad wheels. The other improperly functioning wheel will be held in reserve. Additionally, new software was uploaded which would allow *Rosetta* to function with only two active reaction wheels if necessary.[103][105]

18.6 Timeline of major events and discoveries

Main article: Timeline of Rosetta spacecraft

2004

- 2 March – ESA's *Rosetta* mission was successfully launched at 07:17 UTC (04:17 local time) from Kourou, French Guiana.

2005

- 4 March – *Rosetta* executed its first planned close swing-by (gravity assist passage) of Earth. The Moon and the Earth's magnetic field were used to test and calibrate the instruments on board of the spacecraft. The minimum altitude above the Earth's surface was 1,954.7 km (1,214.6 mi).[106]
- 4 July – Imaging instruments on board observed the collision between the comet Tempel 1 and the impactor of the Deep Impact mission.[107]

2007

- 25 February – Mars flyby.[25][108]
- 8 November – Catalina Sky Survey briefly misidentified the *Rosetta* spacecraft, approaching for its second Earth flyby, as a newly discovered asteroid.
- 13 November – Second Earth swing-by at a minimum altitude of 5,295 km (3,290 mi), travelling 45,000 km/h (28,000 mph).[109]

2008

- 5 September – Flyby of asteroid 2867 Šteins. The spacecraft passed the main-belt asteroid at a distance of 800 km (500 mi) and the relatively slow speed of 8.6 km/s (31,000 km/h; 19,000 mph).[110]

2009

- 13 November – Third and final swing-by of Earth at 48,024 km/h (29,841 mph).[111][112]

2010

- 16 March – Observation of the dust tail of asteroid P/2010 A2. Together with observations by Hubble Space Telescope it could be confirmed that P/2010 A2 is not a comet, but an asteroid, and that the tail most likely consists of particles from an impact by a smaller asteroid.[113]
- 10 July – Flew by and photographed the asteroid 21 Lutetia.[114]

2014

- May to July – Starting on 7 May, *Rosetta* began orbital correction manoeuvres to bring itself into orbit around 67P. At the time of the first deceleration burn *Rosetta* was approximately 2,000,000 km (1,200,000 mi) away from 67P and had a relative velocity of +775 m/s (2,540 ft/s); by the end of the last burn, which occurred on 23 July, the distance had been reduced to just over 4,000 km (2,500 mi) with a relative velocity of +7.9 m/s (18 mph).[15][115] In total eight burns were used to align the trajectories of *Rosetta* 67P with the majority of the deceleration occurring during three burns: Delta-v's of 291 m/s (650 mph) on 21 May, 271 m/s (610 mph) on 4 June, and 91 m/s (200 mph) on 18 June.[15]
- 14 July – The OSIRIS on-board imaging system returned images of Comet 67P which confirmed the irregular shape of the comet.[116][117]
- 6 August – *Rosetta* arrives at 67P, approaching to 100 km (62 mi) and carrying out a thruster burn that reduces its relative velocity to 1 m/s (3.3 ft/s).[118][119][120] Commences comet mapping and characterisation to determine a stable orbit and viable landing location for *Philae*.[121]

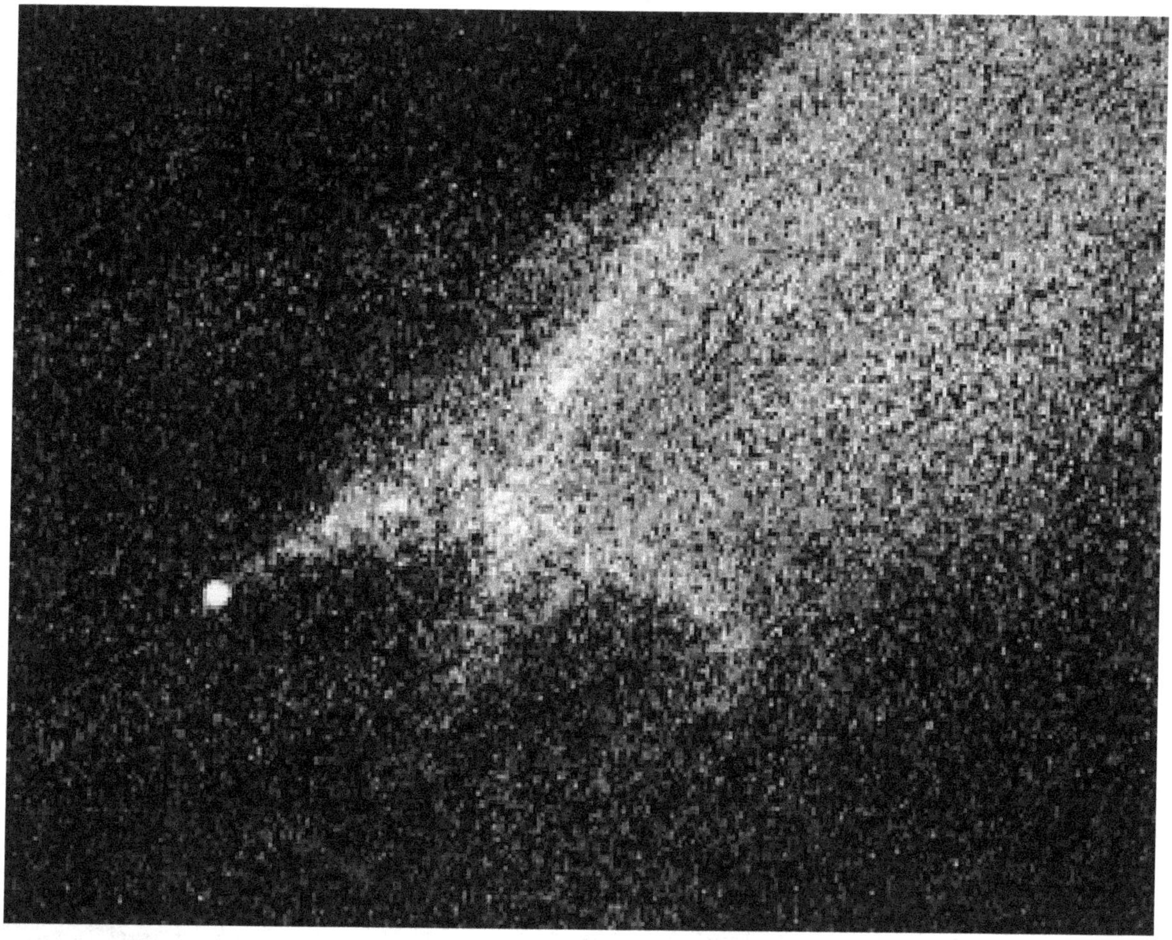

Hubble view of P/2010 A2

- 4 September – The first science data from *Rosetta*'s ALICE instrument was reported, showing that the comet is unusually dark in ultraviolet wavelengths, hydrogen and oxygen are present in the coma, and no significant areas of water-ice have been found on the comet's surface. Water-ice was expected to be found as the comet is too far from the Sun to turn water into vapour.[122]
- 10 September 2014 – *Rosetta* enters the Global Mapping Phase, orbiting 67P at an altitude of 29 km (18 mi).[3]
- 12 November 2014 – *Philae* lands on the surface of 67P.[8]
- 10 December 2014 – Data from the ROSINA mass spectrometers show that the ratio of heavy water to normal water on comet 67P is more than three times that on Earth. The ratio is regarded as a distinctive signature, and the discovery means that Earth's water is unlikely to have originated from comets like 67P.[71][72][73]

2015

- 14 April 2015 – Scientists report that the comet's nucleus has no magnetic field of its own.[69]
- 2 July 2015 – Scientists report that active pits, related to sinkhole collapses and possibly associated with outbursts, have been found on the comet.[123][124]
- 11 August 2015 – Scientists release images of a comet outburst that occurred on 29 July 2015.[125]
- 28 October 2015 – Scientists publish an article in *Nature* reporting high levels of molecular oxygen around 67P.[126][127]

Comet 67P seen from 10 km (6 mi)

Future milestones

- November 2014 to December 2015 – *Rosetta* will escort the comet around the Sun and would perform riskier investigations.[128]
- September 2016 - mission ends by slow crash-landing on the comet. Although *Philae* sent back some data during its descent, *Rosetta* has more powerful and more varied sensors and instruments. The orbiter will also descend much more slowly than *Philae* did, allowing it to gather more data and better images.[129]

18.6.1 Media coverage

The entire mission was featured heavily in social media, with a Facebook account for the mission and both the satellite and the lander having an official Twitter account portraying a personification of both spacecraft. The hashtag "#CometLanding" gained widespread traction. A Livestream of the control centres was set up, as were multiple official and unofficial events around the world to follow *Philae*'s landing on 67P.[130][131]

- Video reports by the German Aerospace Center
- Play media

  About *Rosetta*'s mission
  (9 min., 1080p HD, English)

- Play media

  About *Philae*'s landing
  (10 min., 1080p HD, English)

18.7 See also

- Deep Impact (spacecraft)
- Halley Armada
  - Giotto (spacecraft)
- Hayabusa – successful sample-return mission to an asteroid
- Stardust (spacecraft)
- Timeline of Solar System exploration

18.8 References

[1] "Rosetta at a glance — technical data and timeline". German Aerospace Center. Archived from the original on 8 January 2014. Retrieved 8 January 2014.

[2] "Rosetta timeline: countdown to comet arrival". European Space Agency. 5 August 2014. Retrieved 6 August 2014.

[3] Scuka, Daniel (10 September 2014). "Down, down we go to 29 km – or lower?". European Space Agency. Retrieved 13 September 2014.

[4] "No. 2 — Activating Rosetta". European Space Agency. 8 March 2004. Retrieved 8 January 2014.

[5] "We are working on flight control and science operations for Rosetta, now orbiting comet 67P, and Philae, which landed on the comet surface last week. Ask us Anything! AMA!". Reddit. 20 November 2014. Retrieved 21 November 2014.

[6] Agle, D. C.; Brown, Dwayne; Bauer, Markus (30 June 2014). "Rosetta's Comet Target 'Releases' Plentiful Water". NASA. Retrieved 30 June 2014.

[7] Chang, Kenneth (5 August 2014). "Rosetta Spacecraft Set for Unprecedented Close Study of a Comet". *The New York Times*. Retrieved 5 August 2014.

[8] Beatty, Kelly (12 November 2014). "Philae Lands on Its Comet – Three Times!". *Sky & Telescope*. Retrieved 26 November 2014.

[9] Bibring, Jean-Pierre; Schwehm, Gerhard (25 February 2007). "Stunning view of Rosetta skimming past Mars". European Space Agency. Retrieved 21 January 2014.

[10] Auster, H. U.; Richter, I.; Glassmeier, K. H.; Berghofer, G.; Carr, C. M.; Motschmann, U. (July 2010). "Magnetic field investigations during Rosetta's 2867 Šteins flyby". *Planetary and Space Science* **58** (9): 1124–1128. Bibcode:2010P&SS...58.1124A. doi:10.1016/j.pss.2010.01.006.

[11] Pätzold, M.; Andert, T. P.; Asmar, S. W.; Anderson, J. D.; Barriot, J.-P.; et al. (October 2011). "Asteroid 21 Lutetia: Low Mass, High Density". *Science* **334** (6055): 491–492. Bibcode:2011Sci...334..491P. doi:10.1126/science.1209389.

[12] Sharp, Tim (15 January 2014). "Rosetta Spacecraft: To Catch a Comet". Space.com. Retrieved 25 January 2014.

[13] "Unlocking the secrets of the universe: Rosetta lander named Philae". European Space Agency. 5 February 2004. Retrieved 25 January 2014.

[14] "ESA's Rosetta Probe begins approach of comet 67P". *Long Now.* 6 June 2014. Retrieved 6 August 2014.

[15] Scuka, Daniel (7 May 2014). "Thruster burn kicks off crucial series of manoeuvres". European Space Agency. Retrieved 21 May 2014.

[16] Fischer, D. (6 August 2014). "Rendezvous with a crazy world". The Planetary Society. Archived from the original on 6 August 2014. Retrieved 6 August 2014.

[17] Bauer, M. (6 August 2014). "Rosetta Arrives at Comet Destination". European Space Agency. Archived from the original on 6 August 2014. Retrieved 6 August 2014.

[18] Lakdawalla, Emily (15 August 2014). "Finding my way around comet Churyumov-Gerasimenko". The Planetary Society. Archived from the original on 15 August 2014. Retrieved 15 August 2014.

[19] Algar, Jim (14 October 2014). "Rosetta's lander Philae snaps selfie with comet". *Tech Times.* Retrieved 19 October 2014.

[20] Agle, D. C.; Cook, Jia-Rui; Brown, Dwayne; Bauer, Markus (17 January 2014). "Rosetta: To Chase a Comet". NASA. Retrieved 18 January 2014.

[21] "Rosetta at a glance". European Space Agency. Archived from the original on 14 May 2011. Retrieved 4 October 2010.

[22] Pearson, Michael; Smith, Matt (21 January 2014). "Comet-chasing probe wakes up, calls home". CNN. Retrieved 21 January 2014.

[23] Bauer, Markus (3 September 2014). "RSGS: The Rosetta Science Ground Segment". European Space Agency. Retrieved 20 November 2014.

[24] Gilpin, Lyndsey (14 August 2014). "The tech behind the Rosetta comet chaser: From 3D printing to solar power to complex mapping". *TechRepublic.*

[25] Keller, Uwe; Schwehm, Gerhard (25 February 2007). "Beautiful new images from Rosetta's approach to Mars: OSIRIS Update". European Space Agency.

[26] Glassmeier, Karl-Heinz; Boehnhardt, Hermann; Koschny, Detlef; Kührt, Ekkehard; Richter, Ingo (February 2007). "The Rosetta Mission: Flying Towards the Origin of the Solar System". *Space Science Reviews* **128** (1–4): 1–21. Bibcode:2007SSRv..128....1G.doi:10.1007/s11214-006-9140-8.

[27] Amos, Jonathan (4 October 2010). "Asteroid Lutetia has thick blanket of debris". BBC News. Retrieved 21 January 2014.

[28] Jordans, Frank (20 January 2014). "Comet-chasing probe sends signal to Earth". *Excite News.* Associated Press. Archived from the original on 2 February 2014. Retrieved 20 January 2014.

[29] Morin, Monte (20 January 2014). "Rise and shine Rosetta! Comet-hunting spacecraft gets wake-up call". *Los Angeles Times.* Science Now. Retrieved 21 January 2014.

[30] Agle, D. C.; Webster, Guy; Brown, Dwayne; Bauer, Markus (12 November 2014). "Rosetta's 'Philae' Makes Historic First Landing on a Comet". NASA. Retrieved 13 November 2014.

[31] Chang, Kenneth (12 November 2014). "European Space Agency's Spacecraft Lands on Comet's Surface". *The New York Times.* Retrieved 12 November 2014.

[32] "Rosetta: Comet Probe Beams Back Pictures". Sky News. 12 November 2014. Retrieved 12 November 2014.

[33] "Philae: Its done what it went there to do". BBC News. 15 November 2014. Retrieved 15 November 2014.

[34] "Europe's Comet Chaser-historic mission". European Space Agency. 16 January 2014. Retrieved 5 August 2014.

[35] "Europe's Rosetta probe goes into orbit around comet 67P". BBC News. 6 August 2014. Retrieved 6 August 2014.

[36] "Rosetta". *National Space Science Data Center*. NASA. Retrieved 3 November 2014.

[37] "The Rosetta orbiter". European Space Agency. 16 January 2014. Retrieved 13 August 2014.

[38] D'Accolti, G.; Beltrame, G.; Ferrando, E.; Brambilla, L.; Contini, R.; et al. (2002). *The Solar Array Photovoltaic Assembly for the ROSETTA Orbiter and Lander Spacecraft's*. 6th European Space Power Conference. 6–10 May 2002. Porto, Portugal. Bibcode:2002ESASP.502..445D.

[39] Stage, Mie (19 January 2014). "Terma-elektronik vækker rumsonde fra årelang dvale". *Ingeniøren*. Retrieved 2 December 2014.

[40] Jensen, Hans; Laursen, Johnny (2002). *Power Conditioning Unit for Rosetta/Mars Express*. 6th European Space Power Conference. 6–10 May 2002. Porto, Portugal. Bibcode:2002ESASP.502..249J.

[41] Stramaccioni, D. (2004). *The Rosetta Propulsion System*. 4th International Spacecraft Propulsion Conference. 2–9 June 2004. Sardinia, Italy. Bibcode:2004ESASP.555E...3S.

[42] "No bugs please, this is a clean planet!". European Space Agency. 30 July 2002. Retrieved 7 March 2007.

[43] Gibney, Elizabeth (17 July 2014). "Duck-shaped comet could make Rosetta landing more difficult".*Nature*. doi:10.1038/nature. Retrieved 15 November 2014.

[44] Ulamec, S.; Espinasse, S.; Feuerbacher, B.; Hilchenbach, M.; Moura, D.; et al. (April 2006). "Rosetta Lander—Philae: Implications of an alternative mission".*Acta Astronautica***58**(8): 435–441. Bibcode:2006AcAau..58..435U.doi:10.1016/j.actaastro

[45] "Rosetta correctly lined up for critical Mars swingby". European Space Agency. 15 February 2007. Retrieved 21 January 2014.

[46] "Europe set for billion-euro gamble with comet-chasing probe". Phys.org. 23 February 2007. Archived from the original on 25 February 2007.

[47] Keller, Horst Uwe; Sierks, Holger (15 November 2007). "First OSIRIS images from Rosetta Earth swing-by". Max Planck Institute for Solar System Research. Archived from the original on 7 March 2008.

[48] Lakdawalla, Emily (2 November 2007). "Science plans for Rosetta's Earth flyby". The Planetary Society. Retrieved 21 January 2014.

[49] "M.P.E.C. 2007-V69". Minor Planet Center. Archived from the original on 24 May 2015. Retrieved 6 October 2015.

[50] Sutherland, Paul (10 November 2007). "'Deadly asteroid' is a spaceprobe". *Skymania*. Retrieved 21 January 2014.

[51] Lakdawalla, Emily (9 November 2007). "That's no near-Earth object, it's a spaceship!". The Planetary Society. Retrieved 21 January 2014.

[52] Tomatic, A. U. (9 November 2007). "MPEC 2007-V70: Editorial Notice". *Minor Planet Electronic Circular*. Minor Planet Center. Retrieved 21 January 2014.

[53] "First Asteroid". *Aviation Week & Space Technology* **169** (10): 18. 15 September 2008.

[54] "Rosetta makes final home call". BBC News. 12 November 2009. Retrieved 22 May 2010.

[55] Sierks, H.; Lamy, P.; Barbieri, C.; Koschny, D.; Rickman, H.; et al. (October 2011). "Images of Asteroid 21 Lutetia: A Remnant Planetesimal from the Early Solar System". *Science* **334** (6055): 487–90. Bibcode:2011Sci...334..487S. doi:10.1126/science.1207325.PMID22034428.

[56] Scuka, Daniel (23 July 2014). "Last of the FATties". European Space Agency. Retrieved 31 July 2014.

[57] Agle, D. C.; Brown, Dwayne; Bauer, Markus (25 August 2014). "Rosetta: Landing site search narrows". *NASA*. Retrieved 26 August 2014.

[58] Amos, Jonathan (4 November 2014). "Rosetta comet mission: Landing site named 'Agilkia'". BBC News. Retrieved 5 November 2014.

[59] Bauer, Markus (15 September 2014). "'J' Marks the Spot for Rosetta's Lander". European Space Agency. Retrieved 20 September 2014.

[60] Knapton, Sarah (12 November 2014). "Rosetta mission: broken thrusters mean probe could bounce off comet into space". *The Daily Telegraph*. Retrieved 12 November 2014.

[61] Withnall, Adam; Vincent, James (13 November 2014). "Philae lander 'bounced twice' on comet but is now stable, Rosetta mission scientists confirm". *The Independent*. Retrieved 26 November 2014.

[62] "Rosetta camera captures Philae's descent to the comet". *Spaceflight Now*. 13 November 2014. Retrieved 26 November 2014.

[63] Dambeck, Thorsten (21 January 2014). "Expedition to primeval matter". Max-Planck-Gesellschaft. Retrieved 19 September 2014.

[64] Wall, Mike (30 July 2015). "Surprising Comet Discoveries by Rosetta's Philae Lander Unveiled". *Space.com*. Retrieved 2015-07-31.

[65] Meierhenrich, Uwe (2008). *Amino Acids and the Asymmetry of Life*. Springer-Verlag. doi:10.1007/978-3-540-76886-9. ISBN 978-3-540-76885-2.

[66] Beatty, Kelly (15 November 2014). "Philae Wins Race to Return Comet Findings". *Sky & Telescope*. Retrieved 2 November 2015.

[67] Fessenden, Marissa (12 November 2014). "Comet 67P Has a Welcome Song for Rosetta And Philae". Smart News. Smithsonian.com. Retrieved 26 December 2014.

[68] Edwards, Tim (14 November 2014). "Music emitted from Comet 67P sounds an awful lot like 20th-century harpsichord masterpiece". *Classic FM*. Retrieved 26 December 2014.

[69] Bauer, Markus (14 April 2015). "Rosetta and Philae Find Comet Not Magnetised". European Space Agency. Retrieved 14 April 2015.

[70] Schiermeier, Quirin (14 April 2015). "Rosetta's comet has no magnetic field". *Nature*. doi:10.1038/nature.2015.17327.

[71] Agle, D.C.; Bauer, Markus (10 December 2014). "Rosetta Instrument Reignites Debate on Earth's Oceans". NASA. Retrieved 10 December 2014.

[72] Chang, Kenneth (10 December 2014). "Comet Data Clears Up Debate on Earth's Water". *The New York Times*. Retrieved 10 December 2014.

[73] Morelle, Rebecca (10 December 2014). "Rosetta results: Comets 'did not bring water to Earth'". BBC News. Retrieved 11 December 2014.

[74] Agle, D. C.; Brown, Dwayne; Bauer, Markus (22 January 2015). "Rosetta Comet 'Pouring' More Water Into Space". NASA. Retrieved 22 January 2015.

[75] Agle, D. C.; Brown, Dwayne; Fohn, Joe; Bauer, Markus (2 June 2015). "NASA Instrument on Rosetta Makes Comet Atmosphere Discovery". NASA. Retrieved 2 June 2015.

[76] Feldman, Paul D.; A'Hearn, Michael F.; Bertaux, Jean-Loup; Feaga, Lori M.; Parker, Joel Wm.; et al. (2 June 2015). "Measurements of the near-nucleus coma of comet 67P/Churyumov-Gerasimenko with the Alice far-ultraviolet spectrograph on Rosetta" (PDF). *Astronomy and Astrophysics*. doi:10.1051/0004-6361/201525925.

[77] Stern, S. A.; Slater, D. C.; Scherrer, J.; Stone, J.; Versteeg, M.; et al. (February 2007). "*Alice*: The Rosetta Ultraviolet Imaging Spectrograph". *Space Science Reviews* **128** (1–4): 507–527. arXiv:astro-ph/0603585. Bibcode:2007SSRv..128..507S. doi:10.1007/s11214-006-9035-8.

[78] Stern, S. A.; Slater, D. C.; Gibson, W.; Scherrer, J.; A'Hearn, M.; et al. (1998). "Alice—An Ultraviolet Imaging Spectrometer for the Rosetta Orbiter". *Advances in Space Research* **21** (11): 1517–1525. Bibcode:1998AdSpR..21.1517S. doi:10.1016/S0273-1177(97)00944-7.

[79] Thomas, N.; Keller, H. U.; Arijs, E.; Barbieri, C.; Grande, M.; et al. (1998). "Osiris—The optical, spectroscopic and infrared remote imaging system for the Rosetta Orbiter". *Advances in Space Research* **21** (11): 1505–1515. Bibcode:1998AdSpR..21.1505T. doi:10.1016/S0273-1177(97)00943-5.

[80] Coradini, A.; Capaccioni, F.; Capria, M. T.; Cerroni, P.; de Sanctis, M. C.; et al. (March 1996). "VIRTIS Visible Infrared Thermal Imaging Spectrometer for Rosetta Mission". *Lunar and Planetary Science* **27**: 253. Bibcode:1996LPI....27..253C. doi:10.1109/igarss.1995.521822.

[81] Kofman, W.; Herique, A.; Goutail, J.-P.; Hagfors, T.; Williams, I. P.; et al. (February 2007). "The Comet Nucleus Sounding Experiment by Radiowave Transmission (CONSERT): A Short Description of the Instrument and of the Commissioning Stages". *Space Science Reviews* **128** (1–4): 414–432. Bibcode:2007SSRv..128..413K. doi:10.1007/s11214-006-9034-9.

[82] "RSI: Radio Science Investigation". European Space Agency. Retrieved 26 November 2014.

[83] Balsiger, H.; Altwegg, K.; Arijs, E.; Bertaux, J.-L.; Berthelier, J.-J.; et al. (1998). "Rosetta Orbiter Spectrometer for Ion and Neutral Analysis—ROSINA". *Advances in Space Research* **21** (11): 1527–1535. Bibcode:1998AdSpR..21.1527B. doi:10.1016/S0273-1177(97)00945-9.

[84] Riedler, W.; Torkar, K.; Rüdenauer, F.; Fehringer, M.; Schmidt, R.; et al. (1998). "The MIDAS experiment for the Rosetta mission". *Advances in Space Research* **21** (11): 1547–1556. Bibcode:1998AdSpR..21.1547R. doi:10.1016/S0273-1177(97)00947-2.

[85] Engrand, Cecile; Kissel, Jochen; Krueger, Franz R.; Martin, Philippe; Silén, Johan; et al. (April 2006). "Chemometric evaluation of time-of-flight secondary ion mass spectrometry data of minerals in the frame of future *in situ* analyses of cometary material by COSIMA onboard ROSETTA". *Rapid Communications in Mass Spectrometry* **20** (8): 1361–1368. doi:10.1002/rcm.2448. PMID 16555371.

[86] Colangeli, L.; Lopez-Moreno, J. J.; Palumbo, P.; Rodriguez, J.; Cosi, M.; et al. (February 2007). "The Grain Impact Analyser and Dust Accumulator (GIADA) experiment for the Rosetta mission: design, performances and first results". *Space Science Reviews* **128** (1–4): 803–821. doi:10.1007/s11214-006-9038-5.

[87] Della Corte, V.; Rotundi, A.; Accolla, M.; Sordini, R.; Palumbo, P.; et al. (March 2014). "GIADA: its status after the Rosetta cruise phase and on-ground activity in support of the encounter with comet 67P/Churyumov-Gerasimenko". *Journal of Astronomical Instrumentation* **3** (1): 1–11. doi:10.1142/S2251171713500116.

[88] Trotignon, J. G.; Boström, R.; Burch, J. L.; Glassmeier, K.-H.; Lundin, R.; et al. (January 1999). "The Rosetta plasma consortium: Technical realization and scientific aims". *Advances in Space Research* **24** (9): 1149–1158. Bibcode:1999AdSpR..24.114 9T.doi:10.1016/S0273-1177(99)80208-7.

[89] Glassmeier, Karl-Heinz; Richter, Ingo; Diedrich, Andrea; Musmann, Günter; Auster, Uli; et al. (February 2007). "RPC-MAG The Fluxgate Magnetometer in the ROSETTA Plasma Consortium". *Space Science Reviews* **128** (1–4): 649–670. Bibcode:2007SSRv..128..649G. doi:10.1007/s11214-006-9114-x.

[90] "Rosetta's frequently asked questions". *European Space Agency*. Retrieved 9 August 2014.

[91] Hoover, Rachel (21 February 2014). "Need to Track Organic Nano-Particles Across the Universe? NASA's Got an App for That". NASA.

[92] Chang, Kenneth (18 August 2009). "From a Distant Comet, a Clue to Life". *The New York Times*. Space & Cosmos. p. A18.

[93] Tate, Karl (17 January 2014). "How the Rosetta Spacecraft Will Land on a Comet". Space.com. Retrieved 9 August 2014. A previous sample-return mission to a different comet found particles of organic matter that are the building blocks of life.

[94] Kremer, Ken (6 August 2014). "Rosetta Arrives at 'Scientific Disneyland' for Ambitious Study of Comet 67P/Churyumov-Gerasimenko after 10 Year Voyage". *Universe Today*. Retrieved 9 August 2014.

[95] Baldwin, Emily (26 June 2015). "Rosetta and Philae: Searching for a good signal". European Space Agency. Retrieved 26 June 2015.

[96] Thiemann, Wolfram H.-P.; Meierhenrich, Uwe (February 2001). "ESA Mission ROSETTA Will Probe for Chirality of Cometary Amino Acids". *Origins of Life and Evolution of the Biosphere* **21** (1–2): 199–210. doi:10.1023/A:1006718920805.

[97] Bergeron, Louis (17 October 2007). "William Bonner, professor emeritus of chemistry, dead at 87". *Stanford Report*. Retrieved 8 August 2014.

[98] Capaccioni, F.; Coradini, A.; Filacchione, G.; Erard, S.; Arnold, G.; et al. (23 January 2015). "The organic-rich surface of comet 67P/Churyumov-Gerasimenko as seen by VIRTIS/Rosetta". *Science* **347** (6220). doi:10.1126/science.aaa0628.

[99] Quirico, E.; Moroz, L. V.; Beck, P.; Schmitt, B.; Arnold, G.; et al. (March 2015). *Composition of Comet 67P/Churymov-Gerasimenko Refractory Crust as Inferred from VIRTIS-M/Rosetta Spectro-Imager* (PDF). 46th Lunar and Planetary Science Conference. 16–20 March 2015. The Woodlands, Texas. Bibcode:2015LPI....46.2092Q. LPI Contribution No. 1832, p. 2092.

[100] Rincon, Paul (18 November 2014). "Comet landing: Organic molecules detected by Philae". BBC News. Retrieved 6 April 2015.

[101] Grey, Richard (19 November 2014). "Rosetta mission lander detects organic molecules on surface of comet". *The Guardian*. Retrieved 6 April 2015.

[102] Schulz, Rita; Hilchenbach, Martin; Langevin, Yves; Kissel, Jochen; Silen, Johan; et al. (February 2015). "Comet 67P/Churyumov-Gerasimenko sheds dust coat accumulated over the past four years". *Nature* **518** (7538): 216–218. Bibcode:2015Natur.518..216S. doi:10.1038/nature14159.

[103] "Rosetta's Frequently Asked Questions". European Space Agency. Retrieved 24 May 2014.

[104] Amos, Jonathan (21 May 2014). "Rosetta comet-chaser initiates 'big burn'". BBC News. Retrieved 24 May 2014.

[105] Clark, Stephen (29 January 2014). "ESA says Rosetta in good shape after 31-month snooze". *Spaceflight Now*. Retrieved 29 July 2014.

[106] Montagnon, Elsa; Ferri, Paolo (July 2006). "Rosetta on its way to the outer Solar System". *Acta Astronautica* **59** (1–5): 301–309. Bibcode:2006AcAau..59..301M. doi:10.1016/j.actaastro.2006.02.024.

[107] Schwehm, Gerhard (4 July 2005). "Rosetta camera view of Tempel 1 brightness". European Space Agency. Retrieved 21 January 2014.

[108] Schwehm, Gerhard (25 February 2007). "Rosetta successfully swings-by Mars – next target: Earth". European Space Agency. Retrieved 21 January 2014.

[109] Schwehm, Gerhard; Accomazzo, Andrea; Schulz, Rita (13 November 2007). "Rosetta swing-by a success". European Space Agency. Retrieved 7 August 2014.

[110] "Encounter of a different kind: Rosetta observes asteroid at close quarters". European Space Agency. 6 September 2008. Retrieved 29 May 2009.

[111] "Last visit home for ESA's comet chaser". European Space Agency. 20 October 2009. Retrieved 8 November 2009.

[112] "Rosetta bound for outer Solar System after final Earth swingby". European Space Agency. 13 November 2009. Retrieved 7 August 2014.

[113] Snodgrass, Colin; Tubiana, Cecilia; Vincent, Jean-Baptiste; Sierks, Holger; Hviid, Stubbe; et al. (October 2010). "A collision in 2009 as the origin of the debris trail of asteroid P/2010 A2". *Nature* **467** (7317): 814–816. arXiv:1010.2883. Bibcode:2010Natur.467..814S. doi:10.1038/nature09453. PMID 20944742.

[114] Chow, Denise (10 July 2010). "Mysterious Asteroid Unmasked By Space Probe Flyby". Space.com. Retrieved 10 July 2010.

[115] Scuka, Daniel (20 May 2014). "The Big Burns – Part 1". European Space Agency. Retrieved 21 May 2014.

[116] "The twofold comet: Comet 67P/Churyumov-Gerasimenko". *Astronomy.com*. 17 July 2014. Retrieved 18 July 2014.

[117] Temming, Maria (17 July 2014). "Rosetta's Comet has a Split Personality". *Sky & Telescope*. Retrieved 18 July 2014.

[118] ESA Operations (6 August 2014). "Thruster burn complete". Twitter.com. Retrieved 6 August 2014.

[119] Scuka, Daniel (3 June 2014). "The Big Burns – Part 2". European Space Agency. Retrieved 9 June 2014.

[120] Rkaina, Sam (6 August 2014). "Rosetta probe: Recap updates after spacecraft successfully reached deep space comet orbit". *Daily Mirror*. Retrieved 6 August 2014.

[121] Amos, Jonathan (14 August 2014). "Rosetta: Comet probe gets down to work". BBC News. Retrieved 15 August 2014.

[122] Brown, Dwayne; Agle, A. G.; Martinez, Maria; Bauer, Markus (4 September 2014). "NASA Instrument aboard European Spacecraft Returns First Science Results". NASA. Release 14-238. Retrieved 5 September 2014.

[123] Vincent, Jean-Baptiste; et al. (2 July 2015). "Large heterogeneities in comet 67P as revealed by active pits from sinkhole collapse". *Nature* **523**: 63–66. doi:10.1038/nature14564.

[124] Ritter, Malcolm (1 July 2015). "It's the pits: Comet appears to have sinkholes, study says". Associated Press. Retrieved 2 July 2015.

[125] "PIA19867: Rosetta Comet In Action (Animation)". *NASA*. 11 August 2015. Retrieved 11 August 2015.

[126] Bieler, A.; Altwegg, K.; Balsiger, H.; Bar-Nun, A.; Berthelier, J.-J.; et al. (29 October 2015). "Abundant molecular oxygen in the coma of comet 67P/Churyumov–Gerasimenko". *Nature* **526**: 678–681. doi:10.1038/nature15707.

[127] "Comet gives clues to Earth's beginning". Radio New Zealand. 28 October 2015. Retrieved 29 October 2015.

[128] "Rosetta mission extended". European Space Agency. 23 June 2015. Retrieved 11 July 2015.

[129] Gibney, Elizabeth (4 November 2015). "Historic Rosetta mission to end with crash into comet". *Nature News*. Retrieved 2015-11-05.

[130] "Live updates: *Rosetta* mission comet landing". 12 November 2014.

[131] "Call for Media Opportunities to Follow Rosetta Mission's Historic Comet Landing". European Space Agency. 16 October 2014.

18.9 External links

- *Rosetta* website at ESA.int
- *Rosetta* website at NASA.gov
- *Rosetta* operations site at ESA.int
- *Rosetta* mission profile at NASA.gov
- *Rosetta* Latest science news from the craft at ESA.int

Media

- *Rosetta* outreach image gallery at ESA.int
- *Rosetta* raw image gallery at ESA.int
- *Rosetta* image gallery at Flickr.com
- *Rosetta*'s orbital journey at YouTube.com
- *Rosetta: landing on a comet* at ESA.int
- *How to land on a comet* by Fred Jansen at TED

Chapter 19

Living Interplanetary Flight Experiment

The **Living Interplanetary Flight Experiment**[2] (**LIFE** or **Phobos LIFE**[3]) was an interplanetary mission developed by the Planetary Society. It consisted of sending selected microorganisms on a three-year interplanetary round-trip in a small capsule aboard the Russian Fobos-Grunt spacecraft in 2011, which was a failed sample-return mission to the Martian moon Phobos. The Fobos-Grunt mission failed to leave Earth orbit,[4][5] and was destroyed.

The goal was to test whether selected organisms can survive an as yet undetermined number of years in deep space by flying them through interplanetary space. The experiment would have tested one aspect of transpermia, the hypothesis that life could survive space travel, if protected inside rocks blasted by impact off one planet to land on another.[6][7]

19.1 Precursor

Prior to the *Phobos LIFE* experiment, a precursor *LIFE* prototype was successfully flown in 2011 aboard the final flight of Space Shuttle Endeavour, STS-134. Known as the **Shuttle-LIFE** (also **LIFE**[8]) experiment.[9][10]

19.2 The experiment

The project includes representatives of all three domains of life: bacteria, eukaryota and archaea. The capsule was transporting 10 types of organisms in 30 self-contained samples, i.e., each in triplicate. In addition, one or more natural native soil samples were flown in their own self-contained capsule.[11] The Phobos-Soil sample return mission was the only attempted biological science mission that would have returned to Earth from deep space, far beyond the protection of Earth's magnetic field; sending biological samples through deep space is therefore a much better test of interplanetary survivability than sending the samples on a typical Earth-orbiting flight.[11]

The project was being done in collaboration with the Russian Space Research Institute, the Institute for Biomedical Problems of the Russian Academy of Sciences, the Moscow State University, the American Type Culture Collection (ATCC), and the Institute for Aerospace Medicine in Germany.

19.2.1 Specimens

Three fundamental guidelines governed the selection of the organisms:[12] First, the organisms selected represent the three domains of life – eukaryote, bacteria and archaea. Second, the organisms are very well studied (e.g., having their genome sequenced and studied in many other experiments) to make it possible to accurately assess the effects of the long exposure to space. If they had already been studied in space conditions so much the better, since it would enable researchers to pinpoint precisely how organisms were affected by the years-long exposure to the interplanetary environment. Finally, a strong preference was given to organisms that appear to stand the best chance of surviving the journey. These are extremophiles, organisms that thrive in conditions that would kill the vast majority of Earthly creatures.

The 10 'passenger' organisms selected are listed below:[12]

Bacteria[13]

- *Bacillus safensis*
  - Discovered in JPL's 'clean' room: Spacecraft Assembly Facility.[14] Might already be on Mars with Spirit and Opportunity.[14]
- *Deinococcus radiodurans*
  - Is extremely resistant to radiation, can survive a dose of 5,000 Gy.
- *Bacillus subtilis*, strain MW01
- *Bacillus subtilis*, strain 168
  - Very well known from other experiments. Flew to the Moon with Apollo and had multiyear exposure in low Earth orbit.[14]

Archaea[15]

- *Haloarcula marismortui*
  - If Mars had an ocean, it would have been very salty. *H. marismortui* is halophilic.
- *Methanothermobacter wolfeii*
  - Mars Express has discovered methane in the Martian atmosphere. *M. wolfeii* is a methane-producing organism.
- *Pyrococcus furiosus*
  - *P. furiosus* thrives at about 100 °C, it was supposed to act as a maximum temperature indicator.

Eukaryote[16]

- Fungus – *Saccharomyces cerevisiae* (yeast)[14]
- Plantae – Seeds from *Arabidopsis thaliana* ('mouse-ear cress')
  - Flew to the Moon with Apollo.[14]
- Animalia – Tardigrades ('water bears')
  - Have survived vacuum and radiation in low Earth orbit.[17]

19.2.2 Capsule design

The mass of the Bio-Module on board the Fobos-Grunt spacecraft was 100 grams or less. The design is a short cylinder. The bio-module provided 30 small tubes (3 millimeters in diameter) for individual microbe samples. It also accommodated a native sample of bacteria – derived from a permafrost region on Earth – within a cavity 26 mm in diameter.[11]

19.3 Mission failure

The module passed stress tests including a shake test with vibrations at frequencies to 1,100 Hz and an impact test of 4,000 g, designed to simulate the potential impact of the capsule on Earth.[18] The LIFE experiment was launched on November 8, 2011 on board the Fobos-Grunt, however, the spacecraft failed to depart Earth orbit due to a programming error,[19][20] and fell back to Earth in the Pacific Ocean.[4] The module was not recovered.[21] The team is seeking out future exploratory opportunities.[22]

19.4 Criticism

Barry E. DiGregorio, the director of the International Committee Against Mars Sample Return, criticised the LIFE experiment on the Fobos-Grunt mission as a violation of the Outer Space Treaty of 1967 due to its risk of contaminating Phobos or Mars with the microbial spores and live bacteria it contains. While the mission was supposed to land and return from Phobos, a moon of Mars, the risk to Mars itself would have been from the possibility of Fobos-Grunt losing control and crash landing on the planet.[23] It was speculated that the heat-resistant extremophile bacteria would have been particularly able to survive such a crash, on the basis that heat resistant bacteria *Microbispora* survived the Space Shuttle Columbia disaster.[24]

19.5 See also

- Astrobiology
- Fobos-Grunt
- Panspermia
- Planetary Society

19.6 References

[1] Запуск станции "Фобос-Грунт" к спутнику Марса отложен до 2011 года (in Russian). RIA Novosti. 21 September 2009. Retrieved 14 October 2009.

[2] Asian Scientist, "Phobos-Grunt Mission Carrying Yinghuo-1 Space Probe Suffers Technical Glitch", Srinivas Laxman, 9 November 2011

[3] The Planetary Society, "Phobos LIFE Ready to Launch", Bruce Betts, September 2011 (accessed 11-11-11)

[4] RIA Novosti, "Phobos-Grunt mission 'impossible,' says chief designer", 13 December 2011

[5] Universe Today, "Russian Space Program Prepares for Phobos-Grunt Re-Entry", David Warmflash, 13 December 2011

[6] Warmflash, David; Ciftcioglu, Neva; Fox, George; McKay, David S.; Friedman, Louis; Betts, Bruce; Kirschvink, Joseph (November 5–7, 2007). *Living interplanetary flight experiment (LIFE): An experiment on the survivalability of microorganisms during interplanetary travel* (PDF). Workshop on the Exploration of Phobos and Deimos. Ames Research Center.

[7] Zak, Anatoly (1 September 2008). "Mission Possible". *Air & Space Magazine*. Smithsonian Institution. Retrieved 26 May 2009.

[8] Astrobiology.com, "LIFE Launches Aboard Endeavour's Last Flight", The Planetary Society, 27 April 2011 (accessed 11-11-2011)

[9] SPACE.com, "Salvaging Science from Stricken Mars Moon Probe: A Scientist's View", David Warmflash, 11 November 2011

[10] Astrobiology.com, "LIFE Ready to Launch on Endeavour's Last Flight", The Planetary Society, 16 May 2011 (accessed 11-11-11)

[11] "Frequently Asked Questions". The Planetary Society. Retrieved 2 April 2011.

[12] "Ten Hardy Organisms Selected for the LIFE Experiment: Who will Survive the 3-Year Space Odyssey?". The Planetary Society. Retrieved 2 April 2011.

[13] Alexander, Amir. "The LIFE Organisms: Bacteria". The Planetary Society. Retrieved 2 April 2011.

[14] The Planetary Report, Volume XXIX, number 2, March/April 2009, "We make it happen! Who will survive? Ten hardy organisms selected for the LIFE project, by Amir Alexander

[15] Alexander, Amir. "The LIFE Organisms: Archaea". The Planetary Society. Retrieved 2 April 2011.

[16] Alexander, Amir. "The LIFE Organisms: Eukaryotea". The Planetary Society. Retrieved 2 April 2011.

[17] Jönsson, K. Ingemar Jönsson; Elke Rabbow; Ralph O. Schill; Mats Harms-Ringdahl & Petra Rettberg (9 September 2008). "Tardigrades survive exposure to space in low Earth orbit". *Current Biology* **18** (17): R729–R731. doi:10.1016/j.cub.2008.06.048. PMID 18786368.

[18] Gelfand, Mark (15 October 2008). "LIFE Experiment Module Passes Vibration and Impact Tests". The Planetary Society. Retrieved 2 April 2011.

[19] Clark, Stephen (6 February 2012). "Russia: Computer crash doomed Phobos-Grunt". *Spaceflight Now*. Retrieved 29 February 2012.

[20] "Programmers are to be blamed for the failure of Phobos mission". *ITAR-TASS News Agency*. 31 January 2012. Retrieved 29 February 2012.

[21] Emily Lakdawalla (January 13, 2012). "Bruce Betts: Reflections on Phobos LIFE". *The Planetary Society Blog*. Retrieved March 17, 2012.

[22] "Update on Phobos – Where's a shuttle when you need it?" Victor Grippi – The Atomic Writer. Retrieved 04-10-2012.

[23] DiGregorio, Barry E. (2010-12-28). "Don't send bugs to Mars". New Scientist. Retrieved 8 January 2011.

[24] McLean, R; Welsh, A; Casasanto, V (2006). "Microbial survival in space shuttle crash".*Icarus***181**(1): 323–325. Bibcode:200 doi:10.1016/j.icarus.2005.12.002. PMC 3144675. PMID 21804644.

19.7 External links

- The Planetary Society: Shuttle LIFE
- The Planetary Society: Phobos LIFE
- The Planetary Society: LIFE

19.8 Text and image sources, contributors, and licenses

19.8.1 Text

- **Tardigrade** *Source:* https://en.wikipedia.org/wiki/Tardigrade?oldid=689183497 *Contributors:* AxelBoldt, Magnus Manske, Matthew Woodcraft, Derek Ross, Mav, The Anome, Eclecticology, Josh Grosse, Rmhermen, PierreAbbat, Hephaestos, Quercusrobur, Stevertigo, Shyamal, Kku, Pcb21, Goatasaur, Jimfbleak, Kingturtle, Glenn, Ciphergoth, Cimon Avaro, GregRobson, Emperorbma, N-true, Quoth-22, Eugene van der Pijll, Robbot, Astronautics~enwiki, Schutz, Sam Spade, Babbage, Hadal, UtherSRG, Ruakh, Dave6, Jeremiah, Ancheta Wis, Bogdanb, Jjamison, Azu, Kpalion, Peter Ellis, Gyrofrog, OldakQuill, Andycjp, Gdr, Eep^2, Discospinster, Rich Farmbrough, PezHacker, Vsmith, JPX7, Pavel Vozenilek, Bender235, Mauler, Evand, Meggar, .:Ajvol:., Cwolfsheep, Foobaz, Defsac, Arcadian, Alansohn, NyaR, Andrewpmk, Andrew Gray, Riana, Axl, Barrkel, SidP, Cfrjlr, Pauli133, Dan East, Kazvorpal, Kitch, Adrian.benko, Mcsee, Stemonitis, Isfisk, Gmaxwell, OwenX, Mindmatrix, Uncle G, TomTheHand, BillC, Jacobolus, Benbest, Dayv, Sholtar, GregorB, Macaddct1984, Waldir, RichardWeiss, Graham87, BD2412, Josh Parris, Drbogdan, Rjwilmsi, Nightscream, Koavf, Omicron91, Theinsomniac4life, Salleman, Miserlou, Boccobrock, Yamamoto Ichiro, FlaBot, Crazycomputers, GT, Vdogamez, Fosnez, Intgr, Chobot, Gdrbot, Hall Monitor, Wjfox2005, YurikBot, Wavelength, RussBot, Ivirivi00, Hellbus, Shawn81, Mythsearcher, Quadraxis, NawlinWiki, DavidConrad, Dysmorodrepanis~enwiki, Dumoren, Trovatore, Joelr31, Apokryltaros, Moe Epsilon, TDogg310, Bota47, Blue Danube, Lycaon, Waqcku, Closedmouth, Arthur Rubin, LeonardoRob0t, Allens, Katieh5584, John Broughton, Yakudza, SmackBot, Derek Andrews, InverseHypercube, KnowledgeOfSelf, Grye, Wegesrand, Jacek Kendysz, Ganty, EncycloPetey, Chronodm, Gaff, Skizzik, Chris the speller, Josh215, RDBrown, OrangeDog, Papa November, Hibernian, Droll, George Church, Primacag, WDGraham, Jefffire, Abyssal, Onion~enwiki, Mwinog2777, Rainmonger, Triachus, VMS Mosaic, Seduisant, Leveretth, Edema756, SeanAhern, LavosBaconsForgotHisPassword, DMacks, Kendrick7, Shrumster, Felic~enwiki, Rory096, Kimmylee, Disavian, John Cumbers, Tktktk, Mgiganteus1, IronGargoyle, Bella Swan, Smith609, Optakeover, Sausagerooster, Myopic Bookworm, Grothmag, Paul venter, J Di, Courcelles, PaddyM, Steve64, Chamberlian, JForget, Stacecom, FredWallace18@yahoo.com, CuriousEric, DanielRigal, WeggeBot, ONUnicorn, Jonsinger, Cydebot, Kupirijo, 2 of 8, Hebrides, Tenbergen, JohnInDC, Thijs!bot, Youanden, Calaka, QuiteUnusual, Shirt58, Danger, RedCoat10, Sluzzelin, Janbollaert, JAnDbot, DuncanHill, Aderksen, Mark Shaw, Steveprutz, WolfmanSF, Bongwarrior, VoABot II, Gang14, Catgut, Branka France, 245891, BatteryIncluded, Zorque, Sorie, Allstarecho, Philg88, Robin S, Peter coxhead, NatureA16, Darth.maul.is.alive, Flowanda, Rettetast, Drewmutt, Anaxial, Dilzi, Vorratt, Slugger, J.delanoy, Philcha, Justin, OttoMäkelä, Rufous-crowned Sparrow, LordAnubisBOT, NewEnglandYankee, Nwbeeson, Ef3, 83d40m, Tigerdg, Imagi-King, SirBob42, DorganBot, Mike V, Pdcook, Marcin Suwalczan, VolkovBot, Jmrowland, Streetlightspark, Chienlit, Gmischke, Philip Trueman, TXiKiBoT, Technopat, Anonymous Dissident, Telespiza, John Carter, Bainemo, JimBurd, Greswik, DrMikeF, Traiharder, Leafnode, Newsaholic, Wikilias, Sduibek, SieBot, LarsHolmberg, Work permit, WereSpielChequers, ToePeu.bot, JazzLOCO, RJaguar3, Toghome, Grundle2600, Mrengy, Flyer22 Reborn, Hxhbot, Yerpo, Aperseghin, Antonio Lopez, PhilMacD, Rpgch, Jmesches, Escape Orbit, Wasami007, The Thing That Should Not Be, Rodhullandemu, Microshaw, EoGuy, WDavis1911, Boing! said Zebedee, MichaelAmongMichaels, Michaplot, Magic Player911, ?, Deselliers, Kitsunegami, Excirial, M4gnum0n, The Winged Yoshi, SpikeToronto, Happyguy1000, Nickw96, Jotterbot, 7, Good little chicken, BlueDevil, XLinkBot, Dthomsen8, Ost316, Trabelsiismail, Yuvn86, Spoonkymonkey, Whatelseisthere, Addbot, Mr0t1633, DOI bot, Lisapaloma, Wickey-nl, Bioapt~enwiki, Fluffernutter, Glane23, Borsodas, 84user, Tide rolls, Lightbot, Luckas-bot, TheSuave, Yobot, Ptbotgourou, II MusLiM HyBRiD II, Annehammel, Plazmatyk, Ganesh.rao, AnomieBOT, Metalhead94, Jim1138, JackieBot, AdjustShift, Henry Godric, Kingpin13, Tardidude, Citation bot, MithrasPriest, Verb47WRDC, ArthurBot, Xqbot, Wrightkids, Sionus, Capricorn42, Dtree16, Anna Frodesiak, Mlpearc, JCrue, Pereant antiburchius, Mandrake76, Ernsts, Thehelpfulbot, Izvora, FrescoBot, LucienBOT, Mknote, Lothar von Richthofen, TheArticleBurner, Umawera, Citation bot 1, Pinethicket, Jsrgls, Tom.Reding, Serols, JamesGrimshaw, Full-date unlinking bot, Jauhienij, Animalparty, Arctifox, Skakkle, Spabule, Suffusion of Yellow, Innotata, Jynto, Onel5969, RjwilmsiBot, Lbarquist, AShadowed, Ienpw III, Sirawitp, Androstachys, Skamecrazy123, CanadianPenguin, EmausBot, Koszmonaut, GoingBatty, Tommy2010, Lucas Thoms, ZéroBot, Dense4, Medeis, Donner60, EvenGreenerFish, Orange Suede Sofa, Roto2esdios, Kinkreet, Kleopatra, ClueBot NG, Satellizer, Name Omitted, Mikejamesshaw, Celestechang, Infinifold, Krollo, Widr, Helpful Pixie Bot, Badandy2021, Wbm1058, DBigXray, BZTMPS, BG19bot, Penguinmustard, M0rphzone, Darwin.kim, Sivler, MusikAnimal, Mark Arsten, Bradlake10, NotWith, Docdocdoc01, Mateom28, TBrandley, Vanischenu, K!r!93, Mrt3366, ChrisGualtieri, TheJJJunk, IjonTichyIjonTichy, Kissmegoodbyeandsleep, Dexbot, Crispy00001, Materiapensante, Koeniglra42, Imtanner138, Webclient101, Tyler Hruby, Dragonbladeaidan, Cor Ferrum, VanishedUser 2313214sad1, Shurakai, BreakfastJr, Howicus, Baneful Mauler, Asonofomegasupreme, Dert567, Kovl, Kharkiv07, Jeffroz, Biogerontology, Syedisreallycool, Berwyngio, Johnegbert1995, Ml912013, 7Sidz, ThatRusskiiGuy, Marlaina.vela, Monkbot, AKS.9955, Asteroidenbergbauer, Beachd1, Jaemes123, Cutofmyjib, Richard Yin, Feroshus11, Stacie Croquet, Signedzzz, Tiffanylittleboo, Daltonsexton, Bvansanten08, Religious king, Melbourne9, Loricochran, SandKitty256, IEditEncyclopedia, GoldCoastPrior, Danielfire20, Blub512, Lyricon, KasparBot, Ichtious, QuincyR13, Honkey32, Liam2000848, John Brosin, LexN88, Tardigrade man and Anonymous: 528
- **Micro-animal** *Source:* https://en.wikipedia.org/wiki/Micro-animal?oldid=672974564 *Contributors:* TheRealFennShysa, WadeSimMiser, Zotel, E Wing, Harej bot, AshLin, Crum375, AndrewDressel, TimVickers, 55david, Steveprutz, Outcomer, Justin, Hqb, Dendodge, 12 Noon, Sneep712, InternetMeme, Wikipedian Penguin, ClueBot NG, Theopolisme, Calabe1992, NotWith, Eosar, DoctorCartiledge, Eenth78, Username5373 and Anonymous: 8
- **Ecdysozoa** *Source:* https://en.wikipedia.org/wiki/Ecdysozoa?oldid=686088346 *Contributors:* Josh Grosse, JohnOwens, Shyamal, Stan Shebs, Evercat, MichaK, Wiwaxia, Robbot, Nilmerg, UtherSRG, Eequor, Gdr, DanielCD, Rich Farmbrough, Kwamikagami, Arcadian, House of Shin, Kazvorpal, April Arcus, Adrian.benko, Stemonitis, 2004-12-29T22:45Z, Karmosin, Palica, RichardWeiss, Rjwilmsi, FlaBot, Eubot, Chobot, Gdrbot, YurikBot, Hairy Dude, NTBot~enwiki, Jimp, Dysmorodrepanis~enwiki, Voyevoda, Johndburger, Mike Dillon, Donald Albury, SmackBot, Enlil Ninlil, EncycloPetey, Durova, Polyhedron, Jefffire, Abyssal, Cephal-odd, JorisvS, Smith609, Newone, Eluchil404, Tau'olunga, Alexei Kouprianov, A876, Shall I Make Wikipedia A Boom Town?, Thijs!bot, NJPharris, Kronos, GideonF, Danger, JAnDbot, Deflective, WolfmanSF, CommonsDelinker, Nono64, Petter Bøckman, Justin, Sholt.60, Jwasmuth, VolkovBot, Jalwikip, Dendodge, AllemborgoBot, Kidd Loris, BartekChom, Mild Bill Hiccup, Addbot, DOI bot, Ayrenz, AndersBot, ChenzwBot, Lightbot, Zorrobot, Luckas-bot, Yobot, Kjaer, AnomieBOT, Citation bot, Xqbot, Ernsts, Citation bot 1, DrilBot, Full-date unlinking bot, Trappist the monk, EmausBot, ZéroBot, AManWithNoPlan, ClueBot NG, Helpful Pixie Bot, Trypanosomes, Deeptime000, LastSeaOtter, BattyBot, SantoshBot, Saehry, Monkbot and Anonymous: 31
- **Cambrian** *Source:* https://en.wikipedia.org/wiki/Cambrian?oldid=686036545 *Contributors:* Mav, Bryan Derksen, Zundark, Tarquin, Ed Poor,

Youandme, Hephaestos, Olivier, Michael Hardy, Kudz, GUllman, Ahoerstemeier, Glenn, Emperorbma, Tarosan~enwiki, The Anomebot, Marshman, Dragons flight, Joy, Wetman, DLR (usurped), AnthonyQBachler, Phil Boswell, Romanm, Flauto Dolce, Rursus, Jsonitsac, Casito, Smjg, Graeme Bartlett, Jyril, StevenS757, Barbara Shack, Gilgamesh~enwiki, Jason Quinn, Mboverload, Antandrus, Quota, Imroy, DanielCD, Discospinster, Rich Farmbrough, Pmsyyz, Vsmith, Florian Blaschke, Ascánder, Dlloyd, Mani1, Bender235, JoeSmack, Neko-chan, El C, Pjrich, Kwamikagami, Bobo192, Robotje, I9Q79oL78KiL0QTFHgyc, Grutness, Siim, Alansohn, PAR, EagleFalconn, Dinoguy2, Versageek, Stemonitis, Nuno Tavares, Angr, Kelly Martin, Woohookitty, Kurzon, Zzyzx11, M Alan Kazlev, Prashanthns, Bluesmoon, Rjwilmsi, Koavf, Marasama, Vuong Ngan Ha, FlaBot, Godlord2, Chobot, The Rambling Man, YurikBot, Wavelength, RobotE, Peter G Werner, Pip2andahalf, Stco23, NawlinWiki, Lsisson, Bucketsofg, Wknight94, Richardcavell, Leptictidium, Deville, Rhynchosaur, Pb30, LeonardoRob0t, SmackBot, Hu Gadarn, Bomac, Yamaguchi??, Ohnoitsjamie, Andy M. Wang, Douglas Wilhelm Harder, Keegan, MalafayaBot, Fredvanner, DHN-bot~enwiki, A. B., Abyssal, AMK152, Addshore, Bejnar, Ohconfucius, SashatoBot, MrDarwin, JorisvS, Diverman, A. Parrot, Smith609, RJE42, Ceraurus, Hypnosifl, Jon186, Geologyguy, Qyd, KurtRaschke, Paul venter, P.Geol, StephenBuxton, Boreas74, IvanLanin, Tawkerbot2, The Letter J, Poolkris, Rambam rashi, Woudloper, Dr.Bastedo, CuriousEric, Moreschi, Peripitus, Arskoul, Michael C Price, RandomOrca2, JamesAM, Thijs!bot, Parsa, Marek69, Moon&Nature, AlefZet, AntiVandalBot, Spril4, Res2216firestar, Mikenorton, Plantsurfer, Sophie means wisdom, Pedro, VoABot II, Twsx, Fang 23, NatureA16, MartinBot, Kostisl, CommonsDelinker, J.delanoy, Numbo3, Peter Chastain, Crocadog, Clerks, Janet1983, Pdcook, Dorftrottel, Idioma-bot, Spellcast, Philip Trueman, Zamphuor, Ulioceras~enwiki, KAUclan, Kitty's little helper, Clarince63, Autodidactyl, Maxim, Gorank4, Dessymona, Logan, NHRHS2010, EmxBot, Red, Kennethcgass, Arturo zuniga, SieBot, WereSpielChequers, ToePeu.bot, RJaguar3, Brianchasejared, Happysailor, RadicalOne, Toddst1, Oda Mari, Wilson44691, Jc-S0CO, Oxymoron83, Lightmouse, BenoniBot~enwiki, OKBot, Capitalismojo, Spotty11222, ImageRemovalBot, ClueBot, The Thing That Should Not Be, T.Neo, Niceguyedc, Blanchardb, Harland1, Alxwht, The 888th Avatar, Boneyard90, Awickert, Excirial, BBKoVI, Pmronchi, Jo Weber, Kaiba, Wikimedes, Pieking44, Amaltheus, Darkicebot, Cameta, Qfl247, Culturenut, Facts707, Brianfreud, Badgernet, JBA2008, Aunt Entropy, Kacha94, Roentgenium111, DOI bot, Betterusername, Ronhjones, Lets Enjoy Life, CanadianLinuxUser, Daicaregos, 37ophiuchi, Debresser, Gsusgotmyheart, Favonian, Kyle1278, Tide rolls, Zorrobot, Legobot, Luckas-bot, Yobot, Apollonius 1236, Alnagov, AnomieBOT, DemocraticLuntz, Galoubet, Materialscientist, Are you ready for IPv6?, Citation bot, LilHelpa, Xqbot, Tylersquare, C+C, Omnipaedista, Geopersona, Spesh531, DyingDingo, Samwb123, FrescoBot, Coroboy, Citation bot 2, Drew R. Smith, Citation bot 1, Isolatedconductor, Pinethicket, TobeBot, Trappist the monk, Fama Clamosa, Dinamik-bot, Tobias1984, Lucius Winslow, Weedwhacker128, Censoredchinese, Jfmantis, Obsidian Soul, TjBot, Osephjay1997, Retallack, EmausBot, Orphan Wiki, Domesticenginerd, Look2See1, Chermundy, Tommy2010, Josve05a, Érico, Bamyers99, AManWithNoPlan, Dohn joe, L Kensington, Flightx52, RockMagnetist, Jaw2jaw, Herk1955, Petrb, ClueBot NG, Kenyikpefan, Guy123321, Gilderien, O.Koslowski, Widr, Gob Lofa, Bibcode Bot, Swads, BG19bot, Zollo9999, Hza a 9, MusikAnimal, Flyingtomato1000, Cadiomals, Gorobay, Zedshort, Justanonymous, Shirudo, BattyBot, ChrisGualtieri, Ekren, Froggydoodle, Dexbot, Leo Fyllnet, Lugia2453, Graphium, BetterSkatez, Tamarack velocity, Epicgenius, Loganfalco, Monkbot, Filedelinkerbot, Vieque, TropicalCyclones243, Julietdeltalima, Daniel Patterson98, VexorAbVikipædia, Johaniseenkneus, Rhumidian, KasparBot, A3X2, Odawgl01 and Anonymous: 352

- **Parthenogenesis** *Source:* https://en.wikipedia.org/wiki/Parthenogenesis?oldid=688738550 *Contributors:* AxelBoldt, Magnus Manske, SimonP, Frecklefoot, Shyamal, Tannin, Graue, Paul A, SebastianHelm, Ahoerstemeier, Kingturtle, Aarchiba, Evercat, Samw, Jacquerie27, Emperorbma, David Latapie, David Shay, Val42, Floydian, Pollinator, Fredrik, Zandperl, Altenmann, Lzur, Xanzzibar, MPF, Maha ts, Gilgamesh~enwiki, Crag, Pgan002, Andycjp, Calm, Latitudinarian, Knutux, Antandrus, Phe, Sajb, PDH, B.d.mills, Adashiel, DanielCD, Discospinster, Will2k, Lulu of the Lotus-Eaters, JPX7, Zaslav, Karmafist, Lima, Guettarda, Bobo192, Kghose, Viriditas, Scott Ritchie, Anthony Appleyard, Axl, Tchalvak, Seaboniks, Wtmitchell, SidP, Japanese Searobin, Stemonitis, Simetrical, Scjessey, WadeSimMiser, Chochopk, Damicatz, Marudubshinki, RichardWeiss, WBardwin, Edison, Mlewan, Rjwilmsi, Bensin, Acullum, FlaBot, Spudtater, Pete.Hurd, Daycd, Chobot, Sasoriza, Gwernol, Jayme, YurikBot, Wavelength, Jimp, Acefox, Chris Capoccia, YEvb0, Hydrargyrum, Grafen, Welsh, CPColin, DRosenbach, Deeday-UK, Scheinwerfermann, Karuma, Lt-wiki-bot, Nikkimaria, Arundhati bakshi, Nickybutt, Zvika, Jodarom, KnightRider~enwiki, SmackBot, ZorkFox, WookieInHeat, Pecosdave, EncycloPetey, Eskimbot, Kintetsubuffalo, Giandrea, NCurse, Thumperward, Ehrbar, WikiFlier, Sadads, Colonies Chris, Zazpot, JesseRafe, Aldaron, Jalirpa, Freemarket, Drphilharmonic, Adrigon, Lisasmall, Ojophoyimbo, DavidGC, Acidburn24m, Mgiganteus1, TastyPoutine, Ryulong, Novangelis, Robin Chen, Beefyt, Shoeofdeath, IvanLanin, Coffee Atoms, Dlohcierekim, CmdrObot, Calibanu, Glarbex, Dogman15, Mikebrand, Reywas92, Flowerpotman, Llort, Padent, Brendan, Doug Weller, Dyanega, Epbr123, Missvain, I do not exist, Z10x, Mailseth, Sad mouse, Saimhe, Just Chilling, Vendettax, Narfil Palùrfalas, Ioeth, JAnDbot, Deflective, QuantumEngineer, Lucy1981, Yurei-eggtart, Paradoxtwin, Homunq, WhatamIdoing, Adrian J. Hunter, Zfaulkes, Mnc4t, Atulsnischal, Sagqs, Petter Bøckman, J.delanoy, Hans Dunkelberg, Tomallen23, Maurice Carbonaro, DarkFalls, Interscan, Mikael Häggström, Cypherrage, EMT1871, Thesis4Eva, 83d40m, Nadiatalent, Id711, CardinalDan, VolkovBot, GianMarco Tavazzani, Kriis~enwiki, Philip Trueman, Tcaruso2, Jalwikip, Tameeria, Bspeakmon, Martin451, Rastrojo, RadiantRay, Zain Ebrahim111, One Elephant went out to play..., Avi saig, SieBot, JediRogue, Permacultura, Yahastu, Wilson44691, Yerpo, Garydale, AMbot, OKBot, AlanUS, JL-Bot, Fangjian, Dlrohrer2003, Bookish.blogger, ClueBot, User765, Papblak, Eric Wester, Awesomebitch, A754390463, Gherson2, Ribbon Salminen, Enthusiast01, Pete unseth, Stuthomas4, Excirial, Gnome de plume, Bruceanthro, XCalPab, Vivio Testarossa, Estirabot, Tervan, Sun Creator, SchreiberBike, Dylan38, Teleomatic, DumZiBoT, XLinkBot, Libcub, Alexius08, Menthaxpiperita, Tayste, Intlstemcell, Addbot, DOI bot, Jojhutton, Mr. Wheely Guy, LaaknorBot, Bernstein0275, ChenzwBot, TStein, Tayzhian, Justpassin, Prim Ethics, Tide rolls, Zorrobot, Yobot, Ptbotgourou, Bremenjock, Beeswaxcandle, SafeBoot, AnomieBOT, Piano non troppo, Ambaryer, Materialscientist, Citation bot, FHankFreeman, Xqbot, Thijhgf, Capricorn42, Arojr, FortyYrOldUniverse, Pmlineditor, Shadowjams, Joaquin008, Valentino76, GliderMaven, FrescoBot, BabyGirl3468, Deltik, Lonaowna, FaceInTheSand, Citation bot 1, Citation bot 4, Pinethicket, HRoestBot, Jonesey95, Helios13, Trappist the monk, Diblidabliduu, Animalparty, Falkybassist, Lotje, Miracle Pen, Suffusion of Yellow, Obsidian Soul, Cathardic, Onel5969, RjwilmsiBot, TjBot, Regancy42, EmausBot, John of Reading, Gfoley4, GoingBatty, Ryktz8, Scott9999999, TonyMath, Harshmits, Vogelabv, Spinifer, Staticd, WikiHotep, ClueBot NG, Macarenses, Argalite, Nivektrah, CopperSquare, Chitt66, DrChrissy, RafikiSykes, Helpful Pixie Bot, Maculosae tegmine lyncis, BG19bot, Gurt Posh, W-Blog, Gymnodactylus, Weirdtheory, Solarwind71, NotWith, Mike.BRZ, Morning Sunshine, Gibbja, 618116yfp, BattyBot, DanGuerrero, Meengrn, Dexbot, Darekk2, Sminthopsis84, David Tar, Lunivore, WolfyFTW, Epicgenius, Lefty50, The striker16, Inezdeborahealtar, Physics,Biology, Chemistry, Alyssa Pawl, Cecilauthor, Xolani90, Ashishnaval, Monkbot, Napisdog, Maxzacher, AlexBroBrown, TranquilHope, Kikaya alvin, SarahTehCat, Knowledgebattle, Dardky, Sennsationalist, Pigmentkleur and Anonymous: 337

- **Oviparity** *Source:* https://en.wikipedia.org/wiki/Oviparity?oldid=689267827 *Contributors:* Pigsonthewing, DocWatson42, B.d.mills, Dr.frog, Arcadian, MoraSique, Ghirlandajo, Bomac, LinguistAtLarge, MalafayaBot, Tigerhawkvok, Krashlandon, Courcelles, Ozzieboy, JAnDbot, Kaobear, Steveprutz, Objectivesea, Branka France, LookingGlass, Mike Searson, MartinBot, Wlodzimierz, GDW13, Nadiatalent, Jalwikip,

Haplochromis, AlleborgoBot, SieBot, Iceshark7, Treehill, ClueBot, DragonBot, Sun Creator, Selbymayfair, Addbot, Dawynn, Zorrobot, Jarble, Luckas-bot, Dinesh smita, ArthurBot, Obersachsebot, Matthias Willerich, GrouchoBot, Omnipaedista, FrescoBot, FoxBot, Lotje, Dinamik-bot, EmausBot, WikitanvirBot, Essassian, ZéroBot, Rsrking, MarinaMichaels, Donner60, ClueBot NG, Agent 78787, FamAD123, Missmanasa and Anonymous: 28

- **Cryptobiosis** *Source:* https://en.wikipedia.org/wiki/Cryptobiosis?oldid=671156690 *Contributors:* Nealmcb, Gabbe, Julesd, Bogdangiusca, Auric, Dratman, Andycjp, Khaosworks, Xezbeth, Benbest, Sholtar, Rjwilmsi, Feco, Eubot, YurikBot, Wavelength, Ivirivi00, Annabel, Dysmorodrepanis~enwiki, Daniel Mietchen, Dan Harkless, SmackBot, Giraldusfaber, Bomac, Bluebot, Thumperward, Jagoode, Drphilharmonic, BHC, Mgiganteus1, Tmcw, Stifynsemons, Scirocco6, Christian75, Thijs!bot, Colincmr, Wench999, William E Ostrem, Davidhorman, Smartse, Battlekow, Steveprutz, Magioladitis, DukeTwicep, MarceloB, Wingedsubmariner, Gorank4, SieBot, Happysailor, Eriksiers, Deviator13, Take-away, Ouedbirdwatcher, Vojtěch Dostál, Heavy Seltzer, Addbot, Ronhjones, Eliteramen, Peti610botH, Lightbot, Mfhulskemper, Luckas-bot, AnomieBOT, LilHelpa, Worldoffish, SkyMachine, RjwilmsiBot, Dlambe3, Cihanerkut, ZéroBot, Wingman4l7, EdoBot, ClueBot NG, Frietjes, BG19bot, Nchakrab, Debbie27, BattyBot, Everything Is Numbers, Monkbot, Liance, Drkndm and Anonymous: 46
- **Panspermia** *Source:* https://en.wikipedia.org/wiki/Panspermia?oldid=688834702 *Contributors:* Mav, Bryan Derksen, Timo Honkasalo, Taw, Andre Engels, William Avery, SimonP, Shii, Maury Markowitz, Stevertigo, Edward, Alan Peakall, Alfio, CesarB, JWSchmidt, Whkoh, Kimiko, Timwi, RickK, Rednblu, Abscissa, VeryVerily, Populus, Paul-L~enwiki, Omegatron, Samsara, Warofdreams, Proteus, Tonderai, Jeffq, RadicalBender, Northgrove, Chris Roy, Postdlf, Sverdrup, Rholton, Rebrane, Wlievens, David Edgar, Jor, Cyberia23, Per Abrahamsen, Nagelfar, Alan Liefting, Ncox, Kbahey, Jyril, Bfinn, Dmb000006, MingMecca, Duncharris, Solipsist, SWAdair, Bobblewik, Pgan002, Alexf, Bcameron54, Piotrus, Melikamp, Csmiller, Taka, Obby~enwiki, Mzajac, Cglassey, Sam, Lacrimosus, Rich Farmbrough, Vsmith, ArnoldReinhold, YUL89YYZ, Dbachmann, BACbKA, RJHall, CanisRufus, RoyBoy, Godfreylouis, Viriditas, Cmdrjameson, Nicke Lilltroll~enwiki, Dejitarob, L33tminion, Hob Gadling, Vicarage, Pschemp, Slipperyweasel, Calebe, Chino, Arthena, JoaoRicardo, Ahruman, Hu, Mnemo, TahitiB~enwiki, Pauli133, Gene Nygaard, Dismas, Stephen, Gmaxwell, Mindmatrix, TotoBaggins, Pictureuploader, DanHobley, Holek, Dbutler1986, Mandarax, Marskell, Drbogdan, Rjwilmsi, Oblivious, Miserlou, SonicSpike, Erkcan, Krash, Dionyseus, FlaBot, Ian Pitchford, John Baez, Papacha, Diza, SteveBaker, Gurubrahma, Chobot, Voodoom, Bgwhite, Wavelength, Acefox, Chris Capoccia, Akamad, Dysmorodrepanis~enwiki, Icelight, Anetode, Wangi, Emdx, DeadEyeArrow, Skepticsteve, Elkman, Pegship, Noosfractal, 2over0, Jules.LT, Chase me ladies, I'm the Cavalry, Abune, SMcCandlish, Reyk, Petri Krohn, Pádraic MacUidhir, Brentt, SmackBot, Judith.d, Melchoir, Elfsareus, Kintetsubuffalo, Portillo, Ohnoitsjamie, Hmains, Jushi, Kinhull, Qwasty, Sumthingweird, RDBrown, Thumperward, Nemodomi, Silly rabbit, Hibernian, Zephyr707, WDGraham, Jefffire, Frap, OrphanBot, Nixeagle, Thomqi, Soosed, Wen D House, Tsop, Iamdaniel, Kismetmagic, Wizardman, Bklyce, StN, Virago, BrownHairedGirl, Eaglecros, John, Siddharth srinivasan, John Cumbers, Neodarksaver, ISoron, Extremophile, Skymist, Novangelis, Corykoski, Jason.grossman, FelisSchrödingeris, Greygirlbeast, Courcelles, Brainbark, Centered1, Banedon, Kylu, RagingR2, JFreeman, Daniel J. Leivick, Dancter, Michael C Price, DumbBOT, Iliank, Robertinventor, Arb, UberScienceNerd, Thijs!bot, JAF1970, Mpallen, Headbomb, John254, Second Quantization, Z10x, Iulius, Morgana The Argent, The Hams, NeilEvans, Gnixon, Yellowdesk, Rothorpe, LittleOldMe, Magioladitis, Professor marginalia, Yakushima, Theroadislong, Gabriel Kielland, BatteryIncluded, Joe hill, Thibbs, DerHexer, Lelek, Waninge, Urco, Jim.henderson, Rettetast, Ulisse0, AlphaEta, Trusilver, Terrek, Nsande01, Ian.thomson, ABVS1936, Skier Dude, Kukec, RenniePet, Davy p, M-le-mot-dit, EyeRmonkey, Diletante, Idioma-bot, Evolvearth, Soliloquial, Dom Kaos, Philip Trueman, JayEsJay, TXiKiBoT, Mathwhiz 29, Cloudswrest, Stickyhammer, Agmon, AlleborgoBot, Hrafn, SieBot, Coffee, Hugh16, Arbor to SJ, Tiki 92090, Mimihitam, Excutio, LSmok3, Sunrise, Anchor Link Bot, Mnmautner, Martarius, Doyee5, ClueBot, CarolSpears, Foxj, Wanderer57, DrFO.Jr.Tn~enwiki, 1111news, Niceguyedc, Nanobear~enwiki, Sid-Vicious, Sjdunn9, Deselliers, Scog, BSmith821, HRosenberg, Fitzburgh, Unmerklich, Aitias, Apparition11, Crowsnest, Mrmpsy, DumZiBoT, Hotcrocodile, Rror, Oogaboogabooga, Fabio6043, BrucePodger, Addbot, Basilicofresco, DOI bot, Potentialten, Ronhjones, Chamberlain2007, Tdehel, MrOllie, Download, Glane23, SamatBot, Scienceislife, Beren, MuZemike, Cannizzaro S, Legobot, Luckas-bot, Yobot, Wikipedian Penguin, Untrue Believer, AnomieBOT, Taylordw, Rubinbot, Citation bot, Donkyhotay, Eumolpo, Quebec99, LilHelpa, Xqbot, Silver Spoon Sokpop, Albalma, Luis Felipe Schenone, GrouchoBot, Knightofcydonia49, Omnipaedista, N419BH, Hdrosenberg, Unused0011, A.amitkumar, Nagualdesign, FrescoBot, LucienBOT, Styxpaint, Citation bot 1, Solarflaredigital, Pinethicket, Jonesey95, Tom.Reding, Fumitol, SkyMachine, IVAN3MAN, Trappist the monk, Jonkerz, RjwilmsiBot, Damaavand, Ansaazi, Androstachys, EmausBot, Clive tooth, Emmajanej, Evanh2008, JacobSheehy, Kp grewal, ZéroBot, AManWithNoPlan, David J Johnson, Actsmart, Orange Suede Sofa, Spicemix, Grapple X, ClueBot NG, Njh321, Kikichugirl, Liveintheforests, RocketLauncher2, Old wombat, BrigKlyce, Russellml, Widr, Secret of success, Helpful Pixie Bot, Curb Chain, Gob Lofa, Bibcode Bot, Lilman0509, BG19bot, Mariansavu, Chemistryfan, Walterfarah, Kooky2, BattyBot, Arodr451, MichaelEF71, ChrisGualtieri, Khazar2, Rchouake, Dexbot, SummerWillow, Adam2828, Abepeace, Tomdarkblade, Melonkelon, KingSupernova, AlexeiSharov, Praemonitus, Syd Menon, Npascucci01, Someone not using his real name, HarbingerOfLunch, Monkbot, Denny123123, Formuse, BicelPhD, Mapsfly, HiBlueSky, Cspoleta, ComicsAreJustAllRight, Dsmith125, Alexis Gervais, Jnav7, Dutral and Anonymous: 316
- **Abiogenesis** *Source:* https://en.wikipedia.org/wiki/Abiogenesis?oldid=689226463 *Contributors:* Damian Yerrick, AxelBoldt, Joao, Bryan Derksen, The Anome, Sjc, -- April, Ed Poor, SimonP, Maury Markowitz, AdamRetchless, Zadcat, Mjb, Heron, Someone else, Lexor, Gabbe, Martin BENOIT~enwiki, Bobby D. Bryant, Ixfd64, Cyde, Sannse, Mcarling, Ihcoyc, Ellywa, Mdebets, Cyp, JWSchmidt, Julesd, Raven in Orbit, Norwikian, Ec5618, Charles Matthews, Timwi, Steinsky, Foodman, Maximus Rex, David Shay, Populus, Omegatron, Samsara, Jackson~enwiki, Raul654, Johnleemk, Finlay McWalter, Skaffman, Twang, Jason Potter, Robbot, Fredrik, Goethean, Altenmann, Nurg, Rursus, Rebrane, Sheridan, Hadal, Wereon, Raeky, Xanzzibar, Xyzzyva, Giftlite, Mshonle~enwiki, Polsmeth, Pretzelpaws, Everyking, Curps, Solipsist, Bobblewik, Pgan002, Andycjp, Keith Edkins, Sonjaaa, Quadell, Beland, Onco p53, Nograpes, Savant1984, JohnArmagh, Deglr6328, Flex, Lacrimosus, Mike Rosoft, Ta bu shi da yu, Rfl, Discospinster, Rich Farmbrough, Vsmith, ArnoldReinhold, Dave souza, Paul August, Bender235, ESkog, Srbauer, RJHall, Mr. Billion, Crunchy Frog, José Gnudista, Lycurgus, Kwamikagami, Liberatus, Sietse Snel, Art LaPella, RoyBoy, Fufthmin, Guettarda, Causa sui, Bobo192, John Vandenberg, Enric Naval, Viriditas, .:Ajvol:., ZayZayEM, I9Q79oL78KiL0QTFHgyc, VBGFscJUn3, Sulai~enwiki, Hob Gadling, A Karley, Orangemarlin, Marwood, DanielVallstrom, Darrelljon, Psychofox, SlimVirgin, Ferrierd, Kocio, InShaneee, Wtmitchell, Velella, Darco, XB-70, Knowledge Seeker, Pauli133, Tainter, BerndH, Bdrasin, Linas, Mindmatrix, Anilocra, LOL, Rocastelo, Schultz.Ryan, Tabletop, Grace Note, Sadettin, GregorB, CharlesC, Wdanwatts, Essjay, Palica, Gerbrant, GSlicer, RichardWeiss, Alienus, V8rik, BD2412, Rkevins, Sjö, Drbogdan, Rjwilmsi, Mayumashu, Nightscream, Koavf, Zbxgscqf, OneWeirdDude, Bob A, XP1, Crazynas, Mikedelsol, Bfigura, SLi, Duagloth, Margosbot~enwiki, Nihiltres, Alhutch, Geologist~enwiki, Vanished user psdfiwnef3niurunfiuh234ruhfwdb7, WhyBeNormal, Knoma Tsujmai, Truthteller, Chobot, DVdm, Bgwhite, Poorsod, YurikBot, Spacepotato, RadioFan2 (usurped), GPS Pilot, The Hokkaido Crow, NawlinWiki, Rick Norwood, DragonHawk, Dysmorodrepanis~enwiki, Uberisaac,

Dtrebbien, Seirscius, Zarel, SAE1962, RecSpecz, Apokryltaros, Nick, Kdbuffalo, E rulez, Crasshopper, Kortoso, Stefan Udrea, WAS 4.250, 2over0, Encephalon, Bhumiya, Smoggyrob, Davril2020, Petri Krohn, Fram, DisambigBot, JDspeeder1, NeilN, CIreland, Victor falk, NetRoller 3D, Quadpus, KnightRider~enwiki, SmackBot, Eperotao, PiCo, John Croft, Rtc, TestPilot, David Shear, Lankenau, Bmearns, BiT, Edgar181, Yamaguchi⍰⍰, Macintosh User, Gilliam, Portillo, Betacommand, Skizzik, Eloy, Chris the speller, Kaylus, RDBrown, Davep.org, Jprg1966, Thumperward, Silly rabbit, Hibernian, Complexica, Jeff5102, Scwlong, John Hyams, JoelWhy, Jefffire, Viperphantom, Vanished User 0001, Avb, Cfassett, Ines it, Khukri, John D. Croft, Richard001, Archgoon, Smokefoot, Greg.collver, The PIPE, DMacks, Sammy1339, Daniel.Cardenas, Denise from the Cosby Show, Alan G. Archer, Ohconfucius, SashatoBot, Danielrcote, Technocratic, Gloriamarie, Attys, Atkinson 291, Khazar, John, Writtenonsand, Butko, JoshuaZ, JorisvS, Robert Stevens, Mgiganteus1, Olin, Scetoaux, Fig wright, Extremophile, 041744, Robbins, A. Parrot, Tarcieri, Smith609, Makyen, Stevebritgimp, Tac2z, Mr Stephen, Xiaphias, Larrymcp, NJA, Novangelis, LenW, Dan Gluck, Nehrams2020, Clarityfiend, Twas Now, Lent, The Letter J, George100, Chris55, VinnieCool, DangerousPanda, CRGreathouse, Ale jrb, Memetics, BeenAroundAWhile, Runningonbrains, RoliSoft, ButFli, WeggeBot, Moreschi, Richard Keatinge, Nnp, Myasuda, Ciyean, Abeg92, Peterdjones, Cyhawk, Hughgr, Michael C Price, Doug Weller, DumbBOT, Narayanese, DnimrevO, Ebyabe, Crum375, PKT, Thijs!bot, Barticus88, Ryansca, Pstanton, Mojo Hand, Mungomba, Headbomb, James086, Astrobiologist, Davidhorman, Chandler, Gossamers, AntiVandalBot, Luna Santin, Guy Macon, Dbrodbeck, Gnixon, TimVickers, Cstreet, Smartse, Fluffy654, Danny lost, Princeofexcess, JAnDbot, XyBot, GromXXVII, MER-C, The Transhumanist, Matthew Fennell, Mildly Mad, Andonic, Xeno, Panarjedde, TAnthony, Tstrobaugh, Rothorpe, Kornbelt888, Magioladitis, Carlwev, Sushant gupta, JNW, CattleGirl, Harelx, Trishm, Hubbardaie, Mark PEA, Recurring dreams, Zephyr2k~enwiki, Theroadislong, Cgingold, BatteryIncluded, Allstarecho, Lyonscc, DerHexer, Edward321, Urco, JohanViklund, Mdsats, Robin S, Drm310, Tsinoyboi, Keith D, R'n'B, CommonsDelinker, Verdatum, Leyo, Mzaki, Player 03, PhageRules1, Ulisse0, Ifomichev~enwiki, AstroHurricane001, Avkulkarni, Rlsheehan, Hans Dunkelberg, Sidhekin, AmagicalFishy, Dispenser, It Is Me Here, Enuja, McSly, Tarotcards, Janet1983, Davy p, RobinGrant, Lbeaumont, Jorfer, Cmichael, KylieTastic, AzureCitizen, IceDragon64, Funandtrvl, Novernae, Jamiejoseph, Speaker to wolves, Philip Trueman, Sub-life, Vipinhari, GcSwRhIc, Charlesdrakew, Matthewrossing, Littlealien182, Steven J. Anderson, Awl, AllGloryToTheHypnotoad, Noformation, MacFodder, Mannafredo, Mishlai, Gibson Flying V, Wikiisawesome, Maxim, Shanata, WinTakeAll, Distinguisher, SheffieldSteel, Wolfrock, Lamro, Synthebot, Zarcoen, Omermar, Northfox, Rep07, Planet-man828, Hrafn, Nachohosking, EGMAG, Tczuel, Macdonald-ross, Gnocchi, Carny, KatieandHandy, Nihil novi, ToePeu.bot, Meldor, Dawn Bard, ConfuciusOrnis, Odd nature, Yintan, 0xFFFF, Abhishikt, Chhandama, Oda Mari, Jc-S0CO, Oxymoron83, Lightmouse, Helikophis, RW Marloe, Manifolds~enwiki, Jruderman, RyanParis, Sunrise, Diego Grez-Cañete, Skeptical scientist, StaticGull, Mos bratrud, Tesi1700, Hamiltondaniel, Driftwood87, Kalidasa 777, Marmenta, Lucius Sempronius Turpio, Twinsday, Sfan00 IMG, ClueBot, Tmol42, Fyyer, The Thing That Should Not Be, AstroMark, Sexiestjen4u, Desoto10, Pi zero, Unbuttered Parsnip, Jumacdon, Canopus1, Polyamorph, Timberframe, Tfpsly, Niceguyedc, Baegis, Alexis Brooke M, Rotational, Jandew, Paulcmnt, Excirial, Gustavocarra, Winston365, Vital Forces, Shinkolobwe, Abeo iniuria, Sun Creator, Eznight, Coinmanj, NuclearWarfare, SchreiberBike, Audaciter, BOTarate, Truth is relative, understanding is limited, Thusled, Thingg, Aitias, AC+79 3888, Johnuniq, Egmontaz, Editor2020, Goodvac, Bentheadvocate, Darkicebot, CaptainVideo890, XLinkBot, Roxy the dog, Jytdog, Jovianeye, Rror, Bradv, Elfgeek, Ost316, Jungfruchallan, Aloboof123, Opaq87, Aunt Entropy, Virajelix, Thatguyflint, Janisterzaj, Addbot, Roentgenium111, DOI bot, Landon1980, Swissmeister, Ronhjones, CanadianLinuxUser, Dsmith77, Lindert, Download, Redheylin, Bernstein0275, Camedit, Blade13125, Wildreceleste, Polyp2, LinkFA-Bot, Quietmarc, Partofwhole, U3190, Tide rolls, TL782, Romaioi, Nase, Yobot, StarTroll, Scepticus2, Yngvadottir, The Earwig, Punu, 489thCorsica, Cseppala, CinchBug, Dr.Buttons, AnomieBOT, Brroga, Mike Hayes, Kerfuffler, JWSurf, Trabucogold, Csigabi, Mann jess, Materialscientist, Citation bot, Quebec99, Romandoggie, LilHelpa, FreeRangeFrog, Xqbot, Sventington the Second, Blorblowthno, Wapondaponda, Mnnlaxer, Δζ, Nasnema, Mononomic, Turk oğlan, J JMesserly, Crzer07, 7h3 3L173, DerryTaylor, ProtectionTaggingBot, Gui le Roi, Conquistador, Sophus Bie, Ramssiss, Shadowjams, Methcub, Eugene-elgato, Joaquin008, Biem, FrescoBot, Finstergeist, Yanima, Hoffmannrungethailand, BKMBC3, Machine Elf 1735, Trkiehl, Citation bot 1, Redrose64, ANDROBETA, DrilBot, Winterst, Gravityguy, WaveRunner85, Gamocamo, Jonesey95, Helzrule19, Tom.Reding, Deleteduser2015, Hoo man, SpaceFlight89, FormerIP, Tanzania, Jerrywickey, Mikespedia, Jandalhandler, Kibi78704, MichaelExe, Fartherred, SkyMachine, IVAN3MAN, Trappist the monk, Silenceisgod, MEPK, Fama Clamosa, Comet Tuttle, Mcfl116, Vrenator, Victorfrogg, Jimmetry, Clarkcj12, Bcoolsdad, Diannaa, 564dude, Jynto, Gregrutz, Myrmidon1, DARTH SIDIOUS 2, Tor1714, Onel5969, RjwilmsiBot, Apotheosa, Hppa, Plommespiser, WildBot, Tesseract2, I belong to Jesus Christ, EmausBot, Jeffhughes22, Immunize, Dominus Vobisdu, Niluop, Dewritech, Ibbn, RespoonsibileSQ, Tamtrible, Jmv2009, Pboehnke, Slightsmile, Tommy2010, Kiran Gopi, Mmeijeri, Solomonfromfinland, Ofekalef, H3llBot, Wayne Slam, David J Johnson, Ksarasofi, Korztin, Jesanj, Brandmeister, L Kensington, Scientific29, Ego White Tray, Tanoan, Renji911, SemanticMantis, Dr. Hipopotamo, JanetteDoe, Sven Manguard, JonRichfield, Ldvhl, Zuky79, Gary Dee, AUN4, ClueBot NG, Don Para, E3cubestore, Afterrock81, Colin Fredericks, Rainbowwrasse, Jorge 2701, Joefromrandb, Sketchup123, DonaldRichardSands, DS Belgium, Sjmantyl, Asukite, Telpardec, Keenedged, Wikiwiki180, MerllwBot, Lotterox, Michaeltdeans, Helpful Pixie Bot, Elefnose, Cinnaplum, Anentiresleeve, Curb Chain, Bibcode Bot, Mwregehr, BG19bot, Lebs27, Expewikiwriter, Vevanpelt, Karmstrong909, Knowledge Examiner, Halstedcw, Mark Arsten, IraChesterfield, Drewrainey, Dkspartan1, Cauhtcoatl, Գաբրիելյան Աշոտ, MLearry, Cadiomals, Mthoodhood, Ghostsarememories, Harizotoh9, Blackstar167, HMman, Dontreader, Zedshort, Zetazeros, Dontshootimgay, Benyboy2, BattyBot, Decruft, Sfarney, Hghyux, Marc Tessera, Jimw338, SkepticalRaptor, David B Stephens, Soulbust, TheJJJunk, Garamond Lethe, Tanookiinashu, Khazar2, Ekren, Nathanielfirst, Elfinanciero222, Cmw255, Солярист, Pterodactyloid, RGA1980, Dexbot, Webclient101, Jinx69, Cerabot~enwiki, TippyGoomba, Mbreht, TheTahoeNatrLuvnYaho, CuriousMind01, Saehry, Leptus Froggi, 93, TruthOrTruthy, Corinne, Frivolous Consultant, HerbertHuey, FlaviusFerry, Tjmiler, Reatlas, Anastronomer, Bret palmer, Faizan, ICameHereToEdit, Surfer43, KnowledgeIncreases07, Analiticus, StewartGriffiths, Nirendeka, DavidLeighEllis, Ronaldo Laranja, Nigellwh, AbioScientistGenesis, SzostakJack, Mj12hoaxwriter, MDPub13, PubMed2015, Andreas.Geisler, EunuchRU, PrivateMasterHD, SpazAbiogenesis, Leptinresistinadiponectin, NottNott, SuperFreakCell, Anrnusna, Stamptrader, Sstur, Suelru, Chaya5260, Inphynite, Kkosman, Baltazorgue, Johngraybosch, Monkbot, BethNaught, Acagastya, Garfield Garfield, Shandck, Signedzzz, Brianbleakley, Pombrand, Ruwdaman, Fried Vegetables, BicelPhD, Yazan atheos, BlueFenixReborn, Strongjam, Sarr Cat, Imradinmyownway, Washington Charter, Chemistryorigin, Michaelo1019, One sanguin, KasparBot, Fernando orrego, Atchoum, Ktns, Paula NK, Joholub123, Shadowblade001 and Anonymous: 714

- **Biotic material** *Source:* https://en.wikipedia.org/wiki/Biotic_material?oldid=687601269 *Contributors:* Altenmann, HaeB, Pengo, Rich Farmbrough, Paleorthid, Drbogdan, Chris the speller, Lsjzl, Ventifact, Brewhaha@edmc.net, BatteryIncluded, MartinBot, Sheep2000, AzureIcicle, DRTllbrg, Anxietycello, Erik9bot, Look2See1, Ego White Tray, Mark Arsten, NottNott, Kevt2002, Nahid monavarian and Anonymous: 7

- **Directed panspermia** *Source:* https://en.wikipedia.org/wiki/Directed_panspermia?oldid=684009535 *Contributors:* Bearcat, Graeme Bartlett,

Solipsist, RHaworth, Rjwilmsi, Kolbasz, Bgwhite, Kintetsubuffalo, BatteryIncluded, Dom Kaos, Wingedsubmariner, Eeekster, SchreiberBike, Paper45tee, AnomieBOT, LilHelpa, Safiel, Jonesey95, Tom.Reding, Rushbugled13, RjwilmsiBot, John of Reading, Yiosie2356, EvenGreenerFish, ClueBot NG, AbrahamDavidson, Bibcode Bot, DrCruse, Adam2828, Abepeace, Reatlas, Maplestrip, Tetra quark, Koenprins and Anonymous: 12

- **List of interstellar and circumstellar molecules** *Source:* https://en.wikipedia.org/wiki/List_of_interstellar_and_circumstellar_molecules?oldid=688991332 *Contributors:* Bryan Derksen, CBDunkerson, Dbenbenn, Graeme Bartlett, Dratman, BrendanRyan, Jorge Stolfi, Foobar, Darrien, Karol Langner, Spiffy sperry, Tompw, RJHall, Huntster, Bnikolic, Pearle, Snarfevs, John Coupe, EagleFalconn, Shoefly, ThomasWinwood, Drbogdan, Rjwilmsi, Marasama, Mike s, Mike Peel, HappyCamper, Takometer, Bgwhite, Spacepotato, Deville, Reyk, Poulpy, Tropylium, SmackBot, Edgar181, Chris the speller, Bduke, Modest Genius, Chlewbot, JohnI, AstroChemist, RekishiEJ, CmdrObot, Cydebot, Astrochemist, Rifleman 82, Nikopoley, Headbomb, Pixelface, JAnDbot, Jingxin, Magioladitis, BatteryIncluded, Nono64, Leyo, James McBride, Lightmouse, Nergaal, ImageRemovalBot, NuclearWarfare, Scog, BSmith821, Ost316, Addbot, DOI bot, LinkFA-Bot, Yobot, AnomieBOT, Citation bot, Blundgr2, Hcnhplus, Citation bot 1, Jonesey95, Tom.Reding, Mikespedia, Trappist the monk, InvaderXan, Jynto, RjwilmsiBot, John of Reading, GoingBatty, TuHan-Bot, H3llBot, Whoop whoop pull up, Kikichugirl, Frietjes, Polskivinnik, Helpful Pixie Bot, Bibcode Bot, BG19bot, BattyBot, Dexbot, Ruby Murray, Eyesnore, Monkbot, CxHy and Anonymous: 34
- **Extraterrestrial life** *Source:* https://en.wikipedia.org/wiki/Extraterrestrial_life?oldid=689268571 *Contributors:* Damian Yerrick, Dreamyshade, Eloquence, Mav, Bryan Derksen, Robert Merkel, Zundark, The Anome, Ed Poor, TomCerul, Graft, Edward, Patrick, Infrogmation, Bewildebeast, Lexor, Kku, Ixfd64, Gdvorsky, Skysmith, Paul A, Minesweeper, Alfio, Kosebamse, Tregoweth, Ahoerstemeier, Cyp, JWSchmidt, BigFatBuddha, Darkwind, Julesd, Glenn, Andres, Corixidae, Mxn, Guaka, Timwi, RickK, WhisperToMe, Steinsky, Nv8200pa, Martinphi, Thue, Earthsound, Wetman, Proteus, Johnleemk, PuzzletChung, Riddley, Phil Boswell, Nufy8, Moriori, Tlogmer, Goethean, Romanm, Modulatum, Lowellian, PedroPVZ, Rursus, SchmuckyTheCat, Ojigiri~enwiki, Jondel, Rasmus Faber, Hadal, Mushroom, Lupo, Dina, Tobias Bergemann, David Gerard, Giftlite, Dbenbenn, MPF, Lethe, Tom harrison, Ich, Peruvianllama, TomViza, Maha ts, Gracefool, Eequor, Glengarry, Simulcra, Gadfium, Utcursch, J~enwiki, Sonjaaa, Antandrus, Beland, OverlordQ, Robert Brockway, Mark5677, Jossi, Redroach, Latitude0116, Balcer, Kevin B12, Mysidia, Phil1988, Icairns, Sam Hocevar, Arcturus, KeithTyler, Joyous!, Damieng, Adashiel, Valmi, Jimaginator, Eep², Mike Rosoft, Natrij, Freakofnurture, JTN, Gimmick Account, Discospinster, Rich Farmbrough, Rhobite, Cacycle, Vsmith, ArnoldReinhold, Dave souza, LindsayH, Antaeus Feldspar, Dbachmann, Mani1, Grutter, SpookyMulder, Jackqu7, Bender235, MisterBadIdea, Cyclopia, Kaisershatner, Ben Standeven, Pedant, Brian0918, RJHall, CanisRufus, Jpittman, MBisanz, El C, Shrike, Lycurgus, Nonpareility, Kwamikagami, RoyBoy, EurekaLott, Dustinasby, Bastique, Bobo192, Dystopos, Enric Naval, Viriditas, GTubio, Rbj, Angie Y., Mytildebang, Jag123, Forteanajones, La goutte de pluie, Acjelen, Pierre2012, John Fader, Polylerus, Gsklee, Mareino, OGoncho, Storm Rider, Zachlipton, Alansohn, Coma28, Transfinite, Blahma, Neitram, 119, Arthena, Visviva, Keolah, Andrewpmk, Riana, AzaToth, Hinotori, James.england, Fritzpoll, Daniel.inform, Alex '05, Cjnm, Mysdaao, Mrestko, DreamGuy, Wtmitchell, Velella, Max rspct, Docboat, Sudachi, Jesvane, LFaraone, Bsadowski1, Sfacets, Computerjoe, Kusma, Arthur Warrington Thomas, LukeSurl, Kazvorpal, Ceyockey, Centauri, BerserkerBen, Duke33, Njk, Alex.g, WilliamKF, MickWest, Angr, Lincspoacher, Roboshed, OwenX, Woohookitty, TigerShark, Etacar11, Molloy, Pinball22, Mathmo, Jersyko, StradivariusTV, Urod, WadeSimMiser, Jeff3000, MONGO, Moormand, Oreckel, Dmol, Tomlillis, Grika, Optichan, KFan II, Norro, HollyI, CharlesC, Waldir, Zzyzx11, Wayward, Dromedary, Essjay, MarcoTolo, Mandarax, Nivedh, Ashmoo, Graham87, Marskell, WBardwin, Magister Mathematicae, V8rik, BD2412, MikeDockery, OGRastamon, FreplySpang, Quantum bird, Kane5187, Ciroa, Kafuffle, Ketiltrout, Drbogdan, Sjakkalle, Rjwilmsi, Bremen, Jake Wartenberg, Panoptical, Hiberniantears, Linuxbeak, JHMM13, Adamacious, Stardust8212, Seraphimblade, Mred64, Ligulem, Bubba73, Ivan.Romero~enwiki, Bhadani, Yamamoto Ichiro, Scorpionman, FayssalF, Falphin, Lostsocks, Ian Pitchford, SchuminWeb, Opeth, Tom-b, Petruchi41, Doc glasgow, Nihiltres, JdforresterBot, Crazycomputers, Harmil, Bitoffish, MacRusgail, Rune.welsh, RexNL, Wctaiwan, Gurch, Str1977, KFP, Alphachimp, Malhonen, Bmicomp, LeCire~enwiki, Srleffler, Wrightbus, Kri, Spencerk, Startaq, Butros, Nicholasink, Lord Patrick, Deyyaz, Visor, HKT, GangofOne, Jared Preston, DVdm, Mhking, Antiuser, Bgwhite, Sheean, Hall Monitor, Bomb319, Gwernol, Elfguy, UkPaolo, Theymos, Neitherday, Sceptre, Blightsoot, Stan2525, Rtkat3, Pip2andahalf, Phantomsteve, RussBot, Peoplesunionpro, WritersCramp, Muchness, WAvegetarian, Witan, Zafiroblue05, Chris Capoccia, SpuriousQ, Raquel Baranow, Devahn58, BillMasen, Varenius, Stephenb, Cate, Gaius Cornelius, Bovineone, Varnav, Wimt, Cunado19, Shanel, NawlinWiki, Vyran, Wiki alf, Erielhonan, Jaxl, Djadek, Kvn8907, Alisha Gergett, Mersenne, Introgressive, RazorICE, Bmdavll, Lykaestria, Anetode, Banes, Moe Epsilon, Truthdowser, MSJapan, Phaleux, Rdl381, Dbfirs, Adreamsoul, Mysid, Kortoso, Barnabypage, Csobankai Aladar, Elkman, Aaronb1215, Ignitus, Alpha 4615, Bantosh, Nick123, Wknight94, Elysianfields, DocXango, Jangam, Poleary, 2over0, Rudrasharman, Encephalon, Ageekgal, Knotnic, Theda, Closedmouth, Rpvdk, Arthur Rubin, Pb30, Josh3580, Reyk, Sultan of smoov, JQF, Saudade7, Petri Krohn, JoanneB, Peyna, Sitenl, Bagheera, Urocyon, Tyrenius, Nixer, Ilmari Karonen, Neoaeolian, Bluezy, Junglecat, Tzepish, NeilN, CrniBombarder!!!, DVD R W, WesleyDodds, Dragon of the Pants, Hiddekel, Sardanaphalus, Vanka5, Attilios, Tttrung, SmackBot, Aim Here, Brammers, Dubbin, Unschool, Haymaker, Vati115, Zazaban, KnowledgeOfSelf, TestPilot, Hydrogen Iodide, Although, TBH, C.Fred, Vald, Jacek Kendysz, Jagged 85, Watercolour, Davewild, Dims, Jrockley, Delldot, Michaelll, Jab843, Timeshifter, Edgar181, HalfShadow, Alex earlier account, Septegram, Xaosflux, Yamaguchi??, Cuddlyopedia, Linam97, Gilliam, Portillo, Ohnoitsjamie, Hmains, Skizzik, Martial Law, Weirdoactor, Rmosler2100, Chris the speller, Kurykh, Persian Poet Gal, Cattus, Fluri, Timneu22, Hibernian, Dlohcierekim's sock, J. Spencer, Effer, Ai.kefu, Baa, CMacMillan, MaxSem, Salmar, Zsinj, Can't sleep, clown will eat me, RyanEberhart, Jefffire, Abyssal, Kelvin Case, OrphanBot, Sephiroth BCR, Nima Baghaei, TKD, Addshore, RedHillian, Runefurb, Khoikhoi, Soosed, Aldaron, Krich, Ianmacm, CanDo, Downtown dan seattle, Khukri, Decltype, Bowlhover, Nakon, TedE, James McNally, John D. Croft, Richard001, Gauntlet~enwiki, Lpgeffen, Illlaaa, 44Dume, Polonium, Junyor, WoodyWerm, Sigma 7, ElizabethFong, Bejnar, Jagg2499, Ace ETP, Thor Dockweiler, CIS, Cor anglais 16, Cast, The undertow, SashatoBot, Vildricianus, SingCal, Producercunningham, Swatjester, Aximilli Isthill, Xerocs, Kuru, John, J 1982, Heimstern, Katstevens, Almkglor, Perfectblue97, JorisvS, JayMan, Tlesher, JohnWittle, IronGargoyle, Putnamehere3145, 041744, Ckatz, RandomCritic, MarkSutton, Benjaminlobato, Slakr, Hvn0413, Booksworm, Nyarkni, NJMauthor, George The Dragon, Dicklyon, Nephalim, Mets501, Rtkw, Dhp1080, AdultSwim, Anoyce, Midnightblueowl, Ryulong, Purplekitty, Danilot, Dradious, H, Suthaar, Nicolharper, ShakingSpirit, SimonD, Iridescent, Michaelbusch, Jason.grossman, Joseph Solis in Australia, Debeo Morium, Newone, StephenBuxton, Twas Now, Nubzor, TheBreeze, Gerfinch, LordRahl, Beno1000, CapitalR, DavidOaks, Courcelles, Cheeesemonger, Matlefebvre20, PaddyM, Tawkerbot2, The Letter J, Gveret Tered, ChrisCork, TheFloydman, JForget, MrPeabody, Brainbark, JF Mephisto, CmdrObot, Jargon, Ben groulx, Tobes00, Dycedarg, Calibanu, Falconfly, Makeemlighter, Ruslik0, Rasd, KnightLago, Benwildeboer, Evan7257, Dgw, N2e, AshLin, Chmee2, Moreschi, El sand bag57, QuinnJL, Rgonsalv, Dan Fuhry, MrFish, Ebonize christian, Alexignatiou~enwiki, FatBaka, CMG, Xaman, Shanoman, Icek~enwiki, Slazenger, Cydebot, Shitmaster 5, Ik the Toaster, Bluecurio, Steel, Michaelas10, Gogo Dodo, Jkokavec, Crowish, Flowerpotman,

Daniel J. Leivick, Mycroft.Holmes, Gilabarak, Karafias, Tawkerbot4, FDV, Itsfun, Robertinventor, Vanwiek, Dinnerbone, Garik, Har057, Protious, SpK, Omicronpersei8, Shashankgupta, JodyB, Arb, UberScienceNerd, Nol888, Nadirali, Casliber, Epbr123, Casual Moose, Pajz, Marbie, Pstanton, Aaron Pelzer, Indef blocked user 001, PhilMacdonald, Ucanlookitup, Andyjsmith, 24fan24, Headbomb, Simeon H, Rugadh, Marek69, Sean7phil, Timebender13, John254, A3RO, SGGH, Jakerake, Pmrobert49, Kaishininjou, Wildthing61476, Catsmoke, Eljamoquio, Mikeeg555, Whoda, Cooljuno411, FreeKresge, Ericmachmer, FreshFruitsRule, Keyvez, Noclevername, Mmortal03, Mentifisto, Hmrox, Cyclonenim, AntiVandalBot, Majorly, Gioto, Luna Santin, Seaphoto, Opelio, Paul from Michigan, QuiteUnusual, Prolog, TrulyUnusual.com, Pwhitwor, TimVickers, Exteray, Electromagnet, Kevin Nelson, Robsmyth40, Skynet1216, ARTEST4ECHO, AubreyEllenShomo, Robert A. Mitchell, Myanw, Mad Pierrot, Deadbeef, GWhitewood, JackSparrow Ninja, Peter Harriman, JAnDbot, Xhienne, 24630, Husond, Barek, MER-C, Reduxx, Inks.LWC, Ericoides, Hodgetts, Smiddle, Plm209, Andonic, Kerotan, Jarkeld, Y2kcrazyjoker4, Connormah, Bakilas, Pedro, Bongwarrior, VoABot II, BruceDude, James166, StudierMalMarburg, AuburnPilot, JNW, JamesBWatson, TL789, Azznrivera, Swpb, Mutableye, BobTheMad, Violentbob, Jim Douglas, Truancy07, Tonyfaull, Avicennasis, BrianGV, Catgut, Indon, ClovisPt, C.lettinga, Walkerlamond, Sgr927, IkonicDeath, Ali'i, Dinohunter, BatteryIncluded, Beetfarm Louie, Tenjikuronin, Vssun, JoergenB, TehBrandon, Nikolaj Christensen, DerHexer, Hi hi puka puka, Edward321, Saxophlute, NatureA16, Stephenchou0722, Zawarq, Kamikaze, MartinBot, BetBot~enwiki, Arjun01, Zogundar, Sm8900, Mitchhe 91, Thomasshaw, Deepsupport, CommonsDelinker, AlexiusHoratius, Damodar87, LittleOldMe old, Ash, Alex-Parky, Ulisse0, Master of Tofu, J.delanoy, Sasajid, Pharaoh of the Wizards, Mattlewisthepimp, CFCF, Trusilver, ProfButler, Ville V. Kokko, Bogey97, Hans Dunkelberg, Rhinestone K, Uncle Dick, Maurice Carbonaro, All Is One, Jesant13, Harold56, Dada124C41+, NerdyNSK, Ian.thomson, Darth Mike, Gzkn, Acalamari, M C Y 1008, PeterH2, TheChrisD, BrokenSphere, Pegasus1457, Katalaveno, Magiwand, Dark-Falls, Firedraikke, McSly, SpigotMap, Ryan Postlethwaite, Enricorpg, BeŻet, Vanished User 4517, Belovedfreak, NewEnglandYankee, Rominandreu, Rwessel, SJP, G.Batty, AliensPhD, Malerin, Biglovinb, Workofthedevil, Umair82, West109, MetsFan76, Quogud, Donmarkdixon, Chppxbx, Bettyboop2336, Cometstyles, Kenneth M Burke, Myrealana, Nrorres, Vanished user 39948282, Treisijs, Rising*From*Ashes, Mirage GSM, Pdcook, Ja 62, Mwmillar, Trunkalunk, Suuperturtle, CardinalDan, Idioma-bot, Tobynsaunders, Gueedo, Muoi, Wikieditor06, Light of Shadow, Lights, Caribbean H.Q., Char555, Deor, McNoddy~enwiki, TreasuryTag, A.Ou, TallNapoleon, Hersfold, Jeff G., Orthologist, AlnoktaBOT, Maghnus, Newyowker, Boooobzzz, QuackGuru, Mwamanator, Philip Trueman, DoorsAjar, TXiKiBoT, Cosmic Latte, Convolution223, Mrkwtrs, Allbee Honnête, Jkstark, Jacob Lundberg, Alan Rockefeller, Technopat, Malljaja, Ann Stouter, Huwe2, Drestros power, Arnon Chaffin, Someguy1221, Cosmium, Warrush, Supersmarty, Oxfordwang, Anna Lincoln, Cherhillsnow, Wiikipedian, Martin451, BwDraco, LeaveSleaves, Cremepuff222, Mazarin07, Hippy-dippy, Madhero88, Discgolfrules, Tcgriffin, Jdman24, Comrade Tux, Strangerer, Falcon8765, Sardonicone, Scubasteve111, !dea4u, Sesshomaru, SchumiChamp, Insanity Incarnate, Calvin2011, Brianga, Balderunner, Atmgnef, Uranometria, Zungaphile, Logan, Closenplay, Planet-man828, Megasquid500, Daveh4h, Dinofang, NHRHS2010, Gaelen S., Ponyo, Xzazerx, Randommelon, Dusti, Ellio-3.14, Sonicology, Wibubba48, Ziplock74, Zephyrus67, Lemonflash, Dawn Bard, METIfan, Charsniper, JJ Da Kid, Indelible sinn, Yintan, Tataryn, August Dominus, Crash Underride, Dinokid, AlbertHall, Rackshea, Super Sonic Sloth, Flyer22 Reborn, Radon210, Alexbrn, RoGGe, Undead Herle King, Oxymoron83, Faradayplank, Smilesfozwood, AngelOfSadness, Lisatwo, Steven Crossin, Lightmouse, Techman224, Lumentec, Nancy, Spartan-James, StaticGull, Hamiltondaniel, Dust Filter, Dabomb87, Hawkesy12345678, Neo., Kalidasa 777, Lloydpick, Goodergrammar, Gantuya eng, Revelian, Stillwaterising, Kanonkas, Tolerancebelowzero, Sabbath73, Yaanu4, Runn~enwiki, Hpforlife, Loren.wilton, Martarius, Sfan00 IMG, Elassint, ClueBot, Cmmmm, BlazeMaster4R3AL, Foxj, The Thing That Should Not Be, Vikasatkin, Diazenefz, Gaia Octavia Agrippa, R000t, Shade11sayshello, Warder Alduren, Spandrawn, Ben5ive, SuperHamster, JTBX, Boing! said Zebedee, FractalFusion, CounterVandalismBot, Niceguyedc, Blanchardb, Fallenfromthesky, Cavalryman101, Facemuncher, Btroxmysox, Ernstblumberg, Amirreza, Supergodzilla2090, Excirial, -Midorihana-, Jusdafax, Unrealpotboy3154, Panyd, Vimalkdwivedi, NewBeat, SpikeToronto, Brmabepo2ndof4, Fishing ko, Cenarium, Peter.C, TheRedPenOfDoom, Tnxman307, M.O.X, Istok, Razorflame, Redthoreau, Hoth194, Mikaey, Froogle62, Tired time, Thingg, 7, King of the World 5, Footballfan190, Mczack26, WxHalo, Versus22, Porchcorpter, PCHS-NJROTC, Berean Hunter, Docsavage20, SoxBot III, Hattiel, Vanished user uih38riiw4hjlsd, Youhatemeandidontcare, DumZiBoT, Cowardly Lion, BarretB, XLinkBot, Bjrcoolguy210, Tarheel95, Fastily, Pichpich, Gnowor, Wakawaka1, CGaldi, Andrzej Kmicic, 68Kustom, Fullsick, Badgernet, Truthnlove, SelfQ, Herwest299, Thatguyflint, D.M. from Ukraine, Olyus, Coltonmj11, Haqqmisra, Addbot, Xp54321, Cxz111, Basilicofresco, Manwithahat, Gravitophoton, Tcncv, Unforgiven666, Morriswa, Zellfaze, Joegriff4, Ronhjones, Hollando, CanadianLinuxUser, Bnaur, Cst17, Hugh wiki01, MrOllie, Proxima Centauri, Woo united11, Morning277, Chamal N, Glane23, Juneretos, Lollmarofl, AnnaFrance, Quietmarc, Theking9794, 5 albert square, Avs dps, Flump1, Jaguar65, SmartBagel, Tassedethe, DubaiTerminator, Changin history n00bs, Tide rolls, Bfigura's puppy, Lightbot, Totorotroll, Gail, BazzookaJoe, Phantom in ca, Quantumobserver, GiantPea, Bartledan, BlackMarlin, Megaman en m, Alfie66, Twiggfulysmart, Wellminesthesun, Zobango, TheSuave, Yobot, Ultrasquad, Anomalieshunters, Senator Palpatine, Fraggle81, AliDincgor, Cflm001, Legobot II, Akalvi, AdjutantRefluxSoliton, VertigoOne, Texas™, Drcocapepper13, Ningauble, Black vuh jj, Againme, Knoital, MacTire02, Magog the Ogre, AnomieBOT, Angry bee, DoctorJoeE, Jim1138, IRP, Neko85, Zuko117, Piano non troppo, Llebwoc, Kingpin13, Yachtsman1, Flewis, Saidaziz eng, Materialscientist, Cavemanug1001, The High Fin Sperm Whale, Bestie4, Citation bot, UFOForum, E2eamon, Vuerqex, Raphaelzola, Kel12189, Neurolysis, LilHelpa, The Dean Man, Coolboyno1, Vinego, Random astronomer, Sionus, Capricorn42, Newzebras, Bihco, Pontificalibus, ChildofMidnight, Kmcdm, Darkestdeath1, Grim23, Dkelly1966, Tyrol5, Mlpearc, Singster, Gap9551, Bigaireatscheese, Jamesslot3, J04n, GrouchoBot, Ataleh, MrVoodooArmy, Skraddarbacken, Shirik, Italomex, Hoifon, Mathonius, Amaury, Voltorb, Falcon189, Cailex, Mnance1975, Heavenly Chihuahua, Lexy-lou, Shadowjams, Vanished user giiw8u4ikmfw823foinc2, Glump1, Aaron Kauppi, Jacobkailynzeeba, Erik9, Islas93, Bwham, Griffinofwales, Minkala, BoomerAB, MrAwesomeDudePerson, The Skull Fiend, Captain Weirdo the Great, D10hitman, Rock4, Bobwrits, Johncamanley, Hobbit119, Raj6, Bendover61, Exchangor, Nickucla, Banshee lover, Sanpitch, ReneVenegas95, Knubbe kub, Bobri28, Cargoking, ZOLTAN: THE gOd, Mj12cz, HJ Mitchell, Jamsam70, The One & Only Fools and Horses~enwiki, Citation bot 1, Pizik, Dogshitface, GaussianCopula, Amplitude101, Redrose64, Cubs197, JKDw, Pinethicket, I dream of horses, Vicenarian, Flaneursfields, Usukasslikeurmom, Juliaashford, Tom.Reding, Supreme Deliciousness, Calmer Waters, Yahia.barie, Geogene, A8UDI, Jschnur, Jakuplutzen, Xfact, Serols, Militaryman433, BobaTyroneFett, Oooga Booga 50039, Hairy Tom, Milzmilzmilz, SkyMachine, Iokerapid, AATroop, Chachap, Cmdahler, Trappist the monk, Ksanexx, Fuad Thahir, Tycopup, Marvinandmilo, Vrenator, Suomi Finland 2009, MrX, Vancouver Outlaw, Begoon, Fmndu9acgh79q3, Lynn Wilbur, Reaper Eternal, --(Add Name Here)--, Perrielover123, Urface24, Boucher-craig, Tbhotch, Stroppolo, TheMesquito, K602, Marie Poise, Manuron, DARTH SIDIOUS 2, Firemanic9, Svr21121, Davidjess, Onel5969, Mean as custard, The Fudginator, Maryjaneadams, RjwilmsiBot, Dudemister67, SpartanGreg09, Jake576, Ilikecheeseandcereal, MyFellowAmericans234, Vcrmastr, Latinlover2010, DASHBot, Bob512, ImprovingWiki, John of Reading, Orphan Wiki, Acather96, Ghostofnemo, Immunize, Grrow, Heracles31, ScottyBerg, 333beth333, Notanoption, GoingBatty, RA0808, JustinRussoLove, Minimac's Clone, Icantthinkofasuitableusername, Infringement153, The Mysterious El Willstro, Nailer111, Mr Hat 971, Wikipelli, Sinalese, K6ka, WaightZer, Claytonmaner, Solomonfromfinland, Thecheesykid, Sundance218,

Qwertyrandom, John Cline, John Mentor, Fæ, A2soup, DalitDynamite, Mattimole, The next dimensions, Carva01, ElationAviation, Cami37, Reve316, AnArdentReader, Aga1234567, PhreshKiid, Tylerball22, Force92i, Logexp, QEDK, Wayne Slam, David J Johnson, Litjohn, Cit helper, 567895th, JoeSperrazza, TyA, Yotvata, Brandmeister, Coasterlover1994, Angus.hendrick, L Kensington, Societies, Senjuto, Hell of howling, Donner60, I am beast 97, Gorvindo, EvenGreenerFish, Ego White Tray, Bill william compton, Status, Whydoyoucare555, Himan99, ChiZeroOne, Brotherbro, Maxigorilla, Iketsi, Forever Dusk, Haloreach89, TYelliot, Oursana, Rocketrod1960, E.f.harper, Gary Dee, Omni 98, Petrb, Grapple X, ClueBot NG, Ktmuhone, Thilot, NamlessIamnot, Lalalaa12, Giordanobrunoburning, MelbourneStar, This lousy T-shirt, Prathfig, Aarishshaheen, Gilderien, Iritakamas, Marcoboelling, Luckey313, Cntras, Twillisjr, Hazhk, D Quinn42, Kevin Gorman, Sjmantyl, Space1134, Dream of Nyx, Widr, CostaDax, Tr00rle, Caille91, MerllwBot, Helpful Pixie Bot, LORD CAAZE, Daviddwd, Meerta, Bibcode Bot, Nashhinton, Jeraphine Gryphon, Trunks ishida, Lowercase sigmabot, BG19bot, Krenair, Silentkiller101, Billy from Bath, Meepamoop, James350z, Navhus, Qwerty1214, Sailing to Byzantium, Kangaroopower, Hallows AG, Kaspuhler, Superyoshiw, JohnChrysostom, MusikAnimal, Metricopolus, Exobiologist, Purpleaye, Dan653, Petruss, Bugoyito, Kyurem7, Cadiomals, Olev Vinn, Wld555, BurnerHero2, RakezDurban, MordinSolus0, Awright34, Daniellelynettehardy, MrBill3, Pmhorler, Josepheous, Risingstar12, Doowop62, Gwickwire, Fcwmy, Omgdinosaurs111, Academictask, Dvq, Edonnelly527, JENNATIFFANY, Achowat, Chip123456, Lashleymullins, Wer900, Dubstepallen, Anbu121, EricEnfermero, F33mason, Melodychick, BattyBot, Phinizyspalding, Sparsy, Longboardhard, Allylynn9, CrazeGUY1515, Itrollfaceumadbro, EnglishMole0615, Cyberbot II, Bigbottomedgirl, The Illusive Man, ChrisGualtieri, Nick.mon, AshxRoc, Shiltsev, Slordax, EuroCarGT, Gameboy97q, Scoot4life, VoteLobster, Illia Connell, Jmorton2222, Dannyj24, WelshMan1990, Salismasher12, Party102, Qxukhgiels, 123456jp, Stempkik3, Nuf101, Gadgetball, Dexbot, Greggory12, TrollGlaDOS, Anandaraja, Webclient101, Mogism, JimmyTheRacer, Justinwgaines, Lugia2453, 93, Coolbeans443, 069952497a, Wolly123, Reatlas, Erika Leonard, Big boss170, F6Zman, Jubjub01, TishoYanchev, Jamsontoast987, Melonkelon, Eyesnore, Harlem Baker Hughes, Wethar555, Uncledrew, Wikimee34, Doctor-professor-magnificant, Praemonitus, SamoaBot, Purple Door Door, EvergreenFir, Wagnerslove, Worst cooks in America, Adspred, Kentsy, Fatum81, SchroederB, Gruekiller, Ira Bradley, LieutenantLatvia, Andymatches, Legoman 86, Eagle3399, Joppiesaus123, Fatty black man, Spoon in my bellybutton, Aambeienbeffer, NottNott, Jotknktrnfjkrtklg, Johnwatson94, Tjbfl, Submarinefy, MagicatthemovieS, Jose.rivera143, XdoomdragonX, Billy Cuppertino, Meteor sandwich yum, Racer Omega, Fixuture, Man of Steel 85, Keltonmadden, SilvestreRiseguy123, Spaghetti1033, Sofia Lucifairy, Chaya5260, Pjmiller1978, IRock123, Lion buisness, Rosario Berganza, Hohohoinyourbutthole, Fafnir1, Monkbot, DK0010, Ware6561, Mickeyc02, ADHDavid1, MegaGardevoir68, CalebB21, Lagalog, Sofia Koutsouveli, Jack1990tut, Emeryst, GilCarlson, Akanyang, Was an hero, Gurmanmundi, Baconchocolate, Usukwiki, Godzilla1520, Zeus000000, Poiuytrewqvtaatv123321, BicelPhD, AsteriskStarSplat, Amisharajiv, Dr Mullen, Deadjoker27, Marky999, Gamebuster19901, Heuh0, TeaLover1996, Joethethird, Rubbish computer, Sam taylor47, Jjmanofdoom, Lilymary123, Anunaki truth, Gweraituh3wa98th9384, PionLax, Nightwingandbats!, Isambard Kingdom, Hades324, Jiggywithe123, DN-boards1, ANOMALIEN, GeneralizationsAreBad, Medinalori, Jerodlycett, KasparBot, Supman23, Wikichipmunk, CAPTAIN RAJU, Mrryanboogie, Xand123er, Xyndsbjsjdfhbuasdlk.z.kxjvasgvfsadjhf.., Sujai KC, Glawio204, Ehlmao, Prazl001, BluecometFlag, Goofmel, ScottHla2kk3, Leo513, Beatrice is ugly, Julian.Hatcher.1122, Acowmoo, Shebabroward, Prhdbt, ExocticKnowledge01106 and Anonymous: 2100

- **Extremophile** *Source:* https://en.wikipedia.org/wiki/Extremophile?oldid=689256106 *Contributors:* Magnus Manske, Bryan Derksen, Josh Grosse, Rgamble, PierreAbbat, Edward, Patrick, Lexor, Kku, Axlrosen, Skysmith, Ellywa, Mortene, DavidWBrooks, JWSchmidt, Stone, David Latapie, Dragons flight, SEWilco, WormRunner, Rasmus Faber, Jrash, Matthew McVickar, Everyking, JuanitaJP, DragonflySixtyseven, Icairns, Guernica~enwiki, Jmeppley, Julianonions, Pinnerup, Thorwald, CALR, Rich Farmbrough, Guanabot, Vsmith, Kenb215, Chadlupkes, Bender235, Kbh3rd, CanisRufus, Remember, Bobo192, Viriditas, MrTree, Axl, Apoc2400, Katefan0, CaseInPoint, ClockworkSoul, Admiral Valdemar, TShilo12, Stemonitis, JGodman, Ashmoo, Raivein, Drbogdan, Rjwilmsi, Feydey, Nneonneo, Nihiltres, Vossman, Bgwhite, YurikBot, Cathalgarvey, RobotE, Gaius Cornelius, Grafen, Theodolite, Arthur Rubin, Petri Krohn, Fram, Ilmari Karonen, SmackBot, TestPilot, Wschwarz, Clpo13, CMD Beaker, Eskimbot, BiT, Yamaguchi??, Gilliam, Ohnoitsjamie, Skizzik, Mr Beige, Kurykh, Hibernian, WikiPedant, Valenciano, Richard001, DMacks, Cephal-odd, Vina-iwbot~enwiki, Scharks, Loadmaster, JHunterJ, Bendzh, Dr.K., Dl2000, Natronomonas, AlainD, Yuxinzhu93, CRGreathouse, CmdrObot, Jodawi, Cleanr, AshLin, Cheapestcostavoider, Mato, Gogo Dodo, Eric Martz, Rracecarr, JodyB, Thijs!bot, JAF1970, Epbr123, Runch, Deborahjay, MountainBum83, Mailseth, AntiVandalBot, Saimhe, Danger, Sluzzelin, MortimerCat, Tstrobaugh, Bongwarrior, VoABot II, Nyttend, Rich257, Bubba hotep, Cyktsui, BatteryIncluded, DerHexer, Lmchalwell, Qifeng, Fconaway, Lilac Soul, Effectrode, Century0, Naniwako, OAC, Ajtep, 83d40m, STBotD, Philip Trueman, TXiKiBoT, Antoni Barau, Xarelete, Earthdirt, Aedan cunningham, Ninjatacoshell, Sue Rangell, Jsde, Bfpage, SieBot, Jerryobject, Bentogoa, Flyer22 Reborn, Hobartimus, Fratrep, ImageRemovalBot, Touchstone42, ClueBot, Deviator13, PipepBot, The Thing That Should Not Be, Niceguyedc, DragonBot, Shinkolobwe, Estirabot, Thingg, Rror, Avoided, Salam32, Qgil-WMF, RP459, Bioguz, Grayfell, C6541, Gravitophoton, DOI bot, Atethnekos, Fluffernutter, Cst17, LaaknorBot, Bernstein0275, ChenzwBot, Delemon, Legobot, Luckas-bot, Yobot, KamikazeBot, Robert Treat, Campsis, Rubinbot, 1exec1, Hunnjazal, Citation bot, Frankenpuppy, Gifh, LilHelpa, Tzim78, Xqbot, Anna Frodesiak, CeliaRSC, The Wiki ghost, Eugene-elgato, FrescoBot, Akshatrathi294, VI, Doom Order, Citation bot 1, Xirja, Livestronger, Livestrongest, TjBot, Ripchip Bot, Androstachys, Rayman60, Orphan Wiki, WikitanvirBot, Look2See1, Dcirovic, ZéroBot, Wintermomi, Miguelmote, Crm1003, ClueBot NG, Horoporo, Russellml, HMSSolent, Bibcode Bot, Skullcandle, Sam48823, Thegreatgrabber, Jonadin93, EuroCarGT, Nugget00, GoneMoot00, Kyletheowl, TheGuy520, Nickosr.p, Iztwoz, Dustin V. S., Mudcrab11, Ginsuloft, Fewaut001, Officialdrgamer, Monkbot, DeadCygnus, Dr John Martin Wright, WaffleMan19 and Anonymous: 207
- **Exobiology Radiation Assembly** *Source:* https://en.wikipedia.org/wiki/Exobiology_Radiation_Assembly?oldid=660660556 *Contributors:* Chris Capoccia, BatteryIncluded, Citation bot, Amkilpatrick and Bibcode Bot
- **BIOPAN** *Source:* https://en.wikipedia.org/wiki/BIOPAN?oldid=677517794 *Contributors:* Rjwilmsi, BatteryIncluded, Atethnekos, Plazmatyk, Fotaun, FabC, Josve05a, Lizia7, Tetra quark and Anonymous: 1
- **Long Duration Exposure Facility** *Source:* https://en.wikipedia.org/wiki/Long_Duration_Exposure_Facility?oldid=682249313 *Contributors:* SimonP, Mulad, Ke4roh, Bobblewik, The Singing Badger, Hooperbloob, Ashley Pomeroy, Kitch, Bricktop, Rjwilmsi, Rillian, SchuminWeb, TheDJ, Whosasking, SmackBot, Hibernian, WDGraham, Andy120290, Mr.Z-man, SalopianJames, N2e, Thijs!bot, Owenozier, Swpb, Jatkins, BatteryIncluded, Drm310, RockMFR, Everything counts, Pomona17, Addbot, KSC-on-orbit, Yobot, Ville Siliämaa, Xosema, Fotaun, Torquemada082, Full-date unlinking bot, Trappist the monk, Tbhotch, EmausBot, Odyssomay, Chihuahua State, Northamerica1000, Tony Mach, Jupiter-4 and Anonymous: 7
- **Rosetta (spacecraft)** *Source:* https://en.wikipedia.org/wiki/Rosetta_(spacecraft)?oldid=689197772 *Contributors:* Bryan Derksen, Rmhermen, Matusz, Heron, Edward, Nealmcb, Patrick, Karada, (, Ahoerstemeier, Julesd, Sunbeam60, Epo~enwiki, Wikiborg, Stone, Dtgm, Tpbrad-

bury, SEWilco, Wetman, Jerzy, Phil Boswell, AlexPlank, Robbot, Paranoid, Robinh, Kent Wang, Davidcannon, Alan Liefting, Jacoplane, Awolf002, Michael.chlistalla~enwiki, Elinnea, Perl, Curps, Kuuks, Python eggs, Bobblewik, Christopherlin, Peter Ellis, TerokNor, Keith Edkins, GeneralPatton, Doug Huffman, Urhixidur, Hellisp, KeithTyler, Peter bertok, Janneok~enwiki, Generica, Vsmith, Naive cynic, Bender235, Helldjinn~enwiki, RJHall, Huntster, Kwamikagami, Susvolans, Lunaverse, Iamunknown, 23skidoo, BrokenSegue, Giraffedata, Supersexyspacemonkey, Hektor, Dodonov, Snowolf, Wtmitchell, Suruena, Geraldshields11, Ceyockey, Zntrip, Siafu, Jamsta, Oliphaunt, Danno612, RedBLACKandBURN, Ashmoo, Dwarf Kirlston, Edison, Drbogdan, Rjwilmsi, Tim!, Alaney2k, Georgelazenby, Bubba73, Marsbound2024, FloK, Patrick1982, Jmc, Kolbasz, Bgwhite, Agamemnon2, The Rambling Man, YurikBot, RussBot, Arado, Hydrargyrum, Shell Kinney, Gaius Cornelius, Eleassar, Los688, Nowa, Ravedave, Tony1, Rdl381, Kkmurray, Chesnok, Arthur Rubin, Pietdesomere, Nolanus, Smurrayinchester, Philip Stevens, Serendipodous, Cmglee, The Yeti, Crystallina, KnightRider~enwiki, SmackBot, Giraldusfaber, Vald, MichaelSH, Nickst, Srnelson, Chris the speller, PrimeHunter, Miquonranger03, I7s, Hibernian, Sbharris, WDGraham, Andy120290, Zirconscot, Dragon695, Mwtoews, Ericl, Suthers, Vina-iwbot~enwiki, Ohconfucius, CFLeon, John, Catapult, Mauriciobenes, JorisvS, RandomCritic, Dicklyon, Xiaphias, FredTheDeadHead, Joseph Solis in Australia, Eluchil404, Acom, CmdrObot, Unit~enwiki, ThreeBlindMice, N2e, CKozeluh, Devjanrag, Necessary Evil, Cydebot, Kanags, Headbomb, Dtgriscom, Sobreira, Oosh, Dawnseeker2000, Nicholas0, Widefox, Ricnun, Smartse, Ritwik mango, TuvicBot, JAnDbot, Leuko, MER-C, Planetary, IanOsgood, Rothorpe, Rockower, IanShazell, WolfmanSF, Hroðulf, Nyq, EFletcher, J.P.Lon, Ajsteffl, Ckruetze, BatteryIncluded, A3nm, Dkriegls, Superkuh, Geboy, Jim.henderson, R'n'B, CommonsDelinker, Tinwelint, Rod57, Mottelg, Rwessel, Garafatea, 83d40m, Cs302b, Porckmaster, VolkovBot, Sdsds, Rei-bot, JhsBot, Raymondwinn, JavierCastro, Djmckee1, Truthanado, SieBot, KGyST, Rubble pile, Dzid, Lightmouse, Murlough23, RSStockdale, Nergaal, TSRL, ImageRemovalBot, Sfan00 IMG, MBK004, Anonymous799, Kumuty, AstroMark, Dj.antix, Der Golem, Frmorrison, Consert, Mild Bill Hiccup, Com4space, Pointillist, Hseehtar, Excirial, AssegaiAli, Tokyoguy999, L.tak, Nickispeaki, SchreiberBike, BOTarate, Another Believer, Chaosdruid, Rishi.bedi, DumZiBoT, Ost316, Skarebo, WikHead, PenComputingPerson, Kbdankbot, Pline, Addbot, DOI bot, Non-dropframe, Jncraton, Gregouille, Proxima Centauri, Wilnap, AndersBot, Favonian, LinkFA-Bot, Krano, A-Damage1, Luckas-bot, Yobot, Aldebaran66, AnomieBOT, DoctorJoeE, Jim1138, Wiki007wiki, Hunnjazal, Informationtheory, Citation bot, Jzandin, ArthurBot, DirlBot, Xqbot, Haarnaald, Mlpearc, GrouchoBot, RibotBOT, GreekAlexander, Rainald62, Fotaun, Spazturtle, ???, John85, Citation bot 1, I dream of horses, Jonesey95, Skyraider1, Meaghan, Spaluch1, Primaler, Tlhslobus, Full-date unlinking bot, Horst-schlaemma, IVAN3MAN, Kgrad, Jedi94, Yellowdesk60, Lattas, Earthandmoon, Sideways713, RjwilmsiBot, TGCP, EmausBot, Pete Hobbs, Mrericsully, CaptRik, Jmencisom, Winner 42, ZéroBot, DavidMCEddy, A2soup, Pippo skaio, Kiwi128, IIIraute, Шуряан, Rcsprinter123, Donner60, Anonimski, Ontyx, SkywalkerPL, ChiZeroOne, BlueBreezeWiki, Llightex, ClueBot NG, Michaelmas1957, Jakewright007, Kikichugirl, Bibcode Bot, BG19bot, ThunderThighs97, Dr. Whooves, Infometric21, Altaïr, MrBill3, Diamondandgoldore, Daneon2, BattyBot, Teammm, CrunchySkies, TheJJJunk, Mastros, Khazar2, Dexbot, CuriousMind01, TwoTwoHello, JakobSteenberg, Jc86035, Jamesx12345, Muchotreeo, Joeinwiki, Rfassbind, LudwigSebastianMicheler, Zlelik2000, Msbentley, Ihjdekeijzer, Batyu, MKUV, Barbara.menti, Markh89, Mfb, DPRoberts534, Nilstycho, Neptunia, Lagoset, Filedelinkerbot, Alien Putsch resistant, Susannacb, Andrew.Mclaughlin75, ValeMercury, The Average Wikipedian, WyattAlex, DCLukas, Jchrishall2000, Wikiwhittaker, Silenj, Stir564, ISpacex, Isambard Kingdom, Prinsgezinde, KasparBot, Huritisho, G-dac, Outedexits and Anonymous: 183

- **Living Interplanetary Flight Experiment** *Source:* https://en.wikipedia.org/wiki/Living_Interplanetary_Flight_Experiment?oldid=655027434 *Contributors:* Ouro, Florian Blaschke, Marasmusine, Vegaswikian, Jevon, Wavelength, Episiarch, Chris Capoccia, Awiseman, SmackBot, Skizzik, Hibernian, Acdx, J Milburn, Maolmhuire, Necessary Evil, Cydebot, D4g0thur, OhanaUnited, BatteryIncluded, R'n'B, ImageRemovalBot, Mild Bill Hiccup, Nanobear~enwiki, Addbot, Roentgenium111, Ashanda, LaaknorBot, 84user, Lightbot, AnomieBOT, Jim1138, Materialscientist, Citation bot, Fotaun, DrilBot, Tom.Reding, ChiZeroOne, Bibcode Bot, TheGeneralUser, Mogism, Frinthruit, Monkbot, Tetra quark and Anonymous: 26

19.8.2 Images

- **File:10_small_subunit.gif** *Source:* https://upload.wikimedia.org/wikipedia/commons/3/3d/10_small_subunit.gif *License:* Public domain *Contributors:* <a data-x-rel='nofollow' class='external text' href='http://www.pdb.org/pdb/static.do?p=education_discussion/molecule_of_the_month/pdb10_1.html'>*Molecule of the Month*</a> at the RCSB Protein Data Bank *Original artist:* Animation by David S. Goodsell, RCSB Protein Data Bank
- **File:1990_s32_LDEF_stow.jpg** *Source:* https://upload.wikimedia.org/wikipedia/commons/0/0d/1990_s32_LDEF_stow.jpg *License:* Public domain *Contributors:* NASA *Original artist:* NASA/exploitcorporations
- **File:ALH84001_structures.jpg** *Source:* https://upload.wikimedia.org/wikipedia/commons/a/a8/ALH84001_structures.jpg *License:* Public domain *Contributors:* http://web.archive.org/web/2/curator.jsc.nasa.gov/antmet/marsmets/alh84001/ALH84001-EM1.htm *Original artist:* NASA
- **File:Acetaldehyde-3D-vdW.png** *Source:* https://upload.wikimedia.org/wikipedia/commons/8/8e/Acetaldehyde-3D-vdW.png *License:* Public domain *Contributors:* ? *Original artist:* ?
- **File:Acetic-acid-3D-vdW.png** *Source:* https://upload.wikimedia.org/wikipedia/commons/d/db/Acetic-acid-3D-vdW.png *License:* Public domain *Contributors:* ? *Original artist:* ?
- **File:Aleksandr_Oparin_and_Andrei_Kursanov_in_enzymology_laboratory_1938.jpg** *Source:* https://upload.wikimedia.org/wikipedia/commons/f/f4/Aleksandr_Oparin_and_Andrei_Kursanov_in_enzymology_laboratory_1938.jpg *License:* Public domain *Contributors:* ? *Original artist:* ?
- **File:Animation2.gif** *Source:* https://upload.wikimedia.org/wikipedia/commons/c/c0/Animation2.gif *License:* CC-BY-SA-3.0 *Contributors:* Own work *Original artist:* MG (talk · contribs)
- **File:Archeocyathids.JPG** *Source:* https://upload.wikimedia.org/wikipedia/commons/d/d4/Archeocyathids.JPG *License:* CC BY-SA 3.0 *Contributors:* I (Qfl247 (talk)) created this work entirely by myself. (Original uploaded on en.wikipedia) *Original artist:* Qfl247 (talk) (Transferred by Citypeek/Original uploaded by Qfl247)
- **File:Arecibo_message.svg** *Source:* https://upload.wikimedia.org/wikipedia/commons/5/55/Arecibo_message.svg *License:* CC-BY-SA-3.0 *Contributors:* Own drawing, 2005 *Original artist:* Arne Nordmann (norro)

- **File:Asteroid_P-2010_A2.jpg** *Source:* https://upload.wikimedia.org/wikipedia/commons/0/08/Asteroid_P-2010_A2.jpg *License:* Public domain *Contributors:* Hubblesite (STScI-2010-07 c) *Original artist:* Credit: NASA, ESA, and D. Jewitt
- **File:Blacksmoker_in_Atlantic_Ocean.jpg** *Source:* https://upload.wikimedia.org/wikipedia/commons/6/6f/Blacksmoker_in_Atlantic_Ocean.jpg *License:* Public domain *Contributors:* NOAA Photo Library *Original artist:* P. Rona
- **File:Buckminsterfullerene-perspective-3D-balls.png***Source:* https://upload.wikimedia.org/wikipedia/commons/0/0f/Buckminsterfullerene png *License:* Public domain *Contributors:* Own work *Original artist:* Benjah-bmm27
- **File:Cambrian_Rocks_map.png** *Source:* https://upload.wikimedia.org/wikipedia/commons/6/65/Cambrian_Rocks_map.png *License:* Public domain *Contributors:* 1911 Encyclopædia Britannica *Original artist:* John Allen Howe
- **File:Cambrian_explosion_taskforce_logo.svg***Source:* https://upload.wikimedia.org/wikipedia/commons/b/b0/Cambrian_explosion_task logo.svg *License:* CC BY-SA 3.0 *Contributors:*
- Burgess_scale2.png *Original artist:*
- derivative work: Smith609 (talk)
- **File:Carbon-monoxide-3D-vdW.png** *Source:* https://upload.wikimedia.org/wikipedia/commons/a/a7/Carbon-monoxide-3D-vdW.png *License:* Public domain *Contributors:* ? *Original artist:* ?
- **File:Central_fusion_and_terminal_fusion_automixis.svg** *Source:* https://upload.wikimedia.org/wikipedia/commons/e/ed/Central_fusion_and_terminal_fusion_automixis.svg *License:* GFDL *Contributors:* No machine-readable source provided. Own work assumed (based on copyright claims). *Original artist:* No machine-readable author provided. Staticd assumed (based on copyright claims).
- **File:Champagne_vent_white_smokers.jpg** *Source:* https://upload.wikimedia.org/wikipedia/commons/a/aa/Champagne_vent_white_smokers.jpg*License:* Public domain*Contributors:* http://oceanexplorer.noaa.gov/explorations/04fire/logs/hirez/champagne_vent_hirez.jpg*Originalartist:* NOAA
- **File:Chirality_with_hands.svg** *Source:* https://upload.wikimedia.org/wikipedia/commons/e/e8/Chirality_with_hands.svg *License:* Public domain *Contributors:* Chirality with hands.jpg *Original artist:* Chirality with hands.jpg: Unknown
- **File:Cnemidophorus-ThreeSpecies.jpg** *Source:* https://upload.wikimedia.org/wikipedia/commons/3/36/Cnemidophorus-ThreeSpecies.jpg *License:* Copyrighted free use *Contributors:* Transferred from en.wikipedia to Commons by Innotata using CommonsHelper. *Original artist:* Photograph by Alistair J. Cullum (Acullum at en.wikipedia) Email: acullum@creighton.edu
- **File:Comet_67P_on_19_September_2014_NavCam_mosaic.jpg** *Source:* https://upload.wikimedia.org/wikipedia/commons/8/81/Comet_67P_on_19_September_2014_NavCam_mosaic.jpg *License:* CC BY-SA 3.0-igo *Contributors:* http://www.esa.int/spaceinimages/Images/2014/09/Comet_on_19_September_2014_NavCam (image link)
  http://sci.esa.int/rosetta/54679-comet-67p-on-19-september-2014-navcam-mosaic/
  http://blogs.esa.int/rosetta/2014/09/19/cometwatch-19-september/
  Original artist: ESA/Rosetta/NAVCAM, CC BY-SA IGO 3.0
- **File:Comet_West.png** *Source:* https://upload.wikimedia.org/wikipedia/commons/8/87/Comet_West.png *License:* Public domain *Contributors:* This image was obtained from the NASA StarChild informative page on Comets [1] on December 31, 2011. *Original artist:* NASA, STScI
- **File:Commons-logo.svg** *Source:* https://upload.wikimedia.org/wikipedia/en/4/4a/Commons-logo.svg *License:* ? *Contributors:* ? *Original artist:* ?
- **File:Crab_Nebula.jpg** *Source:* https://upload.wikimedia.org/wikipedia/commons/0/00/Crab_Nebula.jpg *License:* Public domain *Contributors:* HubbleSite: gallery, release. *Original artist:* NASA, ESA, J. Hester and A. Loll (Arizona State University)
- **File:Cyanooctatetrayne-3D-vdW.png** *Source:* https://upload.wikimedia.org/wikipedia/commons/f/f7/Cyanooctatetrayne-3D-vdW.png *License:* Public domain *Contributors:* Derived from File:Acetylene-3D-vdW.png and File:Hydrogen-cyanide-3D-vdW.png. *Original artist:* Ben Mills and Jynto
- **File:Darwin_restored2.jpg** *Source:* https://upload.wikimedia.org/wikipedia/commons/b/b6/Darwin_restored2.jpg *License:* Public domain *Contributors:* Library of Congress[1] *Original artist:* Elliott & Fry
- **File:Desiccation-Tolerance-in-the-Tardigrade-Richtersius-coronifer-Relies-on-Muscle-Mediated-Structural-pone.0085091.s001.ogv** *Source:* https://upload.wikimedia.org/wikipedia/commons/8/87/Desiccation-Tolerance-in-the-Tardigrade-Richtersius-coronifer-Relies-on-Muscle-Mediated-Structural-pone.0085091.s001.ogv*License:* CC BY4.0*Contributors:* Movie S1from"Halberg KA,Jørgensen A,Møbjerg N" (2013). "Desiccation Tolerance inthe Tardigrade Richtersius coronifer Relies on Muscle Mediated Structural Reorganization".*PLOS ONE*.DOI: 10.1371/journal.pone.0085091.PMC:3877342.*Original artist:* Halberg K, J?rgensen A, M?bjerg N
- **File:Diacetylene-3D-vdW-B.png** *Source:* https://upload.wikimedia.org/wikipedia/commons/6/64/Diacetylene-3D-vdW-B.png *License:* Public domain *Contributors:* Derived from File:Acetylene-3D-vdW.png. *Original artist:* Ben Mills and Jynto
- **File:EXPOSE_location_on_the_ISS.jpg** *Source:* https://upload.wikimedia.org/wikipedia/commons/d/da/EXPOSE_location_on_the_ISS.jpg *License:* Public domain *Contributors:* http://spaceflight.nasa.gov/gallery/images/shuttle/sts-124/html/s124e009982.html *Original artist:* NASA
- **File:Earth-moon.jpg** *Source:* https://upload.wikimedia.org/wikipedia/commons/5/5c/Earth-moon.jpg *License:* Public domain *Contributors:* NASA [1] *Original artist:* Apollo 8 crewmember Bill Anders
- **File:Echiniscus_L.png** *Source:* https://upload.wikimedia.org/wikipedia/commons/f/fd/Echiniscus_L.png *License:* CC BY 2.1 jp *Contributors:* http://lifesciencedb.jp/resource_icon/icon.cgi?i=Echiniscus&t=L *Original artist:* DBCLS
- **File:Echiniscus_sp.jpg** *Source:* https://upload.wikimedia.org/wikipedia/commons/8/8f/Echiniscus_sp.jpg *License:* Public domain *Contributors:* unknown, watermark of source of scan says "Schultze, C.A.S., 1861" *Original artist:* Schultze

- **File:PIA01130_Interior_of_Europa.jpg** *Source:* https://upload.wikimedia.org/wikipedia/commons/7/7b/PIA01130_Interior_of_Europa.jpg *License:* Public domain *Contributors:* http://photojournal.jpl.nasa.gov/catalog/PIA01130 *Original artist:* unknown author of the NASA
- **File:Panspermie.svg** *Source:* https://upload.wikimedia.org/wikipedia/commons/4/48/Panspermie.svg *License:* CC BY-SA 3.0 *Contributors:* Own work; For the proto-bacteria I used an adapted version of File:Bacteria-.svg by JrPol and for the DNA I used an adapted version of File:DNA chemical structure.svg by Madprime. Earth from File:Earth Flag.svg by Himasaram *Original artist:* Silver Spoon Sokpop
- **File:Parthenogenesis_-_Bischoff,_Steve_R_2010.tif** *Source:* https://upload.wikimedia.org/wikipedia/commons/a/a8/Parthenogenesis_-_Bischoff%2C_Steve_R_2010.tif *License:* CC BY-SA 3.0 *Contributors:* Own work *Original artist:* Template:Steve R. Bischoff
- **File:People_icon.svg** *Source:* https://upload.wikimedia.org/wikipedia/commons/3/37/People_icon.svg *License:* CC0 *Contributors:* OpenClipart *Original artist:* OpenClipart
- **File:Phospholipids_aqueous_solution_structures.svg** *Source:* https://upload.wikimedia.org/wikipedia/commons/c/c6/Phospholipids_aque solution_structures.svg *License:* Public domain *Contributors:* Own work *Original artist:* Mariana Ruiz Villarreal ,LadyofHats
- **File:Phylogenic_Tree-en.svg** *Source:* https://upload.wikimedia.org/wikipedia/commons/5/58/Phylogenic_Tree-en.svg *License:* CC BY-SA 3.0*Contributors:* This file was derived fromPhylogenic Tree.jpg: <a href='//commons.wikimedia.org/wiki/File:Phylogenic_Tree.jpg' class=' image'><imgalt='Phylogenic Tree.jpg' src='https://upload.wikimedia.org/wikipedia/commons/thumb/d/de/Phylogenic_Tree.jpg/50px-Phylogenic_Tree.jpg'width='50' height='33' srcset='https://upload.wikimedia.org/wikipedia/commons/thumb/d/de/Phylogenic_Tree.jpg/75px-Phylogenic_Tree.jpg1.5x, https://upload.wikimedia.org/wikipedia/commons/thumb/d/de/Phylogenic_Tree.jpg/100px-Phylogenic_Tree.jpg 2x ' data-file-width='1069'data-file-height='713' /></a>*Original artist:* Phylogenic Tree.jpg: John D. Croft
- **File:Plumpollen0060.jpg** *Source:* https://upload.wikimedia.org/wikipedia/commons/3/39/Plumpollen0060.jpg *License:* CC-BY-SA-3.0 *Contributors:* ? *Original artist:* ?
- **File:Polycyclic_Aromatic_Hydrocarbons.png** *Source:* https://upload.wikimedia.org/wikipedia/commons/c/c0/Polycyclic_Aromatic_ png *License:* Public domain *Contributors:* Own work by uploader, Accelrys DS Visualizer *Original artist:* Inductiveload
- **File:Portal-puzzle.svg** *Source:* https://upload.wikimedia.org/wikipedia/en/f/fd/Portal-puzzle.svg *License:* Public domain *Contributors:* ? *Original artist:* ?
- **File:Red_Pencil_Icon.png** *Source:* https://upload.wikimedia.org/wikipedia/commons/7/74/Red_Pencil_Icon.png *License:* CC0 *Contributors:* Own work *Original artist:* Peter coxhead
- **File:RocketSunIcon.svg** *Source:* https://upload.wikimedia.org/wikipedia/commons/d/d6/RocketSunIcon.svg *License:* Copyrighted free use *Contributors:* Self made, based on File:Spaceship and the Sun.jpg *Original artist:* Me
- **File:Rosetta_111106.jpg** *Source:* https://upload.wikimedia.org/wikipedia/commons/6/69/Rosetta_111106.jpg *License:* CC-BY-SA-3.0 *Contributors:* Own work by the original uploader *Original artist:* Garafatea at English Wikipedia
- **File:Rosetta_and_Philae_(crop).jpg** *Source:* https://upload.wikimedia.org/wikipedia/commons/0/04/Rosetta_and_Philae_%28crop%29.jpg *License:* CC BY 3.0 de *Contributors:* File:Rosetta and philae.jpg (https://www.flickr.com/photos/dlr_de/11963777196) *Original artist:* DLR, CC-BY 3.0
- **File:Rosetta_spacecraft_(transparent_bg).png** *Source:* https://upload.wikimedia.org/wikipedia/commons/9/94/Rosetta_spacecraft_%28tra bg%29.png *License:* CC BY-SA 2.0 *Contributors:* https://www.flickr.com/photos/europeanspaceagency/11206647984/ *Original artist:* ESA/ATG medialab
- **File:STS-32_Return_to_KSC_-_GPN-2000-000677.jpg** *Source:* https://upload.wikimedia.org/wikipedia/commons/0/01/STS-32_Return_to_KSC_-_GPN-2000-000677.jpg *License:* Public domain *Contributors:* Great Images in NASA Description *Original artist:* NASA/photographer unknown
- **File:STS-46_EURECA_deployment.jpg** *Source:* https://upload.wikimedia.org/wikipedia/commons/b/bf/STS-46_EURECA_deployment.jpg *License:* Public domain *Contributors:* NASA http://images.jsc.nasa.gov/luceneweb/caption.jsp?photoId=STS046-08-010 *Original artist:* NASA
- **File:Signal_received_from_Rosetta_(12055070794).jpg** *Source:* https://upload.wikimedia.org/wikipedia/commons/1/16/Signal_received_from_Rosetta_%2812055070794%29.jpg *License:* CC BY-SA 3.0-igo *Contributors:* ESA flickr stream: “Signal received from Rosetta”, Credit: ESA - Jürgen Mai *Original artist:* ESA - Jürgen Mai
- **File:Sound-icon.svg** *Source:* https://upload.wikimedia.org/wikipedia/commons/4/47/Sound-icon.svg *License:* LGPL *Contributors:* Derivative work from Silsor's versio *Original artist:* Crystal SVG icon set
- **File:Southern_hawker_ovipositing.png** *Source:* https://upload.wikimedia.org/wikipedia/commons/b/b3/Southern_hawker_ovipositing.png *License:* CC BY-SA 4.0 *Contributors:* Own work *Original artist:* Singinghedgehog
- **File:Stromatolites.jpg** *Source:* https://upload.wikimedia.org/wikipedia/commons/c/c0/Stromatolites.jpg *License:* Public domain *Contributors:* National Park Service - http://www.nature.nps.gov/geology/cfprojects/photodb/Photo_Detail.cfm?PhotoID=204 *Original artist:* P. Carrara, NPS
- **File:Symbol_book_class2.svg** *Source:* https://upload.wikimedia.org/wikipedia/commons/8/89/Symbol_book_class2.svg *License:* CC BY-SA 2.5 *Contributors:* Mad by Lokal_Profil by combining: *Original artist:* Lokal_Profil
- **File:Text_document_with_red_question_mark.svg** *Source:* https://upload.wikimedia.org/wikipedia/commons/a/a4/Text_document_with_red_question_mark.svg *License:* Public domain *Contributors:* Created by bdesham with Inkscape; based upon Text-x-generic.svg from the Tango project. *Original artist:* Benjamin D. Esham (bdesham)
- **File:Tree_of_life.svg** *Source:* https://upload.wikimedia.org/wikipedia/commons/0/09/Tree_of_life.svg *License:* CC-BY-SA-3.0 *Contributors:* No machine-readable source provided. Own work assumed (based on copyright claims). *Original artist:* No machine-readable author provided. Vanished user fijtji34toksdcknqrjn54yoimascj assumed (based on copyright claims).

- **File:Tree_of_life_by_Haeckel.jpg** *Source:* https://upload.wikimedia.org/wikipedia/commons/d/de/Tree_of_life_by_Haeckel.jpg *License:* Public domain *Contributors:* First version from en.wikipedia; description page was here. Later versions derived from this scan, from the American Philosophical Society Museum. *Original artist:* Ernst Haeckel
- **File:Trihydrogen-cation-3D-vdW.png** *Source:* https://upload.wikimedia.org/wikipedia/commons/4/49/Trihydrogen-cation-3D-vdW.png *License:* Public domain *Contributors:* ? *Original artist:* ?
- **File:Varanus_komodoensis5.jpg** *Source:* https://upload.wikimedia.org/wikipedia/commons/a/ab/Varanus_komodoensis5.jpg *License:* CC-BY-SA-3.0 *Contributors:* Own work *Original artist:* Taken by myself (Raul654)
- **File:White_House_position_on_ET.png** *Source:* https://upload.wikimedia.org/wikipedia/commons/4/4a/White_House_position_on_ET.png *License:* CC BY 3.0 *Contributors:* Screenshot of US Government website https://petitions.whitehouse.gov/response/searching-et-no-evidence-yet *Original artist:* US Government (the white house) + Colby Gutierrez-Craybill + NASA
- **File:Wiki_letter_w_cropped.svg** *Source:* https://upload.wikimedia.org/wikipedia/commons/1/1c/Wiki_letter_w_cropped.svg *License:* CC-BY-SA-3.0 *Contributors:*
- Wiki_letter_w.svg *Original artist:* Wiki_letter_w.svg: Jarkko Piiroinen
- **File:Wikinews-logo.svg** *Source:* https://upload.wikimedia.org/wikipedia/commons/2/24/Wikinews-logo.svg *License:* CC BY-SA 3.0 *Contributors:* This is a cropped version of Image:Wikinews-logo-en.png. *Original artist:* Vectorized by Simon 01:05, 2 August 2006 (UTC) Updated by Time3000 17 April 2007 to use official Wikinews colours and appear correctly on dark backgrounds. Originally uploaded by Simon.
- **File:Wikiquote-logo.svg** *Source:* https://upload.wikimedia.org/wikipedia/commons/f/fa/Wikiquote-logo.svg *License:* Public domain *Contributors:* ? *Original artist:* ?
- **File:Wikisource-logo.svg** *Source:* https://upload.wikimedia.org/wikipedia/commons/4/4c/Wikisource-logo.svg *License:* CC BY-SA 3.0 *Contributors:* Rei-artur *Original artist:* Nicholas Moreau
- **File:Wikispecies-logo.svg** *Source:* https://upload.wikimedia.org/wikipedia/commons/d/df/Wikispecies-logo.svg *License:* CC BY-SA 3.0 *Contributors:* Image:Wikispecies-logo.jpg *Original artist:* (of code) cs:User:-xfi-
- **File:Wiktionary-logo-en.svg** *Source:* https://upload.wikimedia.org/wikipedia/commons/f/f8/Wiktionary-logo-en.svg *License:* Public domain *Contributors:* Vector version of Image:Wiktionary-logo-en.png. *Original artist:* Vectorized by Fvasconcellos (talk · contribs), based on original logo tossed together by Brion Vibber
- **File:Wormandwaterbear.jpg** *Source:* https://upload.wikimedia.org/wikipedia/commons/5/59/Wormandwaterbear.jpg *License:* CC BY-SA 3.0 *Contributors:*
- http://tardigrades.bio.unc.edu/ *Original artist:* Bob Goldstein and Victoria Madden, UNC Chapel Hill
- **File:Yellow_mite_(Tydeidae)_Lorryia_formosa_2_edit.jpg** *Source:* https://upload.wikimedia.org/wikipedia/commons/3/3d/Yellow_mite_%28Tydeidae%29_Lorryia_formosa_2_edit.jpg *License:* Public domain *Contributors:* This image was released by the Agricultural Research Service, the research agency of the United States Department of Agriculture, with the ID K9077-22 <a class='external text' href='//commons.wikimedia.org/w/index.php?title=Category:Media_created_by_the_United_States_Agricultural_Research_Service_with_known_IDs,<span>,&,</span>,filefrom=K9077-22#mw-category-media'>(next)</a>. *Original artist:* Photo by Eric Erbe;digital colorization by Chris Pooley. Edited byFir0002

19.8.3 Content license

- Creative Commons Attribution-Share Alike 3.0

CPSIA information can be obtained
at www.ICGtesting.com
Printed in the USA
LVHW061320220421
685237LV00009B/239

9 781519 665898